Advances in Spark-Ignition Engines

Editors

Fabio Bozza
Vincenzo De Bellis
Enrica Malfi

MDPI • Basel • Beijing • Wuhan • Barcelona • Belgrade • Manchester • Tokyo • Cluj • Tianjin

Editors
Fabio Bozza
University of Naples
"Federico II"
Italy

Vincenzo De Bellis
University of Naples
"Federico II"
Italy

Enrica Malfi
University of Naples
"Federico II"
Italy

Editorial Office
MDPI
St. Alban-Anlage 66
4052 Basel, Switzerland

This is a reprint of articles from the Special Issue published online in the open access journal *Energies* (ISSN 1996-1073) (available at: https://www.mdpi.com/journal/energies/special_issues/Spark-Ignition_Engines).

For citation purposes, cite each article independently as indicated on the article page online and as indicated below:

LastName, A.A.; LastName, B.B.; LastName, C.C. Article Title. *Journal Name* **Year**, *Volume Number*, Page Range.

ISBN 978-3-0365-5701-4 (Hbk)
ISBN 978-3-0365-5702-1 (PDF)

Contents

About the Editors

Fabio Bozza

Fabio Bozza is a Full Professor at the Department of Industrial Engineering, University of Naples Federico II, of Fluid Machines, Automotive Propulsion Systems, and Modeling and Optimization of Internal Combustion Engines. He is the author of over 170 papers published in national and international journals. His research activity is primarily concerned with theoretical and experimental studies of internal combustion engines (ICEs). Research topics of particular interest to him are the modeling of turbulence and combustion processes, including innovative combustion modes such as HCCI, RCCI, etc., for ICEs.

Vincenzo De Bellis

Vincenzo De Bellis, Ph.D., is an Associate Professor at the Department of Industrial Engineering, University of Naples Federico II, of Fluid Machines, Automotive Propulsion Systems, and Power and Propulsion Systems for UV. He is the author of over 70 papers published in international journals and conference proceedings. His research activity is primarily concerned with the theoretical and experimental study of internal combustion engines (ICEs). Research topics of interest to him are the modeling of turbulence, combustion, knocking, and pollutant emission in spark-ignition ICEs; advanced combustion modalities, including HCCI, RCCI, pre-chamber ignition; and the turbocharging and energy management of hybrid vehicles.

Enrica Malfi

Enrica Malfi, Ph.D., is a Research Fellow at the Department of Industrial Engineering, University of Naples Federico II. Her research activity concerns the numerical study of internal combustion engines (ICEs) and hybrid vehicles. Research topics of interest to her are the modeling of turbulence and combustion processes, with a particular focus on innovative combustion modes and alternative fuels, and the development of control strategies for hybrid powertrains.

energies

MDPI

Article

An Experimental Investigation of the Impact of Washcoat Composition on Gasoline Particulate Filter (GPF) Performance

Junjun Wang [1,2], Fuwu Yan [1,*], Na Fang [2], Dong Yan [2], Guoqing Zhang [1,2], Yu Wang [1] and Wulin Yang [2]

1 School of Automotive Engineering, Wuhan University of Technology, Wuhan 430070, China; wangjunj@dfmc.com.cn (J.W.); zhanggq@dfmc.com.cn (G.Z.); yu.wang@whut.edu.cn (Y.W.)
2 Powertrain Department, Dong Feng Motor Corporation Technical Center, Wuhan 430058, China; fangn@dfmc.com.cn (N.F.); yand@dfmc.com.cn (D.Y.); yangwulin@dfmc.com.cn (W.Y.)
* Correspondence: yanfuwu@vip.sina.com

Received: 31 December 2019; Accepted: 27 January 2020; Published: 5 February 2020

Abstract: The forthcoming implementation of the China VI emission regulations, which are currently the most stringent around the world targeted at light-duty gasoline engine vehicles, will not only further restrict the emissions of gaseous pollutants, but also put forward, for the first time, the requirements of particulate number (PN) emissions with a limit set at 6×10^{11}#/km. To achieve the stringent emission targets, the ceramic wall-flow gasoline particulate filter (GPF) will be effective to achieve the reduction of the particulate number tailpipe emissions in a way similar to the widely applied diesel particulate filter (DPF) in diesel engines. This paper investigated the effect of a coated gasoline particulate filter (GPF) on the PN emission and engine performance. The effects of two factors, including the washcoat powder material bulk density and type of coating, were studied with regard to three primary performances of GPF, including high three-way catalytic performance, low pressure drop, and high PN filtration efficiency, according to the original equipment manufacturer (OEM) requirements. The outcomes show that the use of high bulk density materials resulted in a low washcoat volume and hence a decrease of flow resistance and backpressure, in addition to high PN filtration efficiency. The type of coating had notable influence on the backpressure and PN filtration efficiency. The coating length and coating amount both had notable influence on the backpressure and PN filtration efficiency.

Keywords: washcoat; powder material bulk density; coating; backpressure; PN filtration efficiency

1. Introduction

In order to further improve engine fuel economy and reduce CO_2 emissions, turbocharged gasoline direct injection (GDI) engines have prevailed as the power source for light-duty vehicles [1]. However, compared to vehicles equipped with traditional port fuel injection engines, GDI engines are known to have higher engine-out soot and particle number emissions, especially under cold start and high-speed, high-load operating conditions [2–4]. The latest China VI light-duty emission regulations will be implemented in 2020. For China VI, the certification test driving cycle will be changed from the New European Driving Cycle (NEDC) to the Worldwide Harmonized Light Vehicles Test (WLTC), which includes more transient operation points, leading to an increase of the maximum speed and acceleration, as well as a move to larger loads for engine operating regions [5]. Moreover, the pollutants limits will be substantially tighter. The particle number emissions requirement will be introduced in the China VI emission regulations for the first time, with the same limit set of 6×10^{11}#/km in the China VI-a and VI-b stages. In addition, the real driving emissions (RDE) certification test will be implemented in China VI-b stage in 2023, in which the conformity factor (CF) of the particulate number (PN) and NOx will be limited to be 2.1 and the CO_2 emissions will be recorded. There is no doubt that the stringent requirements in China VI will be even more challenging for light-duty vehicles equipped with turbocharged GDI engines, especially in terms of meeting the PM/PN limit requirements.

For light-duty gasoline vehicles, in response to the upcoming implementation of the China VI emission regulations, advanced exhaust gas aftertreatment technologies (e.g., GPF) must be applied to reduce the vehicle tailpipe emissions in addition to the adaptation of engine internal parameters, such as air/fuel ratio, fuel injection strategy, and combustion phasing, to further optimize the combustion system in order to reduce the engine-out raw emissions [6–9]. The technology of DPF to solve the problem of diesel particulate emissions has been applied for many years, and its effectiveness on the reduction of particulate emission has been well confirmed [10,11]. It is considered that the use of GPF has the potential to effectively reduce gasoline particulate emissions over a wide range of engine operating conditions [12]. However, employing GPF as a solution to reduce particulate emissions also results in an increase of the pressure drop of the entire exhaust aftertreatment system. Consequently, the drop increases the backpressure of the exhaust manifold, and hence imposes a negative influence on the engine power output, fuel economy, and CO_2 emissions [13].

In general, there are two types of particulate filters applied to the exhaust aftertreatment system currently: One is bare GPF without any catalyst coated, and the other is cGPF, which is coated with some amount of catalyst and could even further improve the three-way catalytic performance. Compared to the bare GPF, it is considered that adopting the coated GPF (cGPF) has more advantages, such as reducing the cost of substrates, packaging improvements, and an increase in soot regeneration performance under cold running conditions [14]. It is believed that the coated GPF will become the mainstream in auto original equipment manufacturers (OEMs) in the future. Coated GPF should fulfill design requirements in order to achieve sufficient particulate number filtration efficiency to meet the PN regulation limit. In the meantime, the pressure drop must be minimized to reduce its influence on engine power output and fuel economy. Different kinds of coatings with integrated pore characteristics of the filter wall were confirmed to have an evident effect on the reduction of the pressure drop and to increase PN filtration efficiency [15,16]. Consequently, this study concerned the coated GPF and describes the optimization design methodology. The results are discussed in detail in the subsequent sections.

2. Experimental Section

In this study, the influence of the pressure drop of the washcoat on the GPF were tested on a cold flow bench (computerized flowbench SF-1020 Probench), as can be seen from the Figure 1 and Table 1. The pressure drop of the test GPF samples was carried out on the SF-1020 Probench with 25 °C ambient temperature and an atmosphere pressure of 1 atm. Five air flow rates were studied, ranging from 200 kg/h to 600 kg/h at a step of 100. The GPF design was a cylinder with a diameter of 132.1 mm and a length of 127 mm. The cell structure was 300 cpsi/8 mil and the porosity was approximately 65%. Detailed parameters are listed in the Table 2.

Table 1. Measurement conditions of the pressure drop on the gasoline particulate filter (GPF).

Measurement Equipment	SF-1020 Probench
Flow Measurement Accuracy	±0.5% of reading in normal operating ranges
Repeatability	±0.25% of reading
Range	0–1000 cfm
Temp. Measurement Accuracy	±0.3% °C

Figure 1. Measurement equipment SF-1020 Probench.

Table 2. Measurement conditions of the pressure drop on the GPF.

Measurement Equipment	SF-1020 Probench
Temperature	25 °C
Atmosphere Pressure	1 atm
Air flow rate	200 kg/h, 300 kg/h, 400 kg/h, 500 kg/h, 600 kg/h
GPF size	Φ132.1 mm × 127mm 1.74 L
Cell structure	300 cpsi/8 mil
Porosity	Approx. 65%
Mean Pore Size	Approx. 20 um
Catalyst amount	100 g/L

In this study, the influence of the particle number filtration and gas pollutant abatement efficiency of the washcoat on the GPF were tested on a light-duty passenger car, equipped with 1.4-liter displacement turbocharged gasoline direct-injection (TGDI) engine, which was employed for the on-engine experiment at test driving cycle Worldwide harmonized Light vehicles Test Cycle (WLTC) on vehicle emission hub of HORIBA test equipment. The tests used China VI emission regulation 92 # gasoline, as can be seen from the Figure 2. The information of the test engine and the washcoat are shown in Table 3.

Figure 2. Tests of the particulate number filtration and efficiency.

Table 3. Main parameters of the test vehicle.

Vehicle reference mass	1460 kg
Engine	1.4 L TGDI
Test driving cycle	WLTC
Closed-coupled TWC	Φ118.4 mm × L60 mm, 200 g/L 0/47/3 Fresh
Under-floor catalyzed GPF	Φ118.4 mm × L127 mm, 100 g/L 0/7/3 Fresh

3. Results and Discussion

3.1. Washcoat Powder Material Bulk Density

The effects of washcoat powder material bulk density on gasoline particulate filter performance were first studied. In order to study the factor, the catalyst coating amount was kept constant at 100 g/L. In the meantime, the bulk density of powder material of the washcoat was changed by adjusting the percentage of composition of the powder. Two kinds of compositions were employed in the study, namely aluminum oxide and cerium zirconium composite oxides. Two types of powder with different bulk densities, referred to as GPF-A and GPF-B, were investigated. Detailed information of the powders is listed in the Table 4. As listed in Table 4, GPF-A represents the washcoat, which used low-level bulk density material. The percentage of the aluminum oxide was 0.37 g/mL and the cerium zirconium was 0.27 g/mL. On the other hand, GPF-B represents the high-level bulk density material washcoat, with 0.76 g/mL and 0.55 g/mL for the two materials. For the convenience of comparison, GPF-Ref represents a commercial GPF with same catalyst coating amount.

Table 4. Comparison of bulk density of GPF-A and GPF-B (g/mL).

	Aluminum Oxide	Cerium Zirconium Composite Oxides
GPF-A	0.37	0.27
GPF-B	0.76	0.55

The pressure drop of test GPF samples against the air flow rate is plotted in Figure 3. It can be seen from the Figure 3 that the pressure drop of the GPF-B in the whole test flow range was about 34.9% lower than that of the GPF-A, and 11.1% lower than that of the commercial GPF-Ref. As can be seen from Table 2, the bulk density of the powdery material used in the GPF-B was about twice as high as that of GPF-A. Therefore, the washcoat volume of former coating was smaller than that of the latter under the same coating amount. Consequently, the type of washcoat with a smaller volume avoided extensive build-up in the pores of the substrate, and hence reduced flow resistance. Eventually, it led to a decrease in pressure drop.

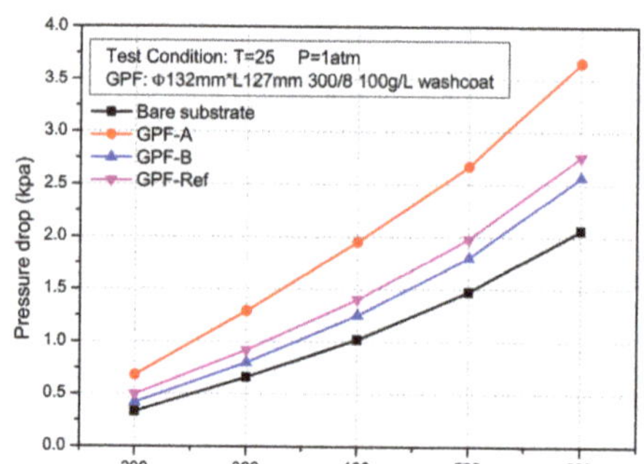

Figure 3. Influence of washcoat material bulk density on pressure drop.

In order to further check the influence of bulk density on the pressure drop, the microstructure of the GPF washcoat was characterized by scanning electron microscope (SEM). The images of the two types of coating are seen in Figure 4. The white part in the figure represents the washcoat, and the dark grey part represents the substrate. As described above, the washcoat in GPF-A showed extensive piling up, especially on one side of the substrate, which had a great hindering effect on the gas flow. As the material bulk density increased, the washcoat of GPF-B had a lower volume and appeared to accumulate very little. Hence, the flow resistance of GPF-B was obviously lower than GPF-A, as confirmed in pressure drop test.

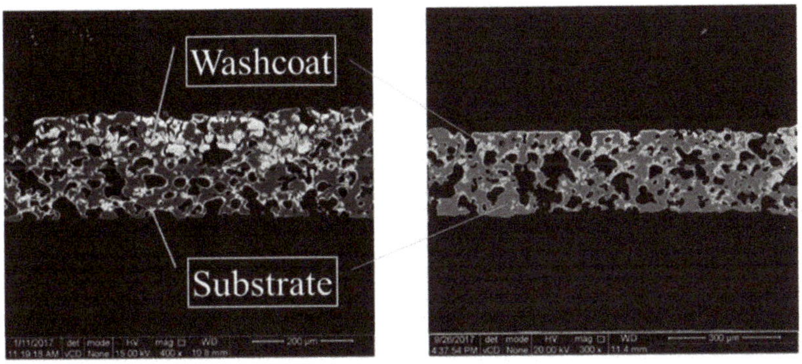

Figure 4. SEM images of the GPF samples of GPF-A and GPF-B.

The results of fresh state PN efficiency by the two types of washcoat are shown in Figure 5. It can be seen that the efficiency was significantly improved by both washcoats, further confirming the effectiveness of the washcoat on the reduction of particle emission. More importantly, compared with GPF-A, the PN filtration efficiency of GPF-B slightly increased by about 2.9% after raising the bulk density of the coating material. More dispersed distribution of the inner coating in the large pores by application of high bulk density coating material is considered to be the reason for the higher PN filtration efficiency.

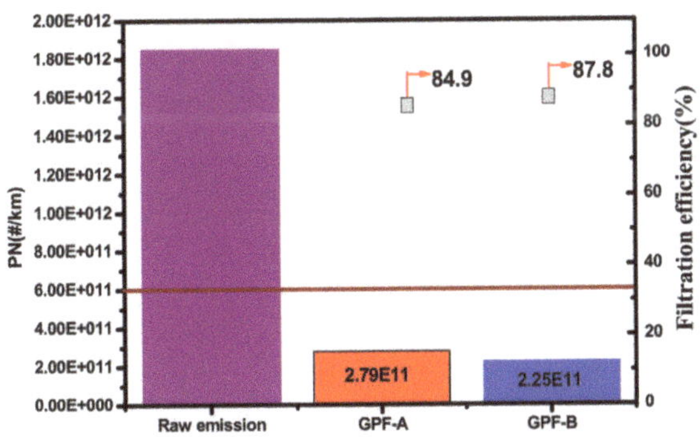

Figure 5. Influence of the washcoat material bulk density on particulate number (PN) filtration efficiency.

In order to check the influence of the bulk density on the light-off temperature on the emission of gasoline engine, the results of CO/HC/NOx light-off temperature (T50) of engine bench test of GPF-A and GPF-B are further shown in Figure 6. It can be confirmed that the coating material bulk density had little impact on CO/HC/NOx light-off temperature.

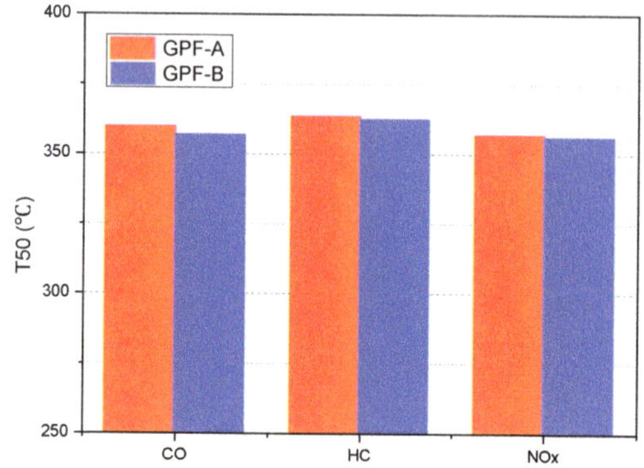

Figure 6. Influence of coating material bulk density on T50 of CO/HC/NOx.

3.2. Catalysts Coating

In addition to the material of the washcoat on GPF performance, the impact of catalyst coating on GPF performance was also studied. Two factors, namely the direction and the length of the coating, were investigated via experiment. The catalyst coating amount was kept constant (approx. 100 g/L) for fair comparison. The test samples were prepared in four different types of coating using the S-75% and S-100% to represent single-side coating with 75% and 100% of substrate length for coating length, respectively, whereas T-50% and T-75% represent two-side coating with 50% and 75% of substrate length for coating length, respectively. The sketches of different catalyst coating are shown in Figure 7.

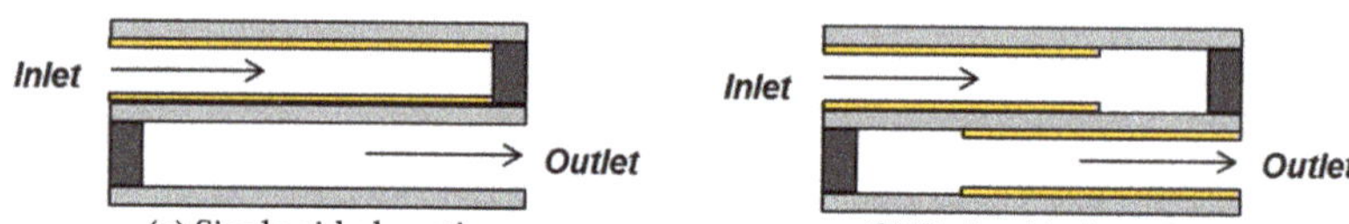

Figure 7. Diagram of two different catalyst coating of GPF.

The experiment boundary conditions for the measurements and procedure are same as Table 1. The experimental test was carried out on the cold flow bench with 25 °C ambient temperature and an atmosphere pressure of 1 atm. The four catalysts coating tested were S-100%, S-75%, T-50%, and T-75%.

The pressure drop, with four different kinds of catalyst coatings against air flow rate, is plotted in Figure 8. It can be seen that catalyst coating had moderate influence on the pressure drop, including the coating direction and coating length. First, all the coatings increased the pressure drop at each flow range. Moreover, the pressure drop of the S-100% was the highest in the whole flow range in the test, followed by T-75%, T-50%, and S-75%. Different coatings had a great influence on the distribution of the coated washcoat in the interior of GPF. The T-75% coating did not only make the washcoat cover the inlet and outlet wall of substrate more uniformly, but also increased the washcoat density per length, resulting in the highest pressure drop. For the S-100% coating, it was not easy to control the coating process operation because it was prevent the penetration of the washcoat throughout the S-100% length-coating process, and it was also difficult to guarantee the uniformity of the coated washcoat. The T-50% coating did not only greatly reduce the pressure drop, but was also easy to be controlled for coating. In addition, better uniformity and distribution of the coated washcoat can also be achieved by this coating. Although the pressure drop was the lowest for the S-75% coating, a blank of 25% length was left to the substrate, which largely affected the PN filtration efficiency.

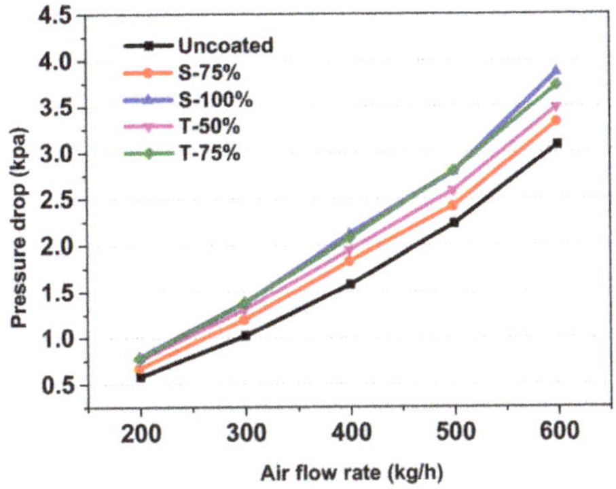

Figure 8. Influence of catalysts coating on GPF cold flow pressure drop.

Furthermore, series of WLTC emission tests were conducted to determine the impact of several different catalyst coatings on PN filtration efficiency. Figure 9 shows the results of the fresh state PN filtration efficiency of four different kinds of catalyst coating. Corresponding to the results of the pressure drop, the T-75% coating had the highest PN filtration efficiency, reaching more than 95%. The PN filtration efficiency of S-100% and T-50% coatings, respectively, were about 92% and 89%, while the S-75% had the lowest PN filtration efficiency with approximately 85%. It can be concluded that the pressure drop is positively correlated with the PN filtration efficiency after further comprehensive analysis in terms of influence of different coatings on pressure drop and PN filtration efficiency. Based on the principle that the pressure drop should be reduced as much as possible on the premise of the PN filtration efficiency to meet development requirements, the T-50% was determined the optimum catalyst coating.

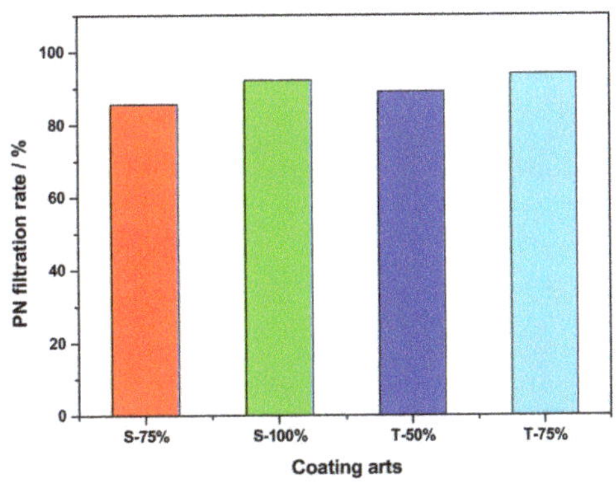

Figure 9. Influence of catalyst coatings on PN filtration efficiency.

4. Conclusions

In this study, the washcoat for a gasoline particulate filter (GPF) was experimentally studied. The influence of material selection and different kinds of coatings on the pressure drop, PN filtration efficiency, and three-way catalytic performance were examined. The two main conclusions were drawn as follows:

1. The bulk density of powder materials employed in GPF washcoat had a great influence on the backpressure. High bulk density materials resulted in a low washcoat volume and hence the decrease of flow resistance and backpressure. The PN filtration can also be improved using high bulk density powder materials due to the more dispersed distribution of the inner coating.

2. The kind of different coatings, which included coating directions, coating length, and coating amount, had notable influence on the backpressure and PN filtration efficiency. T-50% has been experimentally confirmed as the optimum catalyst coating, which meets both the lower backpressure and PN filtration ratio that fulfilled the development requirements.

Author Contributions: Conceptualization, J.W. and F.Y.; Methodology, J.W. and Y.W.; Validation, J.W.; Investigation, N.F and W.Y.; Data Curation, N.F.; Writing—Original Draft Preparation, J.W.; Writing—Review & Editing, N.F., D.Y. and G.Z.; Supervision, F.Y.; Project Administration, F.Y. All authors have read and agreed to the published version of the manuscript.

Funding: This research was funded partially by the National Key R&D Program of China (2017YFC0211201).

Conflicts of Interest: The authors declare no conflict of interest.

Abbreviations

PN	Particulate Number
PM	Particulate Matter
GDI	Gasoline Direct Injection
RDE	Real Driving Emissions
GPF	Gasoline Particulate Filter
PFI	Port Fuel Injection
NEDC	New European Driving Cycle
DPF	Diesel Particulate Filter
MPS	Mean Pore Size
AFR	Air Fuel Ratios
PGM	Platinum Group Metals
TWC	Three-Way-Catalyst
WLTC	Worldwide harmonized Light vehicles Test Cycle
SEM	Scanning Electron Microscope

References

1. Caggiano, M. Technology road map in the next decade for gasoline engine. In Proceedings of the 8th International Conference Powertrain Technologies for CO_2, ATA 2010, Turin, Italy, 17–18 November 2010.
2. Berkemeier, O.; Grieser, K.; Hohenboeken, K.; Karvounis, E.; Springer, K.M. Strategies to Control Particulate Emissions of Gasoline Direct Injection Engines. In Proceedings of the FISITA 2012 World Automotive Congress, Beijing, China, 27–30 November 2012.
3. Chan, T.W.; Meloche, E.; Rosenblatt, D.; Kubsh, J.; Brezny, R.; Rideout, G. Reducing Particulate Emissions for Future Gasoline Direct Injection Vehicles with a Gasoline Particulate Filter. In Proceedings of the ETH Conference on Combustion Generated Nanoparticles, Zürich, Switzerland, 24–27 June 2012.
4. Friedrich, A. Particles from Gasoline Direct Injection Engines: Abatement Options. In Proceedings of the ETH Conference on Combustion Generated Nanoparticles, Zürich, Switzerland, 23–26 June 2013.
5. Limits and Measurement Methods for Emissions from Light-Duty Vehicles (CHINA 6). 23 December 2016. Available online: http://www.mee.gov.cn/ywgz/fgbz/bz/bzwb/dqhjbh/dqydywrwpfbz/201612/W020171207355626647621.pdf (accessed on 31 January 2020).
6. Piock, W.; Hoffmann, G.; Berndorfer, A.; Salemi, P.; Fusshoeller, B. Strategies Towards Meeting Future Particulate Matter Emission Requirements in Homogeneous Gasoline Direct Injection Engines. *SAE Int. J. Engines* **2011**, *4*, 1455–1468. [CrossRef]
7. Heiduk, T.; Stichlmeir, M.; Unselt, F.; Kuhn, M. The New 1.8L TFSI Engine from Audi-Part 2: Mixture Formation, Combustion Method and Turbocharging. *MTZ Worldw. eMag.* **2011**, *72*, 58–64. [CrossRef]
8. Klauer, N.; Kluting, M.; Schunemann, E.; Schwarz, C.; Steinparzer, F. BMW Twin Power Turbo Gasoline Engine Technology-Enabling Compliance with worldwide Exhaust Gas Emission Requirements. In Proceedings of the 34th International Vienna Motor Symposium, Vienna, Austria, 25–26 April 2013.
9. Merdes, N.; Enderle, C.; Vent, G.; Weller, R. The New Turbocharged Four-cylinder Gasoline Engine by Mer-cedes-Benz. *MTZ Worldw. eMag.* **2011**, *72*, 16–23. [CrossRef]

10. Ohara, E.; Mizuno, Y.; Miyairi, Y.; Mizutani, T.; Yuuki, K.; Noguchi, Y.; Hiramatsu, T.; Makino, M.; Takahashi, A.; Sakai, H.; et al. Filtration Behavior of Diesel Particulate Filters (1). *SAE Tech. Pap.* **2007**. [CrossRef]
11. Richter, J.; Klingmann, R.; Spiess, S.; Wong, K. Application of Catalyzed Gasoline Particulate Filters to GDI Vehicles. *SAE Int. J. Engines* **2012**, *5*, 1361–1370. [CrossRef]
12. Saito, C.; Nakatani, T.; Miyairi, Y.; Yuuki, K.; Makino, M.; Kurachi, H.; Heuss, W.; Kuki, T.; Furuta, Y.; Kattouah, P.; et al. New Particulate Filter Concept to Reduce Particle Number Emissions. *SAE Tech. Pap.* **2011**. [CrossRef]
13. Shimoda, T.; Ito, Y.; Saito, C.; Nakatani, T.; Shibagaki, Y.; Yuuki, K.; Sakamoto, H.; Vogt, C.; Matsumoto, T.; Furuta, Y.; et al. Potential of a Low Pressure Drop Filter Concept for Direct Injection Gasoline Engines to Reduce Particulate Number Emission. *SAE Tech. Pap.* **2012**. [CrossRef]
14. Inoda, S.; Nomura, Y.; Ori, H.; Yabuzaki, Y. Development of New Coating Technology Optimized for Each Function of Coated GPF. *SAE Tech. Pap.* **2017**. [CrossRef]
15. Ogyu, K.; Ogasawara, T.; Nagatsu, Y.; Yamamoto, Y.; Higuchi, T.; Ohno, K. Feasibility Study on the Filter Design of Re-Crystallized SiC-GPF for TWC Coating Application. *SAE Tech. Pap.* **2015**. [CrossRef]
16. Ito, Y.; Shimoda, T.; Aoki, T.; Yuuki, K.; Sakamoto, H.; Kato, K.; Thier, D.; Kattouah, P.; Ohara, E.; Vogt, C. Next Generation of Ceramic Wall Flow Gasoline Particulate Filter with Integrated Three Way Catalyst. *SAE Tech. Pap.* **2015**. [CrossRef]

energies

MDPI

Article

Improving Fuel Economy of Spark Ignition Engines Applying the Combined Method of Power Regulation

Yurii Gutarevych [1], Vasyl Mateichyk [2], Jonas Matijošius [3,*], Alfredas Rimkus [4,5], Igor Gritsuk [6], Oleksander Syrota [1] and Yevheniy Shuba [1]

1 Department of Engines and Thermal Engineering, Faculty of Faculty of Automotive and Mechanical Engineering, National Transport University, Mykhaila Omelianovycha-Pavlenka Str. 1, 01010 Kyiv, Ukraine; yugutarevich@gmail.com (Y.G.); cirshu@gmail.com (O.S.); shuba90@i.ua (Y.S.)
2 Department of Ecology and Safety of Vital Functions, Faculty of Faculty of Automotive and Mechanical Engineering, National Transport University, Mykhaila Omelianovycha-Pavlenka Str. 1, 01010 Kyiv, Ukraine; wmate@ukr.net
3 Institute of Mechanical Science, Vilnius Gediminas Technical University, J. Basanavičiaus Str. 28, LT-03224 Vilnius, Lithuania
4 Department of Automobile Engineering, Vilnius Gediminas Technical University, J. Basanavičiaus Str. 28, LT-03224 Vilnius, Lithuania; alfredas.rimkus@vgtu.lt or a.rimkus@vtdko.lt
5 Department of Automobile Transport Engineering, Vilnius College of Technologies and Design, Antakalnio St. 54, LT-10303 Vilnius, Lithuania
6 Department of Technical Exploitation and Service of Cars, Kharkiv National Automobile and Highway University (KhNAHU), Yaroslava Mudrogo St, 25, 61000 Kharkiv, Ukraine; gritsuk_iv@ukr.net
* Correspondence: jonas.matijosius@vgtu.lt; Tel.: +370-684-04-169

Received: 7 January 2020; Accepted: 23 February 2020; Published: 1 March 2020

Abstract: One of the disadvantages of spark ignition engines, whose power is regulated by throttling, is the increased fuel consumption at low loads and when the engine is idle. The combined method of engine power regulation by switching off the cylinder group and throttling working cylinders is one of the effective ways to improve fuel economy in the above-mentioned modes. This article presents the research results of the combined method of engine power regulation which can be realized by minor structural changes in operating conditions. The method implies the following: at low loads and at idle speed of the engine. Fuel supply to the group of cylinders is switched off with the simultaneous increase of the cyclic fuel supply in the working cylinders. The adequacy of the calculated results has been checked by the indication of operating processes in switched off and working cylinders. The research results of a six-cylinder spark ignition engine with the distributed gasoline injection using the combined power regulation system have been shown. The angles of opening the throttle which provides a non-shock transition from the operation with all cylinders to the operation with the cylinder group switched off have been determined.

Keywords: spark ignition engine; combined method of power regulation; fuel economy; the idling mode; switching off the cylinder group

1. Introduction

Many factors influence fuel economy (FE) and the amount of pollution (HS) in the exhaust gas (EG) of cars with internal combustion engines (ICE). Research carried out on automobile engines shows that the modes of car motion have great influence on these indicators.

The movement of a car, in large cities in particular, is accompanied by complicated operating modes of the engine: acceleration, deceleration, constant and unstable modes, and modes of independent and forced idling. In this case, most of the time cars move at a low speed, with the engine running at low loads and at idle speed.

Operating modes of automobile engines under operating conditions have been investigated in many papers. At low speed and low load, high heat transfer and combustion losses occur, directly affecting engine operating conditions [1]. In most cases, these actual engine operating conditions are not reflected in standardized driving cycles and, sometimes, summaries of experimentally-derived cycles are also used to evaluate traffic jams [2]. Design changes, such as ceramic-coated engines, have also been used to reduce heat loss, which, when used in combination with biodiesel as fuel, have shown optimum engine operating conditions [3]. This effect is also noticeable when using triglyceride-based biodiesel as fuel [4]. Hydroxyl gas can also be used as an indicator to improve engine natural conditions, resulting in higher thermal efficiency and improved exhaust emissions [5]. The possible use of biobutanol as another source of biofuels and similar trends in engine operating conditions due to its higher oxidative stability [6] as a spark ignition [7], as a compression ignition engines [8]. It is also known that the use of alcohol additives in petrol improves its properties and achieves an ecological effect [9]. The EGR system shows a different reduction in cycle pressure [10] which is noticeable when operating small engines [11]. Studies [12–15] shows that the basic operating modes of both freight and passenger vehicles moving in cities are part speed and loading modes. They involve acceleration, steady mode, and active and forced idling while driving and braking the car. Most of the time the vehicular engines run the open throttle or with a fuel supply of not more than 50%. The time of operation with a full load, when supplying the fuel is close to the maximum, is not more than 30%. This is particularly emphasized in the context of the development of smart cities, where freight delivery can be modeled on special robotic vehicles [12]. It is important to emphasize that the use of such transport is specialized and more tailored to individual customers, with a higher standardized freight flow due to the limited engine described above and the rail network used [13]. Nevertheless, the combined approach to passenger and freight transport is widely developed [14]. This option offers great advantages especially when considering ecological aspects [15].

In general, it can be concluded that under operating conditions, the main motor vehicle modes of operation are part load speed modes and idling modes. In these modes, the fuel economy of spark ignition engines significantly deteriorates which is one of the main disadvantages of these engines. The reason for this deterioration is the decrease of the indicated efficiency of the actual cycle. It is due to the increase in the relative amount of EG and the deterioration of mixing and combustion conditions [16–23]. Double sparking can cause early release of heat and reduce the knocking effect [16]. Spark shortening phenomena have been observed during the extension of the spark discharge period [21]. What is especially important with the dual fuel system with biogas and ethanol fuel is the reduction in the knocking effect at full engine load [17]. Additionally, exhaust residuals have a direct influence on the combustion process [18]. With the use of hydrogen as an additive in the fuel it is possible to extend the combustion limits of the lean [19]. Another way is to use the thermal efficiency optimization strategy in the case of lean combustion, which also helps to combat combustion variations [20]. The deterioration of mixing and combustion conditions was also addressed using a multi hole injector [22], and in a wide range of engine speed [23]. In addition, the degree of repeatability of successive cycles is increased. Mechanical losses grow which consist of friction losses and pump losses. These losses significantly increase with decreasing load. This is due to the rise of dilution in the inlet line and when throttling can reach 50% of the total mechanical loss [24,25]. Relative friction losses also increase significantly when the load changes. They do not depend on the rotational speed of the crankshaft, which has been proved in many previous papers. This was confirmed by both machine learning model simulations [26], using artificial neural network modeling [27], a zero-dimensional cylinder pressure model [28], the crankshaft rotation measurement experimental method [29], using the crankshaft dynamics research method [30], a comparison of frictional losses research [31], and both by studying the influence of initial starting conditions on the above factors [32].

One of the most effective methods to improve fuel efficiency in the above-mentioned modes is switching off the group of cylinders with the change of the gas exchange system. This method is widely used in modern engines. In fact, it is the combined method of power regulation as it includes switching off the cylinder group and throttling the working cylinders. The homogeneity of the flammable mixture influences the efficiency of disconnecting the cylinders [33]. This is noticeable especially with direct injection spark engines [34]. This also allows for an ecological effect [35] and the fuel economy effect [36], and the issue of tribological performance of piston compression ring and connecting rod bearing is also addressed [37]. According to Ihlemann at al. this is the technology of the future [38].

The use of this method requires significant changes in the design of the engine and is realized in the production process.

The aim of this work is improving fuel economy of spark ignition engines applying the combined method of power regulation.

The combined method, in which the system of gas exchange does not change, has been studied. The power of the engine at small external loads is obtained by switching off the cylinder group when the second group of cylinders works with a large opening of the throttle and higher efficiency. The group of cylinders is switched off by disabling the fuel supply to these cylinders. The position of the throttle valve and the cyclic supply of fuel in the working cylinders is simultaneously increased. It ensures equal power for the operation with all cylinders and with the cylinder group switched off.

2. Theoretical Substantiation of Possible Improvement of Fuel Economy

One of the main reasons for the significant deterioration of fuel economy of spark ignition engines in low load and idling modes is a decrease of the indicated efficiency of the engine and an increase of mechanical losses. Under the combined power regulation method, if the indicated efficiency increases and the mechanical losses decrease, as compared to throttle all cylinders, fuel economy of spark ignition engines is improved.

The comparison has been made for a six-cylinder engine with a distributed gasoline injection system and feedback [35,39]. The dependency of the indicated efficiency on the average effective pressure, which is proportional to the external load, has been determined taking into account the nature of the combustion, the duration of the combustion phases, and the effect of residual gas [10,34,40].

The dependencies of the indicated efficiency "η_i" on the average effective pressure "p_e" from zero to an average effective pressure of 0.35 MPa are shown in Figure 1. It implies throttling and switching off three cylinders at loads where the operation of three and six cylinders is possible [39].

As it can be seen from the shown dependencies, the value of the indicated efficiency when three cylinders are switched off, η_i varies from 0.306 (in idling mode) to 0.345 (before the enrichment of the mixture). When throttling, the indicated efficiency $\eta_{i\ th}$ at this load varies within the range of 0.26, ..., 0.32. Thus, the transition from throttling fuel-air mixture to the combined method at idling modes and low loads by switching off the cylinder group and throttling working cylinders can increase the indicated efficiency of a gasoline engine with an electronic control and a feedback injection system by 13.7–7.8%. With an average effective pressure of more than 0.325 MPa, it is advisable to throttle all cylinders.

The effect of the combined method of power regulation on mechanical losses has been estimated at the same values of the average effective pressure p_e.

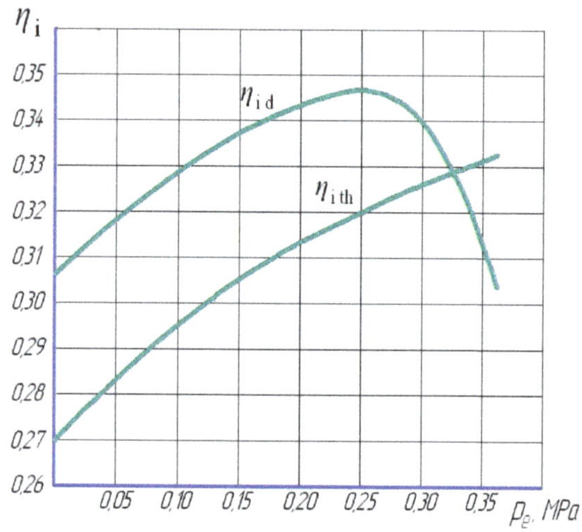

Figure 1. The dependency of the indicated efficiency on the average effective pressure when six cylinders are throttled and when three cylinders are switched off (Opel C30 LE engine). Indices: th, index to indicate throttling; d, index to indicate deactivation; $\eta_{i\,th}$, indicated efficiency of throttling; η_{id}, indicated efficiency when three cylinders are switched off (deactivated).

The average effective pressure $\boldsymbol{p_e}$ required for vehicle motion when throttling and switching off "k" cylinders of the engine with "i" cylinders has been determined by the dependencies:

$$p_{e\,th} = p_{i\,th} - p_{m\,th}, \tag{1}$$

$$p_{ed} = \frac{i-k}{i} p_{id} - p_{md} \tag{2}$$

where $p_{i\,th}$ and p_{id} are the average indicated pressure when throttling and switching off the cylinders; $p_{m\,th}$ and p_{md} are the average pressure of mechanical losses using the combined method of power regulation.

It is known that, during throttling, the average mechanical loss pressure $p_{m\,th}$ includes the average pressure $p_{t\,th}$ used for friction, pressure $p_{g\,th}$ for the gas exchange process, and pressure $p_{d\,th}$ for the camshaft drive, auxiliary equipment and ventilation losses. That is:

$$p_{mth} = p_{tth} + p_{gth} + p_{dth}. \tag{3}$$

When "k" cylinders of an "i"-cylinder engine are switched off, the average mechanical loss pressure $p_{m\,d}$ includes the average pressure $p'_{t\,d}$ used for friction in the working "i-k" cylinders and pressure $p''_{t\,d}$ in "k" switched off cylinders, the loss pressure $p_{g\,d}$ for gas exchange process in the "i-k" working cylinders and $p''_{g\,d}$ in "k" switched off cylinders, the pressure $p_{d\,d}$ for the camshaft drive, auxiliary equipment, and ventilation losses, as well as the average pressure $p_{q\,d}$ for heat losses in the switched off "k" cylinders is:

$$p_{md} = \frac{i-k}{i} p_{td} + \frac{k}{i} p''_{td} + \frac{i-k}{i} p'_{gd} + \frac{k}{i} p''_{gd} + p_{dd} + \frac{k}{i} p_{qd} \tag{4}$$

Components of mechanical losses have been calculated using theoretical and empirical dependencies. Some of them have been partially identified experimentally.

Using these data, the dependence of the engine performance on the average effective pressure has been shown when throttling (Figure 2) and switching off the three cylinders of the engine (Figure 3) for $n = 1800$ rpm.

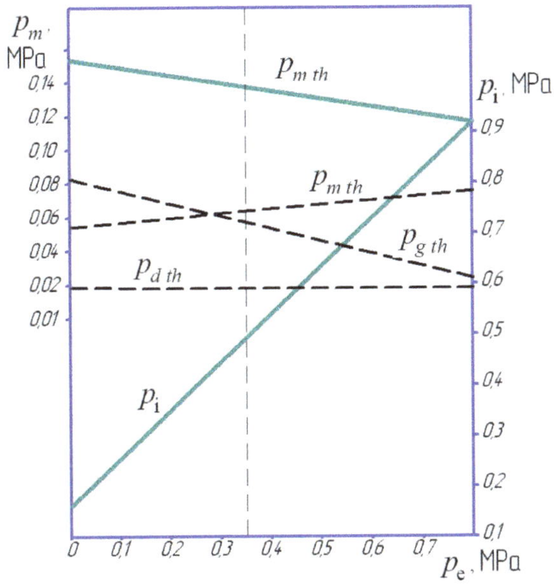

Figure 2. The dependencies of engine parameters on average effective pressure during throttling (n = 1800 rpm).

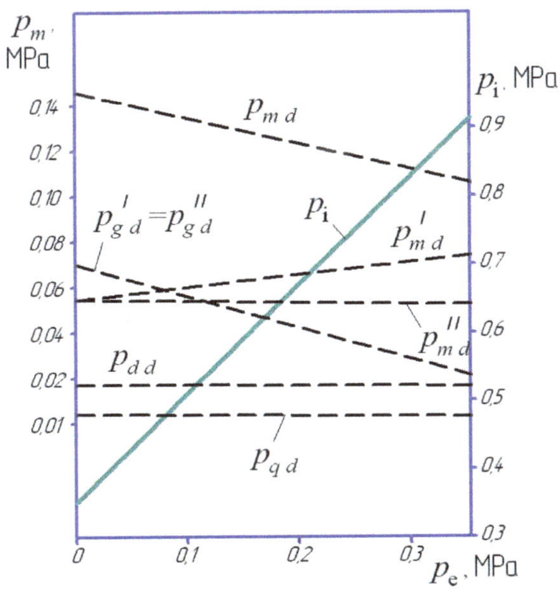

Figure 3. The dependencies of engine parameters on average effective pressure when three cylinders are switched off (n = 1800 rpm).

As can be seen from the comparison of the dependencies of $p_{md}(p_{eth})$ (Figure 3) and p_{mth} (p_{ed}) (Figure 2), the same $p_{e,}$ mechanical losses when the cylinders are switched off are smaller compared with the throttling. For p_e = 0 it has been obtained: p_{mth} = 0.153 MPa, p_{md} = 0.145 MPa, and for p_e = 0.35 MPa: p_{mth} = 0.138 MPa, p_{md} = 0.109 MPa. Thus, with the transition to the combined method of power regulation of gasoline engine, fuel economy improvements can be expected as a result of reducing the internal losses.

Using the method of the volume balance, a mathematical model of the engine operating process with cylinders switched off has been developed [23,41,42]. The adequacy of the calculated results has been checked by the indication of operating processes in switched-off and working cylinders.

The mathematical model to evaluate vehicle fuel consumption and pollution emission in a driving cycle using the combined method of power regulation has been developed in scientific papers. Fuel economy modeling is closely linked to vehicle weight, which is increasing by 4.7% annually on cars sold in the Chinese market since 2015 [43], therefore, modeling of exhaust emissions and energy sustainability is performed [44], which is heavily dependent on the interests of energy distribution companies [45]. The assessment of pollutions emissions is related to the promotion of green vehicle use [46], both with an efficient quantified risk assessment [47] resulting from the use of floating vehicle data [48] and on the optimization of control strategy to reduce fuel consumption [49], fuel consumption software (PΔP) [50], interactions between vehicles and pedestrians [51], and vehicle routing problems [52]. Ehsani et al. provides a clear model for the dependence of fuel consumption on carbon dioxide emissions [53]. Applying such a model would reduce freight transport exhaust and improve fuel economy [54]. However, such links are very much dependent on traffic jams if we look at the urban situation [55].

The theoretical analysis shows that the use of the combined power regulation method will improve fuel economy of spark ignition engines compared with throttling in partial load modes and idling modes.

3. Materials and Methodology

3.1. Working out the Fuel Supply System

The experimental model of a fuel supply system to implement the combined power regulation method is based on a Bosch "Motronik ML 4.1" ("Robert Bosch GmbH", Gerlingen, Germany) fuel supply system in the mechanical drive throttle. The scheme of the experimental system is shown in Figure 4 [20]. The main algorithm of its operation implies the termination of the fuel supply to the switched-off cylinders without changing the gas exchange in small loads and idling modes. The engine is equipped with the new system without structural changes in the electronic control unit of the injection system and engine ignition. It can, therefore, be installed on vehicles in operation. This system of power regulation has five patents in the Ukraine for utility models [35].

The system works as follows: an electromagnet (31), whose operation is controlled by the electronic control unit of the nozzles (11), is added to the fuel supply system of the engine with a mechanical cable throttle actuator. The core of the electromagnet is connected to the tension roller (27) through which the control cable passes to the throttle (28).

With partial loads and idling mode running, the electric current of the electromagnet (30) is activated by the command of the electronic control unit (11) to the coil relay switching the electromagnet. When the electromagnet (31) switches on, its core moves the tension roller (27) to a predetermined value Δl. As a result of it the opening angle of the throttle (17) is changed to a predetermined value according to the engine operation with the cylinders switched off.

At the same time, the electronic control unit of the nozzles (11) receives the signals from the rotational speed sensors (24) and the position of the throttle (16) disables the nozzle assembly (10) by re-directing the electric current to the ballast resistance (9). As a result, the engine does not enter the emergency operating mode.

With an increase in the engine load to a mode where the operation with a subset of cylinders becomes less economical with respect to fuel consumption than with all cylinders, at the same crankshaft rotation frequency, the electronic control unit of the nozzles (11), according to the signal of the position sensor of the throttle valve (16), connects the block of the inactive nozzles (10) to the fuel supply system.

At the same time, electromagnet (31) controls the tension roller (27). The angle of opening the throttle (17) changes to the value corresponding to the engine operation with all cylinders.

When reducing the load of the engine or during transition to the idling mode, when the engine is more efficient at switching off the cylinder group, there is a reverse process of disconnecting the block of injectors and the inclusion in the chain of the ballast resistance. The angle of opening the throttle changes simultaneously to the value corresponding to the engine operation for part of cylinders.

The cold start of the engine and its heating occurs due to the engine operation with all cylinders. This is provided by the air temperature sensor (19) and the temperature sensor of the engine (22), the signal of which comes to the electronic control unit of the nozzles (11). The group of switching off cylinders can also be changed.

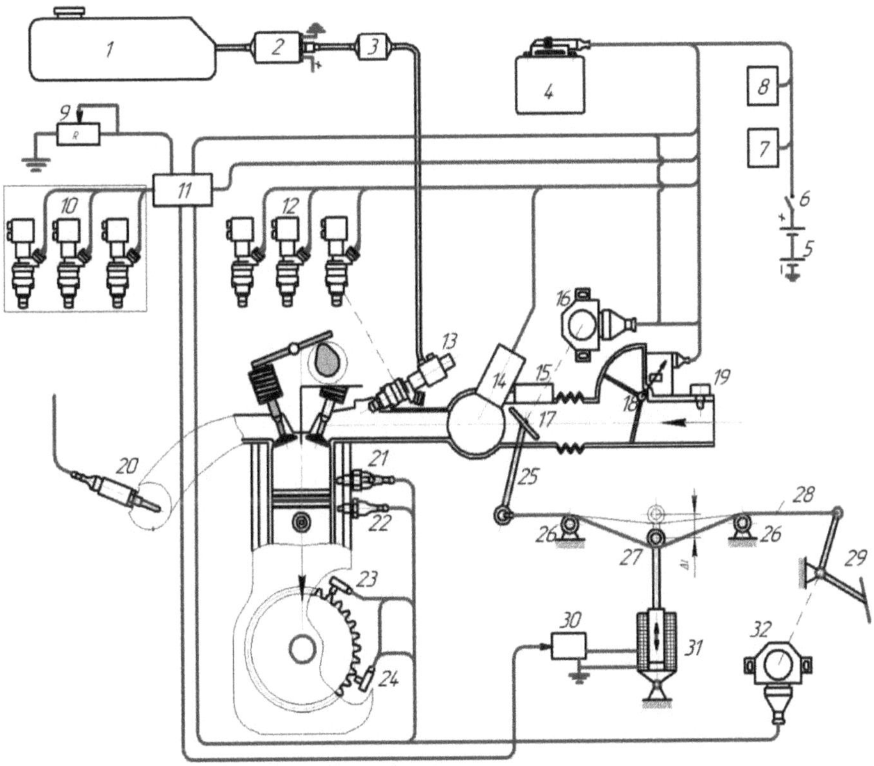

Figure 4. The scheme of experimental power system: 1—fuel tank, 2—electric fuel pump, 3—fuel filter, 4—electronic control device (microprocessor), 5—battery, 6—ignition switch, 7—main relay, 8—pump relay, 9—ballast resistance, 10—disconnected injectors, 11—electronic unit for controlling the nozzles, 12—non-inverting nozzles, 13—fuel distributor, 14—cold start system, 15—idling speed stabilization device, 16—sensor of position and a value of accelerating throttle valve, 17—throttle valve, 18—air vent, 19—air temperature sensor, 20—λ-sensor, 21—thermal timer, 22—engine temperature sensor, 23—crankshaft angle sensor, 24—speed sensor, 25—throttle control lever, 26—locking roller, 27—tension roller, 28—throttle valve drive cable, 29—gas pedal, 30—relay for electromagnet switching, 31—electromagnet, 32—load sensor.

3.2. Working out the Experimental Installation

For experimental research, an experimental installation has been developed consisting of a six-cylinder inline Opel C30 LE engine. The technical characteristic, given in Table 1.

The engine has been installed on a universal brake stand SGEU-100 with an electric brake AKB-92-4, the power of 100 kW and a maximum rotation 3000 rpm. The installation enables testing the engine in wide limits of external loads. The schematic diagram of the experimental installation with measuring devices is shown in Figure 5.

The experimental prototype of the fuel supply system to implement the combined power regulation method is installed on the engine (Figure 6).

Separate exhaust pipes for a group of three cylinders that constantly work and a group of switched-off cylinders have been installed on the engine. In the outlet pipeline of a continuously working cylinder group, a three-component catalytic converter manufactured by the Arvin Tefh company (Madison, IN, USA) has been installed on the 1.5 L engine. This is only the working cylinder group.

Table 1. Technical characteristics of the Opel C30 LE engine.

Parameters	Value
Type and location of cylinders	gasoline, inline
Number of cycles	4-stroke
Order of cylinders	1-5-3-6-2-4
Diameter of cylinder and piston stroke, mm	95 × 69.8
Working volume, cm^3	2969
Compression ratio	8.6
Nominal power kW/at the frequency of rotation, rpm	115/5400
Maximum torque, Nm/at the frequency of rotation, rpm	230/4100
Minimum rotational speed of the crankshaft in idle mode, rpm	750, … , 820
Number of valves per cylinder	2
Fuel injection system	Motronik ML4.1
Fuel	Automotive gasoline A-95

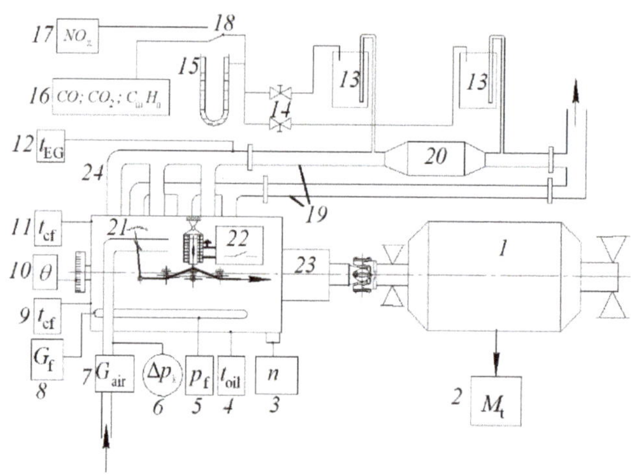

Figure 5. Schematic diagram of the experimental installation: 1—stand SGEU-100 with an electric brake AKB-92-4; 2—weighing head VKM-57; 3—device to measure the frequency of crankshaft rotation of the engine; 4—oil temperature sensor; 5—pressure sensor in the fuel line; 6—dilution sensor in the intake manifold; 7—air flow meter RG-600; 8—weight measuring device for fuel consumption; 9—coolant temperature sensor at the exit from the block of cylinders; 10—device to measure the ignition angle; 11—temperature sensor of the coolant at the inlet to the block of cylinders (temperature sensor for cooling fluid); 12—EG temperature sensor; 13—dryers of VH samples; 14—throttle valve; 15—liquid U-shaped pressure gauge; 16—SINCRO EGA 2001 gas analyzer; 17—344 HL–01 gas analyzer; 18—switch of EG flow between gas analyzers; 19—exhaust pipe of the group of cylinders which are switched off; 20—three-component catalytic converter; 21—device to measure the angle of opening the throttle valve; 22—experimental installation to switch off a group of cylinders; 23—gearbox of the Opel Omega A3.0; 24—the exhaust pipe of a group of cylinders which constantly work.

Figure 6. The fuel supply system for the Opel C30 LE engine.

Measurement equipment accuracy is presented in Table 2.

Table 2. Equipment and measurement accuracy.

Parameter	Dimension	Measuring Equipment	Measurement Error
Engine speed	rpm	Electronic digital frequency meter F 5035	±1
Torque	Nm	Weight head VKM-57	±0.69
Fuel and air consumption time	s	Electronic stopwatch	0.01
Fuel consumption	g	Electronic scales VNM-3/6	±0.5
Air flow	m^3	Gas meter "РГ–600"	0.1
Air pressure	Pa	Barometer, aneroid	±0.5
Air temperature	°C	Mercury thermometer	±0.1
SOI	degree	Strobe DA-5100	±1
Vacuum in intake manifold	kPa	Model vacuum gauge "OBB1–160"	±1
Coolant temperature at the outlet from the cylinder block	°C	Gauge thermometer	±1
Coolant temperature at the inlet to the cylinder block and oil temperature	°C	Logometer	±5
Oil pressure in the lubrication system	kgf/cm^2	Pressure gauge	0.1
Exhaust temperature	°C	Potentiometer EPP-09	±10
Concentrations in the exhaust gas: carbon monoxide CO	%	Infrared type gas analyzer: SINCRO EGA 2001 C	±0.01
carbon dioxide CO_2	%		±0.01
hydrocarbon HC	ppm		±1
Exhaust concentrations of nitrogen oxides NO_x	ppm	Chemiluminescent gas analyzer 344-HL-14	±10

4. Results and Discussion

The efficiency of the combined method as for fuel economy improvement has been assessed on the load characteristics in the load area, where the engine can operate with throttling six cylinders or with throttling three cylinders and three cylinders are switched off.

Figure 7 shows the load characteristics of the engine for its operation with three and six cylinders at n = 2600 rpm. Hence, it is evident that throughout all loads from idling modes and before the operation of the enrichment device, the stoichiometric composition of the fuel and air mixture is maintained, regardless of the number of working cylinders (λ = 1.0). When three cylinders are switched off in this high-speed mode, the maximum torque of the engine M_t decreases from 191 Nm to 84 Nm. For loads up to 65 Nm for three-cylinder engine performance there is an improvement in fuel economy compared to the operation with all cylinders, hourly G_f and brake specific fuel consumption (*BSFC*) is reduced by an average of 13%. At the same time, load reduction fuel economy improves. In idling modes for fuelling a three-cylinder engine, the fuel economy is about 19%. When the load M_t = 65 Nm, fuel consumption for the engine running on six and three cylinders is the same. At a larger load the engine operation with a group of switched off cylinders becomes less economical and therefore inappropriate. The operation of the engine with three cylinders compared to the operation with six cylinders is characterized by a larger opening of the throttle ϕ_{dr}, the lower thinning of Δp_k, and larger volumetric efficiency of engine η_v, with practically the same air flow G_{air}.

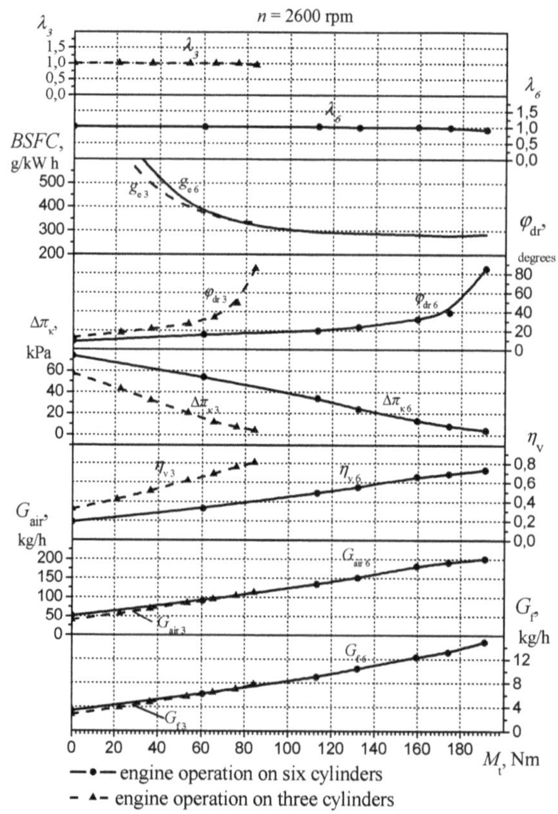

Figure 7. Load characteristics of the Opel C30 LE engine at n = 2600 rpm operating with three and six cylinders.

When determining the load characteristics, the concentration of pollutions in exhaust gases of the engine has been measured before and after the converter. The obtained characteristics are shown in Figure 8.

An analysis of the environmental performance of the engine indicates a significant increase in nitrogen oxide NO_x concentration when the engine runs on three cylinders compared with the engine running on six cylinders. When applying a three-component catalytic converter, the NO_x concentration in both cases is significantly reduced to practically the same level. Similar results were obtained by Ozhan et al. after years of research using multilevel flow modeling [56]. Concentrations of carbon oxide CO, carbon dioxide CO_2, and hydrocarbons *HC* in the operation of the engine with catalytic converters with different numbers of working cylinders are also practically the same.

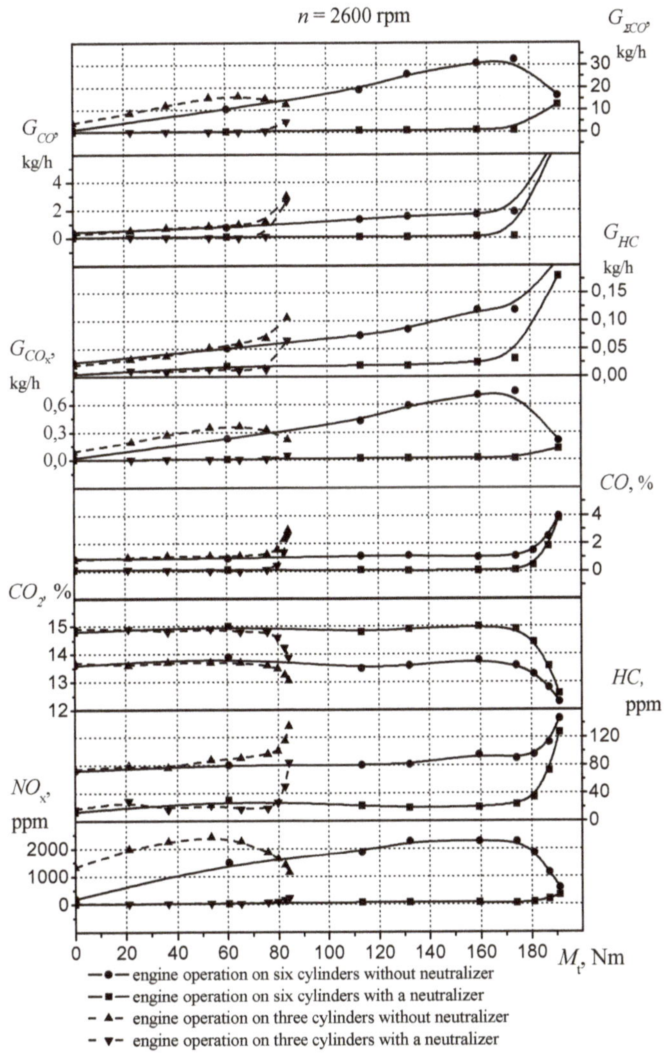

Figure 8. Load characteristics of the Opel C30 LE engine at n = 2600 rpm operating with three and six cylinders (environmental indicators).

The improvement in fuel economy for the operation with three cylinders has been obtained in other high-speed modes (Figure 9); the engine load was low (M_t = 40 Nm).

One of the main issues when adjusting the power of a gasoline engine by the combined method is to determine the rules of changing the position of the throttle valve when switching off and switching on the cylinder group. It provides a non-shock process of changing the number of working cylinders. To establish such a pattern, an experimentally determined series of load characteristics with the measurement of the engine torque, fuel consumption, and the position of the throttle valve for each speed mode, a load mode has been established. The results obtained are similar to those of Fleischman et al. bus tests, where the test was conducted under roadworthiness test conditions [57]. When it is exceeded, the operation of the engine with the switched-off cylinders is inappropriate in terms of fuel economy.

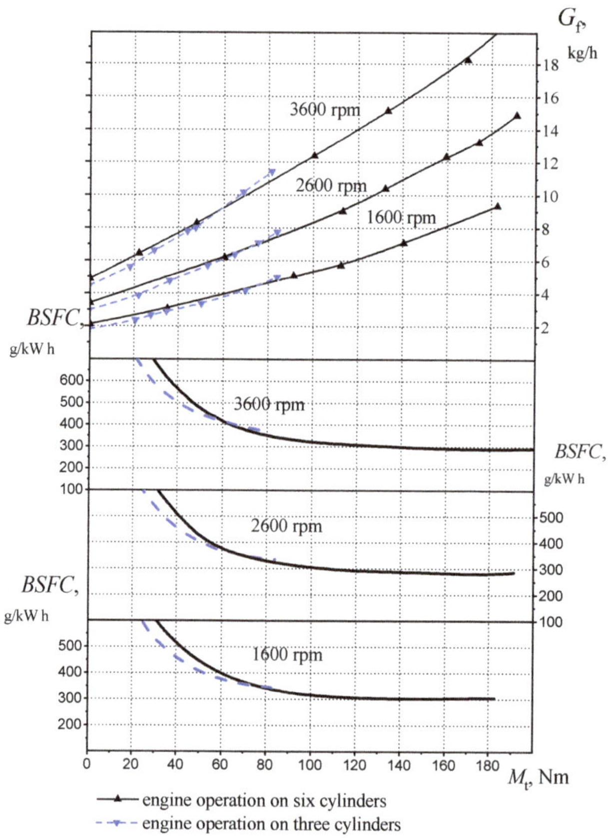

Figure 9. Load characteristics of the Opel C30 LE engine at different speeds operating with three and six cylinders.

Figure 10 shows the experimental dependencies of the hourly flow of petrol G_p and the opening angle of the throttle valve ϕ_{dr} on the torque M_t of a six-cylinder Opel C30 LE petrol engine with a distributed injection system with a feedback with six and three cylinders at different speeds.

The dependencies show that when the group of cylinders is switched off and on, the position of the throttle valve does not change, and the torque will vary several times.

As can be seen from the characteristics shown, the same torque when switching off and on a group of cylinders can be obtained at different ratios of the opening angle of the throttle valve.

The analysis of experimentally determined characteristics (Figure 11) shows that for the engine in all high-speed modes to obtain a practically unbreakable transition from the operation with three cylinders to the operation with six cylinders, and vice versa, it is advisable to do the transition at a torque moment M_t close to 50 Nm and a ratio of the opening angle of the throttle valve of 1.75.

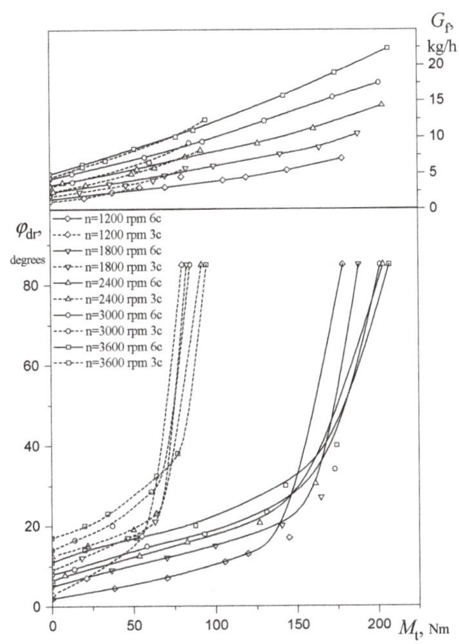

Figure 10. The dependencies of hourly consumption of gasoline and angles of opening throttle flaps on the torque of the Opel C30 LE engine. The 6c shows that engine operation on six cylinders; 3c: engine operation on three cylinders.

One of the main factors influencing the efficiency of the combined method for power regulation of a multi-cylinder petrol engine is the duration of switching off and on the cylinder group.

The exact duration values of the transition process when switching off and on the cylinder group have been determined by indicating the operating process in the cylinder. Figure 11 shows a fragment of oscillograms of the operating process in cylinder A, when fuel is supplied to the cylinder; B, when fuel is not supplied to the cylinder. Oscillograms have been recorded at a load of M_t = 50 Nm, (n = 2000 rpm).

From these fragments it can be seen that both switching on and switching off the cylinder have a transition process in 1–2 operating cycles and almost without gradual increase or decrease of the indicated pressure. When the cylinders are switched on (Figure 11A), the full operating cycle has been restored for approximately 0.15 s. At the same time, the maximum combustion pressure, starting from the second cycle, remains unchanged during all subsequent cycles within the uneven values of cylinder operation.

A similar process occurs with switching off the cylinders (Figure 11B). After the fuel injector stops working, the first operating cycle is without breaking the parameters of the operating process. The next one is without the combustion process. The duration of the transition process when switching off the cylinders lasts for about 0.1 s.

To prove a short-term nature of the transition process, when cylinders are turned on and off, the oscillograms of the operating process during acceleration and deceleration of the engine (Figure 12) at loads have been recorded in unstable modes. The brake stand has created loads according to its speed characteristics (Figure 13).

Additionally, Figure 12 shows the oscillograms of the throttle position change ϕ_{dr}, the dilution in the inlet line Δp_k, the operation of the fuel nozzle, the moment of the TDP, and the moment of a spark jump.

With the gradual opening of the throttle, the rotational speed "n", the maximum pressure in the switched off cylinder "p" and, therefore, the load on the engine increase. At the throttle angle the cylinders are switched off. When the angle is larger, the engine operation with the cylinder part becomes inefficient. At the same time, the angle of opening the throttle is reduced to avoid a spin-like torque change. This causes an increase in the dilution in the inlet pipeline Δp_k, which results in a decrease of the maximum pressure "p" in the next cycle after the cylinders are switched on. In this cycle, the combustion process still does not occur. In subsequent cycles the combustion process is restored and the maximum pressure increases as a result of further opening of the throttle.

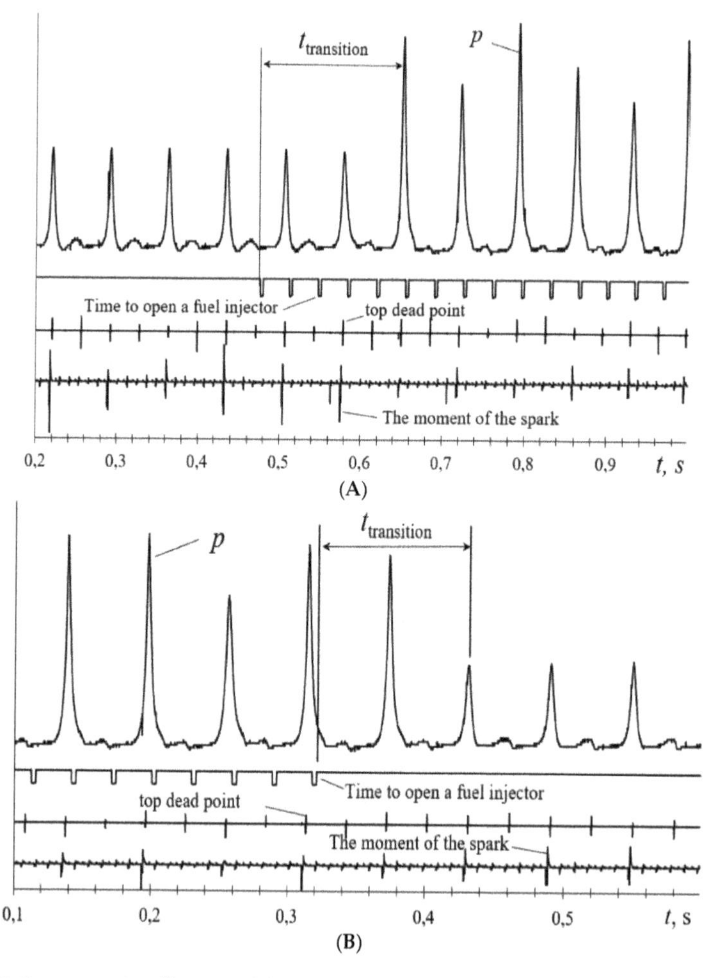

Figure 11. Fragments of oscillograms of the transition process of the Opel C30 LE engine when: (**A**) fuel is supplied to the cylinder, and (**B**) fuel is not supplied to the cylinder.

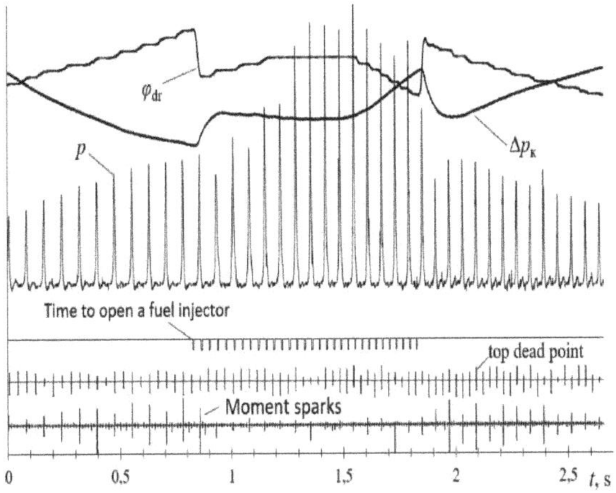

Figure 12. Fragments of oscillograms of the transition processes of the engine when a cylinder is switched on and off during acceleration and deceleration φ_r: the position of the rheostat to control the stand load.

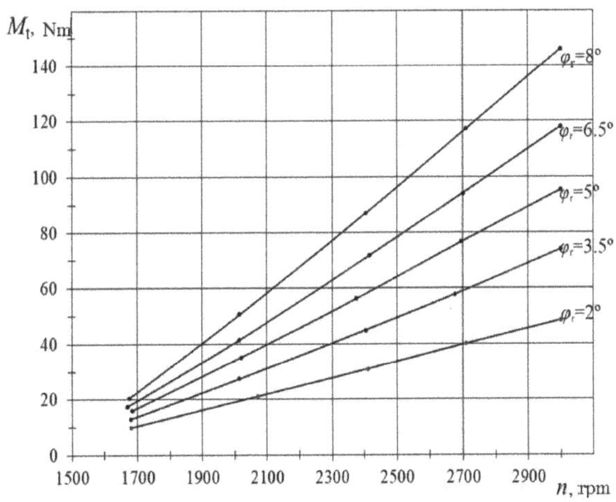

Figure 13. Speed characteristics of the SGEU-100 brake stand.

When the throttle valve closes, a reverse process occurs. In the process of switching the opening angle of the throttle valve ϕ_{dr} sharply decreases. There is a decrease of the dilution in the inlet line. In the next cycle after the cylinders are switched off, the combustion characterized by low pressure in the cylinder does not occur. In the first cycle after the cylinders are switched off, the maximum pressure becomes even lower than in subsequent cycles. It is due to the fact that the pressure reduction wave from the throttle valve has not yet been able to reach the switched off cylinders.

Such a short-term transition process is facilitated by the instantaneous reduction of the dilution when the position of the throttle valve is changed. Due to this all fuel residues from the inlet valve are absorbed into the cylinder.

With further closure of the throttle valve, the dilution increase and the maximum cylinder pressure gradually decreases as a result of volumetric efficiency of engine reduction. The oscillograms recorded with different acceleration and deceleration rates show similar processes of switching the cylinders on and off.

The oscillograms of transition processes recorded in various modes of the engine show that the duration of the transition processes when the cylinder is switched off and on the petrol injection is insignificant and is within 1–2 operating cycles. It enables to increase the speed of the transition process, to improve the engine power regulation. It is one of the reasons to improve fuel economy when applying the combined method of power regulation for a multi-cylinder petrol engine.

5. Conclusions

Experimental studies of a six-cylinder engine have shown that when using the combined method of power regulation. The engine runs steady in all load and high-speed modes. The experimentally established fuel economy at the idling mode is about 19%, and the average economy of loading modes is about 13%.

The application of the combined power regulation method almost does not affect the environmental performance of the engine with a three-component catalytic converter installed in the production system.

The shock-free process of switching off and on a group of cylinders may provide a change in the position of the throttle in these processes with a certain ratio. In this case, the cyclic supply of gasoline also varies with the corresponding value. The composition of fuel and air mixture does not change.

For the Opel C30 LE engine in all high-speed modes to obtain an almost unbreakable transition from the operation with three cylinders to the operation with six cylinders and vice versa, it is advisable to do the transition at a torque close to 50 Nm and the ratio of the throttle opening angle at 1.75.

It has been established that the duration of switching on and switching off the cylinder group does not exceed 1–2 operating cycles. It cannot significantly affect fuel economy.

Author Contributions: Conceptualization: Y.G. and V.M.; methodology: I.G.; software: O.S.; validation: J.M., A.R. and Y.S.; formal analysis: Y.G.; investigation: O.S. and I.G; resources: J.M. and A.R.; data curation: V.M.; writing—original draft preparation: Y.G. and O.S.; writing—review and editing: I.G., V.M., and Y.S.; visualization: A.R.; supervision: Y.G. and V.M.; project administration: J.M.; funding acquisition: J.M. and A.R. All authors have read and agreed to the published version of the manuscript.

Funding: This research received no external funding.

Conflicts of Interest: The authors declare no conflict of interest.

References

1. Fu, J.; Shu, J.; Zhou, F.; Zhong, L.; Liu, J.; Deng, B. Quantitative analysis on the effects of compression ratio and operating parameters on the thermodynamic performance of spark ignition liquefied methane gas engine at lean burn mode. *Fuel* **2020**, *263*, 116692. [CrossRef]
2. Rosero, F.; Fonseca, N.; López, J.-M.; Casanova, J. Real-world fuel efficiency and emissions from an urban diesel bus engine under transient operating conditions. *Appl. Energy* **2020**, *261*, 114442. [CrossRef]
3. Karthickeyan, V.; Thiyagarajan, S.; Ashok, B.; Edwin Geo, V.; Azad, A.K. Experimental investigation of pomegranate oil methyl ester in ceramic coated engine at different operating condition in direct injection diesel engine with energy and exergy analysis. *Energy Convers. Manag.* **2020**, *205*, 112334. [CrossRef]
4. Szabados, G.; Bereczky, Á. Experimental investigation of physicochemical properties of diesel, biodiesel and TBK-biodiesel fuels and combustion and emission analysis in CI internal combustion engine. *Renew. Energy* **2018**, *121*, 568–578. [CrossRef]
5. Kumar Sharma, P.; Sharma, D.; Lal Soni, S.; Jhalani, A.; Singh, D.; Sharma, S. Energy, exergy, and emission analysis of a hydroxyl fueled compression ignition engine under dual fuel mode. *Fuel* **2020**, *265*, 116923. [CrossRef]
6. Kivevele, T.T.; Mbarawa, M.M.; Bereczky, Á.; Zöldy, M. Evaluation of the Oxidation Stability of Biodiesel Produced from Moringa oleifera Oil. *Energy Fuels* **2011**, *25*, 5416–5421. [CrossRef]
7. Zhen, X.; Wang, Y.; Liu, D. Bio-butanol as a new generation of clean alternative fuel for SI (spark ignition) and CI (compression ignition) engines. *Renew. Energy* **2020**, *147*, 2494–2521. [CrossRef]
8. Zoldy, M.; Hollo, A.; Thernesz, A. Butanol as a Diesel Extender Option for Internal Combustion Engines. In Proceedings of the SAE 2010 World Congress & Exhibition, Detroit, MI, USA, 13–15 April 2010.

9. Shirazi, S.A.; Abdollahipoor, B.; Windom, B.; Reardon, K.F.; Foust, T.D. Effects of blending C3-C4 alcohols on motor gasoline properties and performance of spark ignition engines: A review. *Fuel Process. Technol.* **2020**, *197*, 106194. [CrossRef]
10. Pan, M.; Qian, W.; Wei, H.; Feng, D.; Pan, J. Effects on performance and emissions of gasoline compression ignition engine over a wide range of internal exhaust gas recirculation rates under lean conditions. *Fuel* **2020**, *265*, 116881. [CrossRef]
11. Lijewski, P.; Fuc, P.; Dobrzynski, M.; Markiewicz, F. Exhaust emissions from small engines in handheld devices. *MATEC Web Conf.* **2017**, *118*, 00016. [CrossRef]
12. Silvestri, P.; Zoppi, M.; Molfino, R. Dynamic investigation on a new robotized vehicle for urban freight transport. *Simul. Model. Pract. Theory* **2019**, *96*, 101938. [CrossRef]
13. Behiri, W.; Belmokhtar-Berraf, S.; Chu, C. Urban freight transport using passenger rail network: Scientific issues and quantitative analysis. *Transp. Res. Part E Logist. Transp. Rev.* **2018**, *115*, 227–245. [CrossRef]
14. Schröder, S.; Liedtke, G.T. Towards an integrated multi-agent urban transport model of passenger and freight. *Res. Transp. Econ.* **2017**, *64*, 3–12. [CrossRef]
15. Bigazzi, A. Comparison of marginal and average emission factors for passenger transportation modes. *Appl. Energy* **2019**, *242*, 1460–1466. [CrossRef]
16. Chen, L.; Pan, J.; Liu, C.; Shu, G.; Wei, H. Effect of rapid combustion on engine performance and knocking characteristics under different spark strategy conditions. *Energy* **2020**, *192*, 116706. [CrossRef]
17. da Costa, R.B.R.; Valle, R.M.; Hernández, J.J.; Malaquias, A.C.T.; Coronado, C.J.R.; Pujatti, F.J.P. Experimental investigation on the potential of biogas/ethanol dual-fuel spark-ignition engine for power generation: Combustion, performance and pollutant emission analysis. *Appl. Energy* **2020**, *261*, 114438. [CrossRef]
18. Szwaja, S.; Ansari, E.; Rao, S.; Szwaja, M.; Grab-Rogalinski, K.; Naber, J.D.; Pyrc, M. Influence of exhaust residuals on combustion phases, exhaust toxic emission and fuel consumption from a natural gas fueled spark-ignition engine. *Energy Convers. Manag.* **2018**, *165*, 440–446. [CrossRef]
19. Gong, C.; Li, Z.; Yi, L.; Liu, F. Experimental investigation of equivalence ratio effects on combustion and emissions characteristics of an H2/methanol dual-injection engine under different spark timings. *Fuel* **2020**, *262*, 116463. [CrossRef]
20. Xu, Z.; Zhang, Y.; Di, H.; Shen, T. Combustion variation control strategy with thermal efficiency optimization for lean combustion in spark-ignition engines. *Appl. Energy* **2019**, *251*, 113329. [CrossRef]
21. Tsuboi, S.; Miyokawa, S.; Matsuda, M.; Yokomori, T.; Iida, N. Influence of spark discharge characteristics on ignition and combustion process and the lean operation limit in a spark ignition engine. *Appl. Energy* **2019**, *250*, 617–632. [CrossRef]
22. Lee, Z.; Kim, T.; Park, S.; Park, S. Review on spray, combustion, and emission characteristics of recent developed direct-injection spark ignition (DISI) engine system with multi-hole type injector. *Fuel* **2020**, *259*, 116209. [CrossRef]
23. Ghaderi Masouleh, M.; Keskinen, K.; Kaario, O.; Kahila, H.; Karimkashi, S.; Vuorinen, V. Modeling cycle-to-cycle variations in spark ignited combustion engines by scale-resolving simulations for different engine speeds. *Appl. Energy* **2019**, *250*, 801–820. [CrossRef]
24. Heywood, J.B. *Internal Combustion Engine Fundamentals*, 2nd ed.; McGraw-Hill Education: New York, NY, USA, 2018; ISBN 978-1-260-11610-6.
25. Korakianitis, T.; Namasivayam, A.M.; Crookes, R.J. Natural-gas fueled spark-ignition (SI) and compression-ignition (CI) engine performance and emissions. *Prog. Energy Combust. Sci.* **2011**, *37*, 89–112. [CrossRef]
26. Cocco Mariani, V.; Hennings Och, S.; dos Santos Coelho, L.; Domingues, E. Pressure prediction of a spark ignition single cylinder engine using optimized extreme learning machine models. *Appl. Energy* **2019**, *249*, 204–221. [CrossRef]
27. Palomo Guerrero, D.; Jiménez-Espadafor, F.J. Torsional system dynamics of low speed diesel engines based on instantaneous torque: Application to engine diagnosis. *Mech. Syst. Signal Process.* **2019**, *116*, 858–878. [CrossRef]
28. Weißenborn, E.; Bossmeyer, T.; Bertram, T. Adaptation of a zero-dimensional cylinder pressure model for diesel engines using the crankshaft rotational speed. *Mech. Syst. Signal Process.* **2011**, *25*, 1887–1910. [CrossRef]

29. Aono, T.; Saruwatari, M.; Furuya, J. Estimation of Engine Torque and Cylinder Pressure Index Based on Crankshaft Rotation Measurement. *IFAC Proc. Vol.* **2013**, *46*, 360–365. [CrossRef]
30. McKelvey, T.; Andersson, I. System identification of the crankshaft dynamics in a 5 cylinder internal combustion engine. *IFAC Proc. Vol.* **2006**, *39*, 440–445. [CrossRef]
31. Jia, B.; Mikalsen, R.; Smallbone, A.; Roskilly, A.P. A study and comparison of frictional losses in free-piston engine and crankshaft engines. *Appl. Therm. Eng.* **2018**, *140*, 217–224. [CrossRef]
32. Zhong, B.; Hong, W.; Xie, F.; Su, Y.; Li, X. Effect of initial start conditions on combustion and rotation characteristics of first firing cycle during direct-start process for a GDI engine. *Appl. Therm. Eng.* **2017**, *127*, 116–126. [CrossRef]
33. Pham, P.X.; Vo, D.Q.; Jazar, R.N. Development of fuel metering techniques for spark ignition engines. *Fuel* **2017**, *206*, 701–715. [CrossRef]
34. Duronio, F.; De Vita, A.; Montanaro, A.; Villante, C. Gasoline direct injection engines—A review of latest technologies and trends. Part 2. *Fuel* **2020**, *265*, 116947. [CrossRef]
35. Gritsuk, I.V.; Mateichyk, V.; Tsiuman, M.; Gutarevych, Y.; Smieszek, M.; Goridko, N. *Reducing Harmful Emissions of the Vehicular Engine by Rapid After-Start Heating of the Catalytic Converter Using Thermal Accumulator*; SAE Technical Paper; SAE International: Warrendale, PA, USA, 2018.
36. Zhao, J.; Xi, Q.; Wang, S.; Wang, S. Improving the partial-load fuel economy of 4-cylinder SI engines by combining variable valve timing and cylinder-deactivation through double intake manifolds. *Appl. Therm. Eng.* **2018**, *141*, 245–256. [CrossRef]
37. Morris, N.; Mohammadpour, M.; Rahmani, R.; Johns-Rahnejat, P.M.; Rahnejat, H.; Dowson, D. Effect of cylinder deactivation on tribological performance of piston compression ring and connecting rod bearing. *Tribol. Int.* **2018**, *120*, 243–254. [CrossRef]
38. Ihlemann, A.; Nitz, N. *Schaeffler_Kolloquium_Cylinder Deactivation—A technology with a Future or a Niche Application*; Schaeffler Colloquium 2014; SAE International: Warrendale, PA, USA, 2014.
39. Gritsuk, I.; Volkov, V.; Mateichyk, V.; Gutarevych, Y.; Tsiuman, M.; Goridko, N. The Evaluation of Vehicle Fuel Consumption and Harmful Emission Using the Heating System in a Driving Cycle. *SAE Int. J. Fuels Lubr.* **2017**, *10*, 236–248. [CrossRef]
40. Kukharonak, H.; Ivashko, V.; Pukalskas, S.; Rimkus, A.; Matijošius, J. Operation of a Spark-ignition Engine on Mixtures of Petrol and N-butanol. *Procedia Eng.* **2017**, *187*, 588–598. [CrossRef]
41. Zavadskas, E.K.; Čereška, A.; Matijošius, J.; Rimkus, A.; Bausys, R. Internal Combustion Engine Analysis of Energy Ecological Parameters by Neutrosophic MULTIMOORA and SWARA Methods. *Energies* **2019**, *12*, 1415. [CrossRef]
42. Rimkus, A.; Žaglinskis, J.; Stravinskas, S.; Rapalis, P.; Matijošius, J.; Bereczky, Á. Research on the Combustion, Energy and Emission Parameters of Various Concentration Blends of Hydrotreated Vegetable Oil Biofuel and Diesel Fuel in a Compression-Ignition Engine. *Energies* **2019**, *12*, 2978. [CrossRef]
43. Wang, S.; Chen, K.; Zhao, F.; Hao, H. Technology pathways for complying with Corporate Average Fuel Consumption regulations up to 2030: A case study of China. *Appl. Energy* **2019**, *241*, 257–277. [CrossRef]
44. Zheng, Y.; Niu, S.; Shang, Y.; Shao, Z.; Jian, L. Integrating plug-in electric vehicles into power grids: A comprehensive review on power interaction mode, scheduling methodology and mathematical foundation. *Renew. Sustain. Energy Rev.* **2019**, *112*, 424–439. [CrossRef]
45. Zargarnezhad, S.; Dashti, R.; Ahmadi, R. Predicting vehicle fuel consumption in energy distribution companies using ANNs. *Transp. Res. Part Transp. Environ.* **2019**, *74*, 174–188. [CrossRef]
46. Turkensteen, M. The accuracy of carbon emission and fuel consumption computations in green vehicle routing. *Eur. J. Oper. Res.* **2017**, *262*, 647–659. [CrossRef]
47. Huang, W.; Fan, H.; Qian, Y. Modeling and efficient quantified risk assessment of haze causation system in China related to vehicle emissions with uncertainty consideration. *Sci. Total Environ.* **2019**, *668*, 74–83. [CrossRef] [PubMed]
48. Du, Y.; Wu, J.; Yang, S.; Zhou, L. Predicting vehicle fuel consumption patterns using floating vehicle data. *J. Environ. Sci.* **2017**, *59*, 24–29. [CrossRef] [PubMed]
49. Sawulski, J.; Ławryńczuk, M. Optimization of control strategy for a low fuel consumption vehicle engine. *Inf. Sci.* **2019**, *493*, 192–216. [CrossRef]
50. Smit, R. Development and performance of a new vehicle emissions and fuel consumption software (PΔP) with a high resolution in time and space. *Atmos. Pollut. Res.* **2013**, *4*, 336–345. [CrossRef]

51. Li, X.; Sun, J.-Q. Effect of interactions between vehicles and pedestrians on fuel consumption and emissions. *Phys. Stat. Mech. Appl.* **2014**, *416*, 661–675. [CrossRef]
52. Zhang, J.; Zhao, Y.; Xue, W.; Li, J. Vehicle routing problem with fuel consumption and carbon emission. *Int. J. Prod. Econ.* **2015**, *170*, 234–242. [CrossRef]
53. Ehsani, M.; Ahmadi, A.; Fadai, D. Modeling of vehicle fuel consumption and carbon dioxide emission in road transport. *Renew. Sustain. Energy Rev.* **2016**, *53*, 1638–1648. [CrossRef]
54. Malik, L.; Tiwari, G. Assessment of interstate freight vehicle characteristics and impact of future emission and fuel economy standards on their emissions in India. *Energy Policy* **2017**, *108*, 121–133. [CrossRef]
55. Maghrour Zefreh, M.; Torok, A. Theoretical Comparison of the Effects of Different Traffic Conditions on Urban Road Traffic Noise. *J. Adv. Transp.* **2018**, *2018*, 1–11. [CrossRef]
56. Ozhan, C.; Fuster, D.; Da Costa, P. Multi-scale flow simulation of automotive catalytic converters. *Chem. Eng. Sci.* **2014**, *116*, 161–171. [CrossRef]
57. Fleischman, R.; Amiel, R.; Czerwinski, J.; Mayer, A.; Tartakovsky, L. Buses retrofitting with diesel particle filters: Real-world fuel economy and roadworthiness test considerations. *J. Environ. Sci.* **2018**, *67*, 273–286. [CrossRef] [PubMed]

energies

MDPI

Article

Application of a Model-Based Controller for Improving Internal Combustion Engines Fuel Economy

Teresa Castiglione *, Pietropaolo Morrone, Luigi Falbo, Diego Perrone and Sergio Bova

Department of Mechanical, Energy and Management Engineering, University of Calabria, Ponte Bucci, Cubo 46C, 87036 Rende, Italy; pietropaolo.morrone@unical.it (P.M.); luigi.falbo@unical.it (L.F.); diego.perrone@unical.it (D.P.); sergio.bova@unical.it (S.B.)

* Correspondence: teresa.castiglione@unical.it; Tel.: +39-0984-494162

Received: 23 January 2020; Accepted: 20 February 2020; Published: 3 March 2020

Abstract: Improvements in internal combustion engine efficiency can be achieved with proper thermal management. In this work, a simulation tool for the preliminary analysis of the engine cooling control is developed and a model-based controller, which enforces the coolant flow rate by means of an electrically driven pump is presented. The controller optimizes the coolant flow rate under each engine operating condition to guarantee that the engine temperatures and the coolant boiling levels are kept inside prescribed constraints, which guarantees efficient and safe engine operation. The methodology is validated at the experimental test rig. Several control strategies are analyzed during a standard homologation cycle and a comparison of the proposed methodology and the adoption of the standard belt-driven pump is provided. The results show that, according to the control strategy requirements, a fuel consumption reduction of up to about 8% with respect to the traditional cooling system can be achieved over a whole driving cycle. This proves that the proposed methodology is a useful tool for appropriately cooling the engine under the whole range of possible operating conditions.

Keywords: spark-ignition engine; fuel economy; model predictive control; electric pump; engine thermal management

1. Introduction

Steadily increasing challenges for climate change are leading to the development of new technologies for sustainable mobility, which include powertrain electrification like hybrids, plug-in hybrids, and fully electric vehicles. These solutions provide a very interesting opportunity in terms of fuel-saving and air quality, especially in very congested areas; however, they have the drawback of being expensive and some environmental issues seen from a life-cycle assessment (LCA) point of view are still unresolved [1]. In this context, internal combustion engines, and in particular spark ignition (SI) engines, still offer significant developmental margins to meet future requirements, both in terms of CO_2 emissions reduction within the limits prescribed by the regulatory agencies [2,3] and in terms of engine performance and will presumably be widespread in the automotive industry for the next few decades [4].

Several technical solutions have been developed to optimize SI engines in terms of fuel consumption, performance, and emissions. Gasoline engines "downsizing" and supercharging boosting [5], innovative intake strategies [6–8], cooled external exhaust gas recirculation [9], and innovative after-treatment technologies [10,11] have been shown to improve the efficiency of spark-ignited gasoline engines. A further contribution to fuel consumption reduction is given by the adoption of an optimized engine thermal management [12,13]. In fact, with optimized thermal management, a contribution of about 3% over the total CO_2 decrease was estimated [5] and, if compared to different technological options, it is the most cost-effective by accounting for about 200 Euro/(km/L saved) [14].

The primary requirement of optimized thermal management is the reduction of the warm-up time, which, by speeding-up lubricant heating, reduces frictional losses [13–15] and improves engine efficiency as a direct consequence. Lower coolant flow rates than the ones delivered by the traditional engine cooling system are, therefore, needed in this case. On the contrary, higher coolant flow rates than the ones commonly used under low-speed, high-load conditions, can be effective in mitigating knock and, therefore, diminishing the use of more fuel-consuming strategies like spark retard and mixture enrichment [16]. Unfortunately, the regulation capabilities of a standard engine cooling system are restricted by the adoption of a crank-shaft driven pump and a wax thermostat valve. These devices cause engine overcooling for about 95% of its operating time [17], during which the single-phase forced convection takes place. The possibility of optimizing the cooling system relies, therefore, on the removal of the standard crankshaft-driven pump and the inclusion of electrically driven pumps and of advanced electronic actuators, associated with proper control strategies. Solutions, which include an electrically driven pump, whose control is based on empirical methods can be found in [18,19], where improvement in specific fuel consumption of up to 2% is achieved. The adoption of electrically driven pumps and actuators presents a further advantage: the possibility of moving towards nucleate boiling flow regimes and to achieve a precision cooling strategy, which can provide a reduction in warm-up time of about 18% and a remarkable reduction of the power supplied to the coolant pump (up to 1.85 kW) [20]. To this aim, more sophisticated control approaches than the empirical ones are needed for keeping the boiling regime under control.

A control method, which has gained attention in recent years, is model predictive control (MPC), which permits the control of multi-input multi-output (MIMO) systems with constraints in an optimized way [21] and has found wide application in many slow-dynamic areas of engine control [22,23]. MPC was primarily implemented on compression-ignition (CI) engines [24–26] and recent attempts on spark-ignited (SI) engines [27–29] have also been proposed. The working principle of the MPC is based on the definition of a quadratic cost function, which integrates the deviation of an engine parameter from its desired values over a period of time, which is the prediction horizon; the cost function is minimized by manipulating the actuator set-points; in such a way, the optimal control actions are determined. In particular, the MPC algorithm requires the use of an engine model, which determines the current state of the engine from the sensors signals; the model is linearized about the current states and is then used to determine new actuator set-points that meet the cost function requirements within prescribed constraints. Linearization is carried out off-line for a variety of operating points and the results of the optimization problem are stored in the onboard controller.

A systematic control approach for the optimal thermal management of the coolant flow rate in an SI engine, based on MPC, was proposed by Pizzonia et al. [30]. The controller is based on the development of an engine thermal model, which predicts the heat transfer regime occurring between the engine walls and the coolant for a given engine operating condition. The model also predicts the occurrence of nucleate boiling and defines an index, namely the NB_Index, which operates as a virtual sensor for the detection of the boiling intensity. The control actuator is an electric pump, which gives the possibility of operating with reduced coolant flow rates when compared to the ones delivered by the belt-driven pump. The controller can also operate higher coolant flow rates under low-speed, high-load conditions and can keep the pump running after engine switch-off to avoid after-boiling [31–33].

In this work, a simulation tool for the proposed controller is developed with the aim to preliminarily evaluate different control strategies and give an estimation of the controller's effectiveness in terms of constraint requirements and engine efficiency. The proposed engine simulator is validated at the experimental test rig during a standard homologation cycle. Then, a variety of controllers are proposed and the benefits in terms of fuel consumption reduction, when compared to the crankshaft-driven pump, are quantified.

The model developed for the controller design and the controller working principle are summarized in Section 2. A simulation algorithm, which predicts the controller behavior, is validated in Section 3. Different control strategies are developed and simulated during a New European Driving Cycle (NEDC) homologation cycle and the results in terms of fuel consumption reduction and CO_2 emissions are presented in Section 4. Several engine operating conditions different from the NEDC cycle are analyzed in Section 5. Finally, the achieved results are discussed in Section 6.

2. Materials and Methods

The proposed approach is based on the development of a controller, which is able to optimize the coolant flow rate in the presence of nucleate boiling. For this purpose, a lumped-parameters model for the thermal exchange between the coolant and the engine walls is developed. The model predicts the occurrence of nucleate boiling by defining a metric for nucleate boiling detection. An extensive description of the model and controller is included in [30,34]; however, both the model and the controller are briefly described in the following subsections for the sake of clarity.

In the present work, the operating scheme of the controller is implemented numerically, with the aim to preliminarily simulate its behavior under various engine operating conditions, to predict the effects of the control strategy on fuel consumption and eventually to improve the controller design. The controller is also implemented at the experimental test rig and is tested under real engine operation.

2.1. Engine Thermal Model

A zero-dimensional approach is adopted for modeling the heat transfer between the engine walls and the coolant; the model dynamically predicts the occurrence of nucleate boiling [34]. The spatial-averaged wall and coolant temperatures, $\dot{T}_W$ and $\dot{T}_C$ are obtained from the following energy conservation equations:

$$\dot{T}_W \dot{C}_W = \dot{Q}_g - \dot{Q}_C, \tag{1}$$

$$\dot{T}_C \dot{C}_C = \dot{Q}_C - \dot{Q}_r, \tag{2}$$

C_W and C_C are the engine and coolant thermal capacities. $\dot{Q}_g$, $\dot{Q}_C$, and $\dot{Q}_r$ are the heat transfer rates from the combustion gases to the metal, from the metal to the coolant, and from the coolant to the radiator (Figure 1).

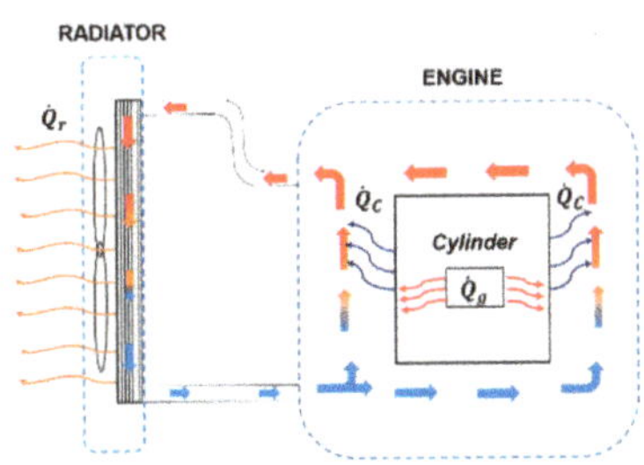

Figure 1. Thermal energy fluxes between the in-cylinder gases and the wall ($\dot{Q}_g$) between the walls and the coolant ($\dot{Q}_C$) and between the coolant and atmosphere ($\dot{Q}_r$) [35].

$\dot{Q}_g$ is estimated by Equation (3) developed in [34], which includes the fuel flow rate, $\dot{m}_f$, the coolant flow rate, $\dot{m}_c$, and the engine speed, N:

$$\dot{Q}_g = c \cdot N^{n1} \dot{m}_c^{n2} \dot{m}_f^{n}. \tag{3}$$

The coefficients c, n, n_1 and n_2 were calibrated at the experimental test rig.

The heat-transfer rate to the coolant is determined by the following correlation:

$$\dot{Q}_C = h_{mac}A(T_W - T_\infty) + h_{mic}A_{nb}(T_W - T_{sat}) \tag{4}$$

which takes into account forced convection and nucleate boiling. The heat transfer coefficient, h_{mac}, due to forced convection is modeled through the well-known Dittus–Boelter correlation [36], while for h_{mic}, due to nucleate boiling, the Chen approach is used [37]. In Equation (4), T_W is the metal temperature, T_∞ is the bulk flow temperature, T_{sat} is the coolant saturation temperature at the given pressure, A is the total heat transfer area and A_{nb} is the percentage of the wall area affected by the nucleate boiling. The latter is calculated as:

$$\begin{cases} A_{nb} = \frac{(q_W - q_{ONB})}{q_W} \cdot A & q_W > q_{ONB} \\ A_{nb} = 0 & q_W < q_{ONB} \end{cases}, \tag{5}$$

where q_W is the actual thermal flux through the engine walls and q_{ONB} is the threshold thermal flux, which determines the nucleate boiling onset [34].

The heat transfer rate from the coolant to the atmosphere through the radiator is computed as:

$$\dot{Q}_r = \dot{m}_c c_p (T_{C,out} - T_{C,in}) \tag{6}$$

where $T_{C,out}$ and $T_{C,in}$ are the coolant temperatures at the radiator inlet (engine-outlet) and radiator outlet (engine-inlet), respectively, and c_p is the coolant specific heat.

Finally, an index, which indicates if boiling occurs, is defined:

$$NB\_Index = \frac{q_W - q_{ONB}}{q_{ONB}}. \tag{7}$$

Hence, if NB_Index ≤ 0 no boiling occurs; if NB_Index ~ 0 nucleate boiling occurs; for NB_Index ~ 1 saturated boiling conditions are reached. Details on the modeling equations for nucleate boiling are included in [34].

2.2. Controller Design

The coolant flow rate is the variable selected for the optimization of the engine thermal management. Other quantities, which define the engine operating point, such as the engine speed and load, are determined by the drivers' needs; therefore, they cannot be known a priori and, from the control point of view, are considered as disturbances. However, these quantities are measured onboard and are useful for the proper functioning of the controller.

Figure 2 shows the working scheme of the controller in real-time operation at the experimental test rig. Measurements of coolant temperature at the engine outlet, of fuel flow rate and engine speed are usually available on modern engines; in addition, the proposed controller requires the measurement of coolant pressure for the detection of nucleate boiling. At each time step, the signals from transducers are sent to the data acquisition system (DAQ) and to the computing facility, where the controller is implemented. The model is used to calculate the wall and coolant temperatures, which, in turn, are used as controller inputs for the calculation of the correction to be applied to the actual coolant flow rate. This correction is then converted into a proper voltage, and hence into a proper rotational speed for the electrically driven pump, which delivers the corrected coolant flow rate to the engine. The working principle onboard a vehicle in real-time is very similar as the computing facility and the DAQ are substituted by the electronic control unit (ECU).

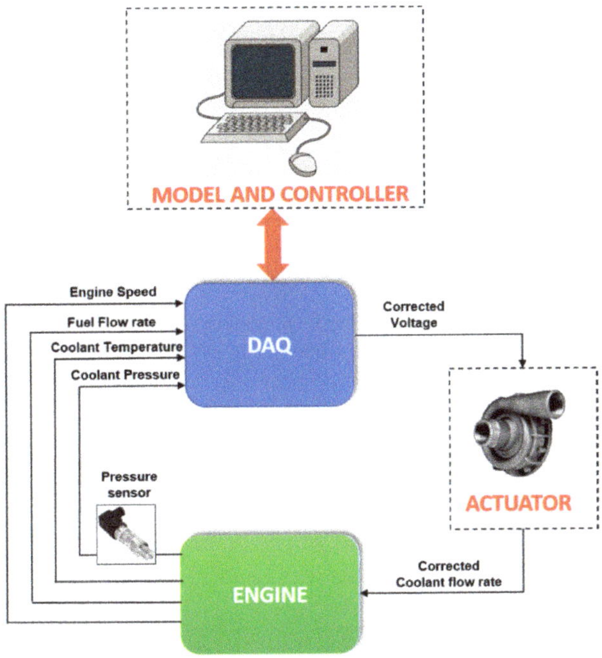

Figure 2. Operating scheme of the controller at the experimental test rig.

Figure 3 enters in more detail regarding the interaction between the model and the controller. The signals, out from engine transducers, are sent to the model, which computes the wall and coolant temperatures (T_W, T_C) for that particular engine operating point (Equations (1) and (2)).

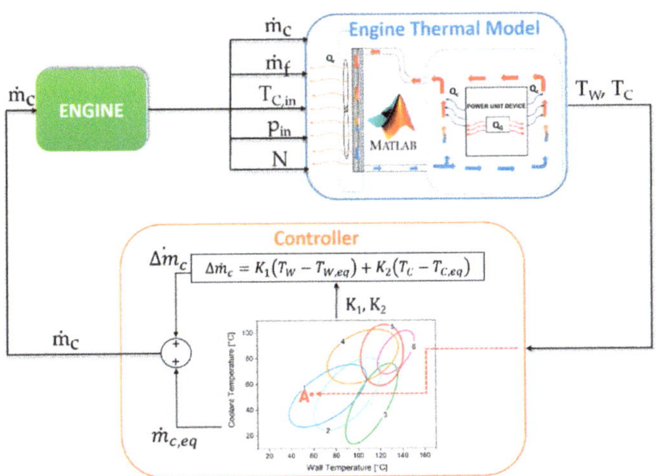

Figure 3. Schematic of the interaction between the model and the controller. The variables computed by the model determine the state of the system and define the control stability region.

These values are used by the controller to select a control region in the plane (T_W, T_C), which defines the state of the system. The controller, in fact, enters the state space with a couple of values (T_W, T_C) (Point A in Figure 3) and selects the control region, which is approximated by an ellipsoid, where the point enters. Each ellipsoid is characterized by an equilibrium coolant flow rate, $\dot{m}_{c,eq}$, an equilibrium coolant temperature, $T_{C,eq}$, an equilibrium wall temperature, $T_{W,eq}$ and two-state feedback control gains, K_1 and K_2. These characteristic values are collected and read from a look-up table and the correction to the equilibrium coolant flow rate is calculated as:

$$\Delta\dot{m}_c = K_1(T_W - T_{W,eq}) + K_2(T_C - T_{C,eq}) \tag{8}$$

The number of control regions (ellipsoids) is determined in order to cover all the possible states of the systems and, hence, the entire range of engine operating conditions and to make sure that the behavior of the linearized model does not differ significantly from the non-linear one. The values of K_1, K_2, and the definition of the equilibrium values, which determine the coolant flow rate correction, are the result of an optimization problem, which is carried out off-line, during the controller design stage. The optimization problem includes the model linearization for an engine operating point, the definition of the constraints on input and output variables and the minimization of a cost function. In particular, constraints are defined for the disturbances and the output variables; for the disturbances, the constraints define the minimum and maximum expected values for engine load and speed; for the output variables, the constraints define the maximum and minimum allowed values of wall temperature, coolant temperature, and boiling level, which guarantee for instance engine reliability. Then, a cost function, which integrates the deviation of the engine parameters from the desired values is defined and the linear objective minimization problem is formulated. The solution returns the state-feedback controller gains, K_1 and K_2 and the controller stability region, whose shape, in the state space, is that of an ellipsoid. In conclusion, the controller acts with the aim to vary the coolant flow rate to guarantee that, as long as the disturbances are contained within prescribed constraints, the engine temperatures are kept inside the stability region of the controller, rejecting the disturbances variations and satisfying, therefore, the constraints on the output variables T_W and T_C and on the NB_Index.

The simulation tool developed in the present work operates following the scheme in Figure 3. In this case, the input data, instead of being read from the transducers, are read from an input file, which includes the fuel flow rate, $\dot{m}_f$, engine speed, N, coolant pressure, p_{in} and coolant temperature, $T_{C,in}$; these data are usually obtained from real engine operating conditions. The coolant flow rate, $\dot{m}_c$, is computed by the controller and is given as an input to the engine thermal model.

3. Experimental Validation

The controller is validated at the experimental test rig. Tests are carried out on a spark-ignition engine, whose displacement is 1.2 dm^3; it delivers a maximum power of 60 kW between 5000 and 6000 rpm. The test bench is equipped with an eddy current engine torque dynamometer, which controls engine speed and allows load variation by setting either torque or throttle position. The fuel consumption is measured through an AVL 733S metering system; coolant pressure and temperatures at engine inlet and exit are measured by means of pressure transducers and temperature sensors (PT100-type), respectively. Figure 4 shows the experimental test rig.

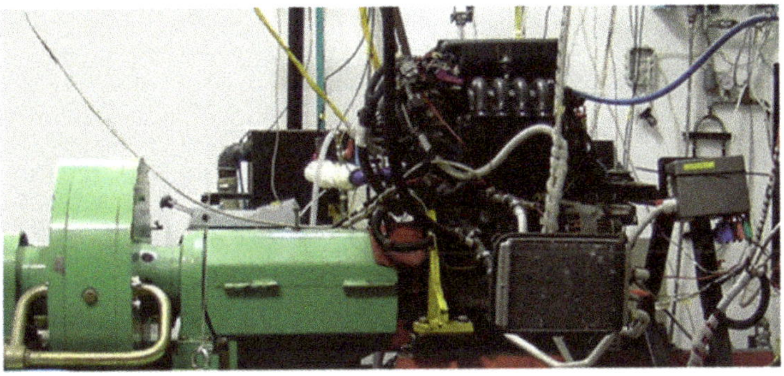

Figure 4. Experimental test rig for the 1.2 dm^3 engine.

The cooling circuit layout includes an electrically driven pump manufactured by OMP, which delivers a maximum coolant flow rate of 240 dm^3/min at 5000 rpm, maximum power of 1000 W, and overall efficiency of 67%. The coolant flow rate is measured by means of an electromagnetic flowmeter (Proline Promag E 100). Finally, K-type thermocouples are positioned in the cylinder head and block (Figure 5). The radiator and pressurized expansion tank are the production ones.

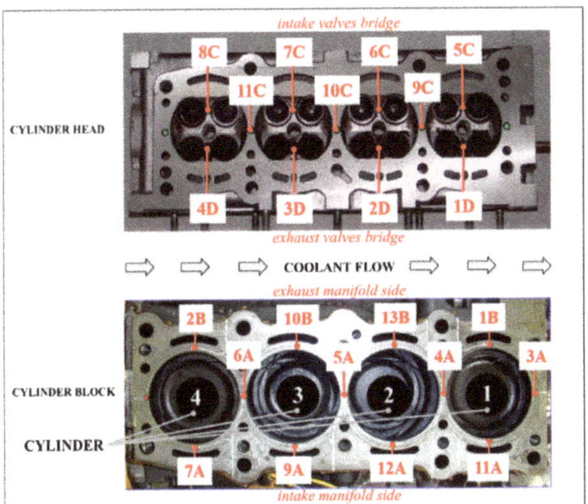

Figure 5. Thermocouple positions in engine head and block.

The NEDC is replicated at the stationary test bench. To this aim, the torque-speed couples of values adopted during the NEDC tests at the rolling bench have been adapted for the stationary test rig tests and have been changed into throttle opening-engine speed series of data, applied to the dyno. The fuel flow rate (in arbitrary units for the privacy policy of the engine constructor) and the engine speed from rolling test bed and stationary test rig are shown in Figure 6, where a high level of similarity between both cases is well visible.

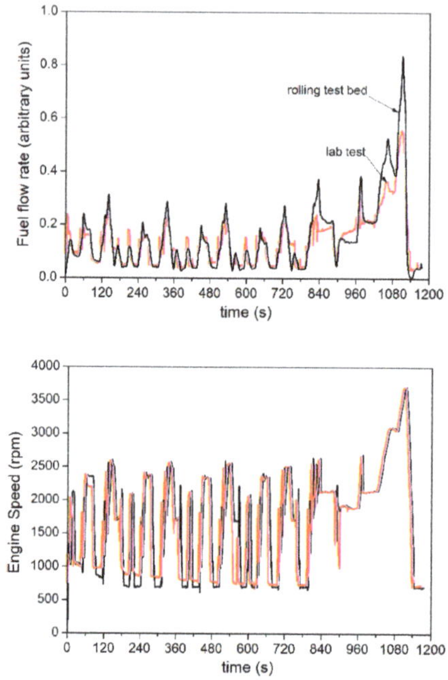

Figure 6. Comparison between the rolling test bed and stationary test rig fuel flow rate (**top**) and engine speed (**bottom**) for the 1.2 dm^3 engine.

Figure 7 shows a comparison between the average wall temperature calculated by the model and the thermocouple measurement in a specific engine head location when the controller is operating.

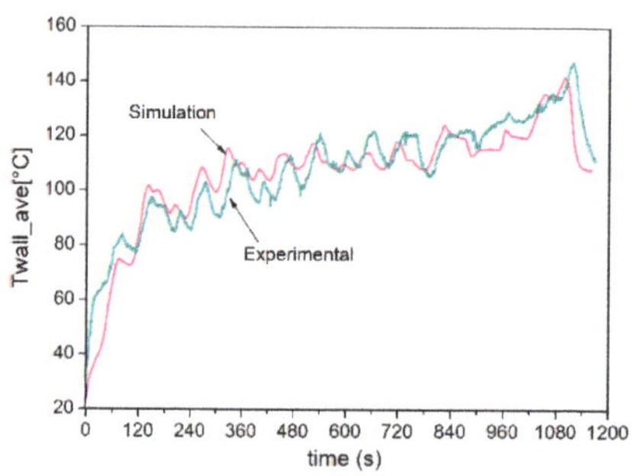

Figure 7. The average metal temperature during the New European Driving Cycle (NEDC); comparison between numerical simulations and experimental tests (the temperature at position 11C of Figure 5).

The agreement is satisfactory; the differences can be explained by considering that simulations are carried out by using engine speed, fuel flow rate, and coolant temperature values given by the constructor as measured at the rolling bed. On the contrary, in the current experiments, the engine speed, fuel flow rate, coolant pressure, and temperature values are those actually measured during the lab tests (Figure 6, top).

Figure 8 displays the NB_Index for both simulated and experimental controller. The results are satisfactory and simulations are in accordance with the experimental data.

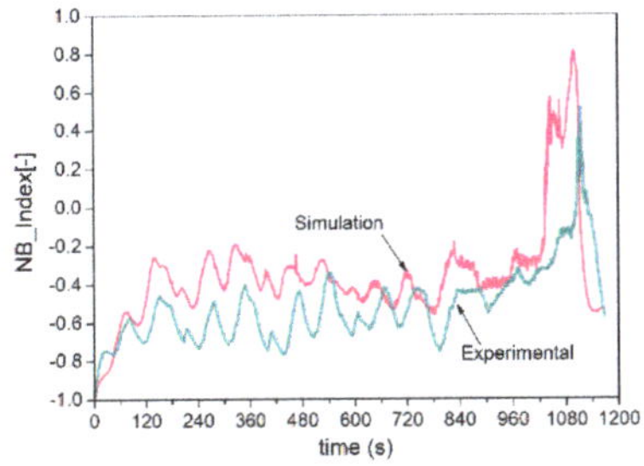

Figure 8. NB_Index during the NEDC; comparison between numerical simulations and experimental tests.

Details on the controller parameters adopted for validation purposes will be described in the next section, where the results of numerical simulations carried out with three different controllers will be presented.

4. Results

The controller is designed to operate at the lowest coolant flow rate as long as the wall temperature is lower than 100 °C. In such conditions, the energy required by the pump is the lowest possible; at the same time, the engine temperature is in the range of metal reliability and the coolant is safe from boiling. After the engine wall reaches the threshold value of 100 °C, the controller adjusts the coolant flow rate determined by the ellipsoid crossed by the trajectory. In the present work, three different control strategies are developed: in the first case, a low nucleate boiling level of the coolant is allowed; in the second case, the allowed boiling level is the same as the one obtained when the standard belt-driven pump operates; finally, a slightly higher boiling level is admitted. The tests are carried out for the New European Driving Cycle (NEDC), for which experimental data of fuel flow rate and engine speed for the current engine are supplied by the manufacturer (Figure 6).

4.1. Controller with Low Boiling Levels

The control region and the evolution of the state of the system during the NEDC are plotted in Figure 9.

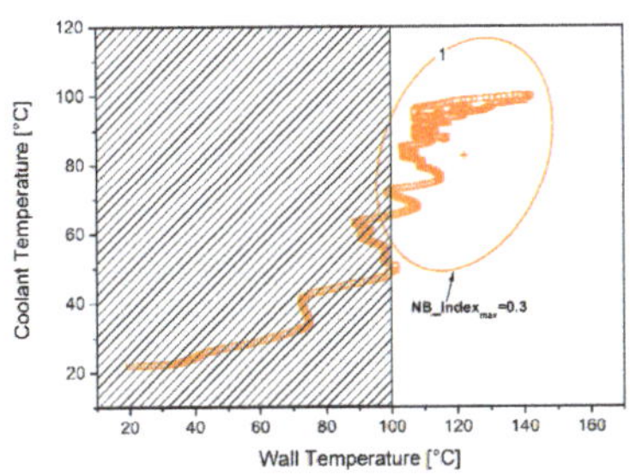

Figure 9. Control region and system trajectory during the NEDC for low boiling levels.

The initial wall and coolant temperature are the ambient values corresponding to 20 °C. In the dashed area, where the wall temperature is lower than 100 °C, the coolant flow rate is the lowest (~180 dm^3/h) and the wall and coolant temperature vary according to the fuel flow rate enforced during the low part of the NEDC. When the wall temperature rises to reach 100 °C, the trajectory crosses the ellipsoid and the coolant flow rate is modified to satisfy the controller constraints for the selected ellipsoid.

The constraints for Ellipsoid No.1 are summarized in Table 1. The constraints on fuel flow rate and engine speed are given by the NEDC; the maximum fuel flow rate, $\dot{m}_f$, is about 9 kg/h and the engine speed can vary in the range 500–4000 rpm. The maximum coolant flow rate, $\dot{m}_c$, is determined by the pump, which delivers about 6000 dm^3/h. Finally, the constraints on the output variables are defined by the constructor. If a moderate boiling level is required (NB_Index$_{max}$ = 0.3), with an average pressure of 1.4 bar, a maximum average coolant temperature of T_C of 110 °C and a maximum average wall temperature T_W of 160 °C can be achieved.

Table 1. Constraints on input and output variables for Ellipsoid No.1.

Input/Output Variable	Equilibrium	Min/Max
$\dot{m}_f$ (kg/h)	3.5	0–9
Eng. Speed (rpm)	2750	500–4000
$\dot{m}_C$ (dm^3/h)	1800	176–6000
T_W (°C)	122.5	85–160
T_C (°C)	83	56–110
NB_Index (-)	−0.1	−0.5–0.3

The resulting coolant flow rate during the NEDC is displayed in Figure 10a. It can be observed that, as long as the wall temperature is lower than 100 °C (Figure 10c), the pump delivers the lowest coolant flow rate and no boiling occurs (Figure 10b) owing to the low metal and coolant temperatures achieved in this operating range (Figure 10c). When the threshold wall temperature is exceeded, the state of the system crosses Ellipsoid No.1 (Figure 9) and the coolant flow rate varies in order to satisfy the controller constraints whilst the fuel flow rate and engine speed vary according to Figure 6. Figure 10c,d shows that both the wall and coolant temperatures are kept within the prescribed ranges. At the same time, the NB_Index is lower than the maximum allowed value (Figure 10b).

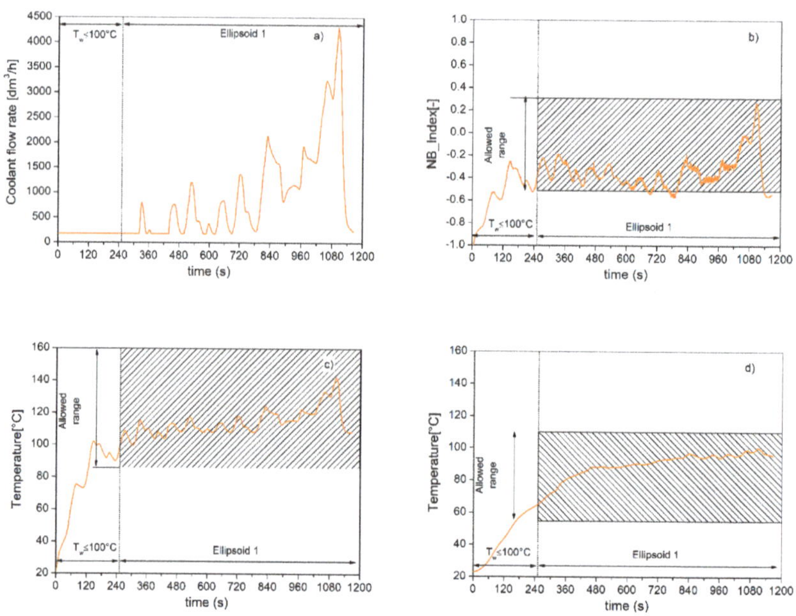

Figure 10. Coolant flow rate (**a**), NB_Index (**b**), wall (**c**), and coolant (**d**) temperature when the controller operates with low boiling levels.

4.2. Controller with Moderate Boiling Levels

If moderate boiling levels are allowed, like the ones that occur when the standard belt-driven pump is adopted (NB_Index ~ 0.5/0.6), a new controller can be adopted. In this case, a new control region with a higher NB_Index is obtained with lower coolant flow rates, which increase the wall and coolant temperatures and move the equilibrium point towards higher values (Ellipsoid No.2). The control region is, however, smaller than the previous one (Ellipsoid No.1), owing to the constraints on the maximum wall and coolant temperatures, which are substantially unchanged when compared to the previous case. As Ellipsoid No.2 is not large enough to cover the gap with the dashed area, a transition region (Ellipsoid No.1) is, however, needed and the controller is made up of two ellipsoids, as displayed in Figure 11. When the system state (T_W, T_C) is in the region, which belongs to both ellipsoids, Ellipsoid No.2 is used.

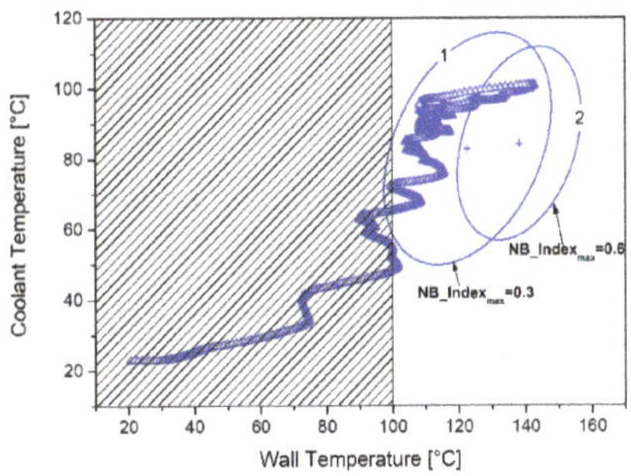

Figure 11. Control region and system trajectory during the NEDC for moderate boiling levels.

The equilibrium values and the constraints for the moderate boiling controller are summarized in Table 2. The results in terms of coolant flow rate, wall and coolant temperature, and NB_Index are displayed in Figure 12. As long as the state of the system lies in Ellipsoid No.1, no differences occur with respect to the previous case. Owing to the fuel flow rate increase, at time 822 s, the state of the system crosses Ellipsoid No.2; as this controller allows higher boiling levels than Ellipsoid No.1, the coolant flow rate decreases, while the wall and coolant temperatures and the NB_Index rise. The state of the system remains in this controller region for about 60 s before returning to Ellipsoid No.1, owing to the reduced fuel flow rate. In this region, the coolant flow rate tends to increase and to lower the temperatures; however, as the fuel flow rate is reduced, no significant variations in coolant flow rate occur. In the final part of the NEDC cycle, where the fuel flow rate increases significantly, the state of the system crosses Ellipsoid No.2 again and the coolant flow rate increases considerably; the wall and coolant temperatures and the boiling levels are kept within the prescribed limits. The effectiveness of the controller is demonstrated for the whole NEDC cycle, where the constraints on the variables are satisfied, as displayed by the dashed areas in Figure 12b–d.

Table 2. Constraints on input and output variables for Ellipsoid No.1 and Ellipsoid No.2.

	Ellipsoid No.1		Ellipsoid No.2	
Input/Output Variable	**Equilibrium**	**Min/Max**	**Equilibrium**	**Min/Max**
$\dot{m}_f$ (kg/h)	3.5	0–9	3.5	0–9
Eng. Speed (rpm)	2750	500–4000	2750	500–4000
$\dot{m}_C$ (dm3/h)	1800	176–6000	1200	176–6000
T_W (°C)	122.5	85–160	138	110–160
T_C (°C)	83	56–110	84.4	70–110
NB_Index (-)	−0.1	−0.5–0.3	0.1	−0.2–0.6

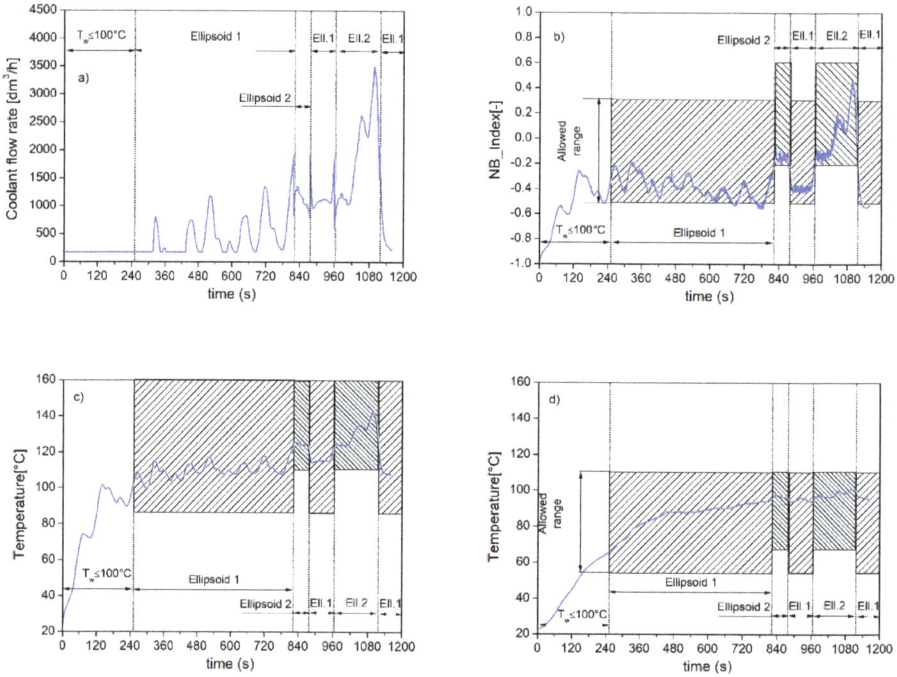

Figure 12. Coolant flow rate (**a**), NB_Index (**b**), and wall (**c**) and coolant (**d**) temperatures for moderate boiling levels.

4.3. Controller with High Boiling Levels

If higher boiling levels are accepted, Ellipsoid No.2 can be further modified and a new equilibrium point with a higher NB_Index can be designed (Table 3).

Table 3. Constraints on input and output variables for Ellipsoid No.1 and Ellipsoid No.2.

	Ellipsoid No.1		Ellipsoid No.2	
Input/Output Variable	Equilibrium	Min/Max	Equilibrium	Min/Max
$\dot{m}_f$ (kg/h)	3.5	0–9	3.5	0–9
Eng. Speed (rpm)	2750	500–4000	2750	500–4000
$\dot{m}_C$ (dm3/h)	1800	176–6000	915	176–6000
T_W (°C)	122.5	85–160	135.5	110–160
T_C (°C)	83	56–110	85.0	70–110
NB_Index (-)	−0.1	−0.5–0.3	0.2	0–0.8

The maximum boiling level considered in the present case is about 0.8; the new controller is displayed in Figure 13 and the corresponding coolant flow rate, wall and coolant temperatures, and NB_Index are shown in Figure 14. As the state of the system crosses Ellipsoid No.2, the coolant flow rate is reduced as higher temperatures are allowed; at the same time the NB_Index increases and, at the highest fuel flow rate, it reaches, without exceeding, the maximum allowed value.

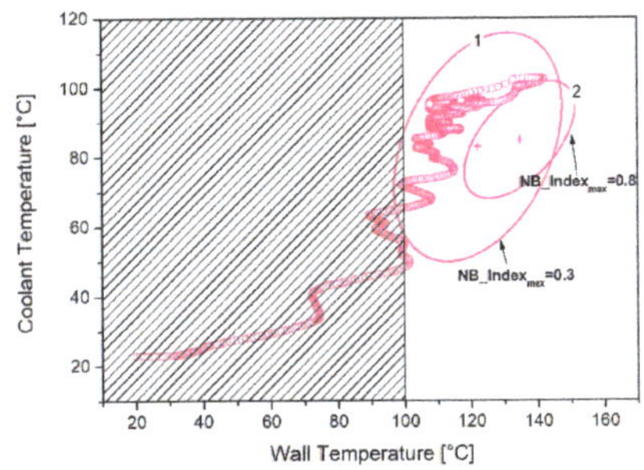

Figure 13. Control region and system trajectory during the NEDC for high boiling levels.

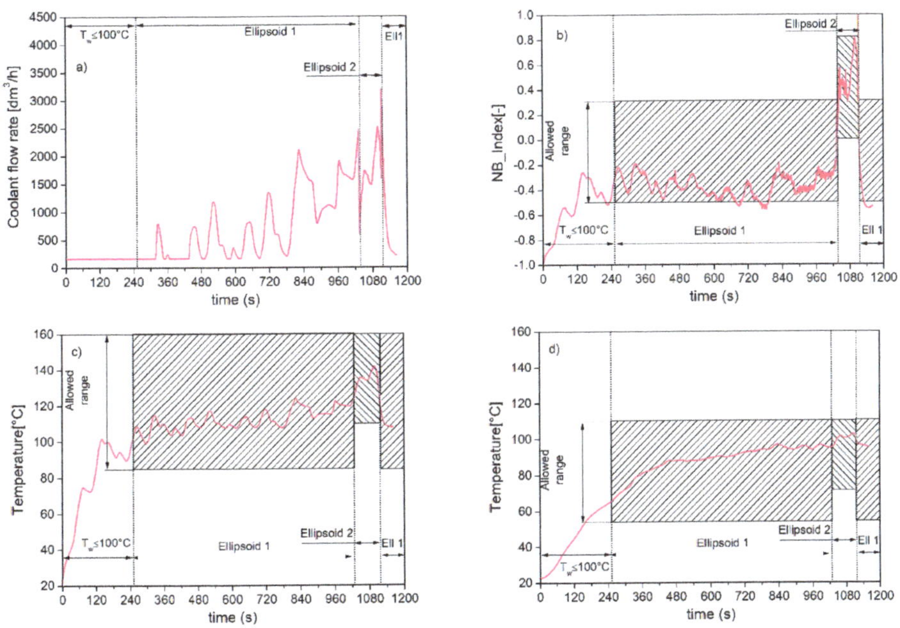

Figure 14. Coolant flow rate (**a**), NB_Index (**b**), and wall (**c**) and coolant (**d**) temperature for high boiling levels.

4.4. Comparison with the Belt Driven Pump

A comparison of the adoption of the standard belt-driven pump and the electrically driven pump with the developed controllers is presented in this section. In the following figures, the controllers with low, medium and high boiling levels will be denoted as Controller A, B, and C, respectively.

Figure 15 displays a comparison between the coolant flow rate as obtained with the crankshaft-driven pump (top) and with the controllers (bottom). In the urban part of the cycle (time ~ 770 s), the standard belt-driven pump delivers a considerably higher coolant flow rate (800 ÷ 2000 dm^3/h) than the controller (176 ÷ 1000 dm^3/h), while no difference occurs among the various controllers. In fact, Ellipsoid No.1, which acts in this part of the cycle, is the same for the three controllers. In the extra-urban part of the NEDC, the coolant flow rate increases for all cases. The highest coolant flow rate, even higher than the belt-driven pump, is delivered by Controller A, which allows low boiling levels, while Controller C, which allows a higher degree of boiling, reduces considerably the coolant flow rates in this part of the cycle.

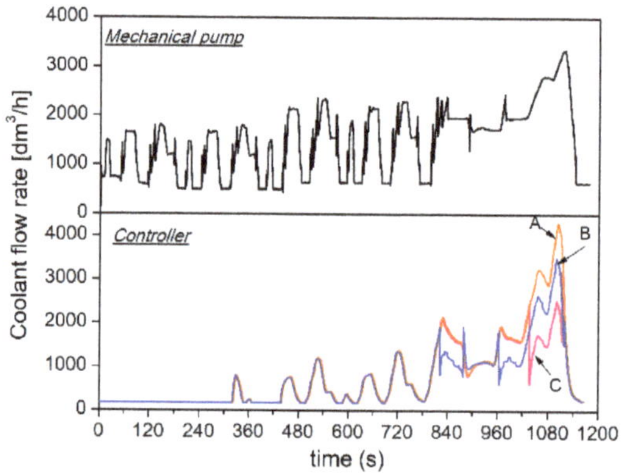

Figure 15. Coolant flow rate delivered by the standard belt-driven pump (**top**) and by the various controllers during the NEDC (**bottom**).

The boiling levels are plotted in Figure 16 for all the cases. The figure shows that, in the urban part of the NEDC (time ~ 770s), the mechanical pump overcools the engine; the lower coolant flow rate delivered by the controller enhances NB_Index, which is, however, negative and the coolant is safe from boiling. In the extra-urban part of the cycle, the boiling level starts to increase and, with the adoption of the mechanical pump, the maximum level of boiling is NB_Index ~ 0.5. Controller A limits the NB_Index at 0.3, Controller B allows the same boiling level as the belt-driven pump and Controller C slightly enhances this limit.

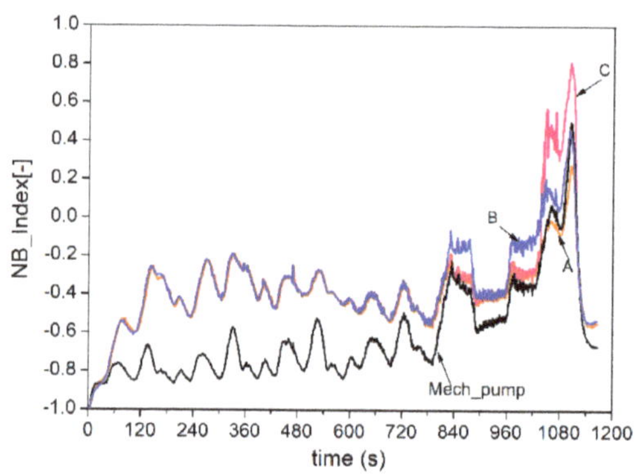

Figure 16. NB_Index for the various cooling strategies.

The boiling levels, however, preserve the engine metal, whose average temperature during the NEDC cycle is plotted in Figure 17. The figure shows that, despite the adoption of lower coolant flow rates, in the hottest part of the cycle the temperatures are well within the reliability limits. The analysis is carried out over the average wall temperature; it is, however, worth pointing out that, from the experimental data, the difference between the average metal temperature and the hottest thermocouple is about 40 °C, as displayed in Figure 18, and is still in the metal reliability range.

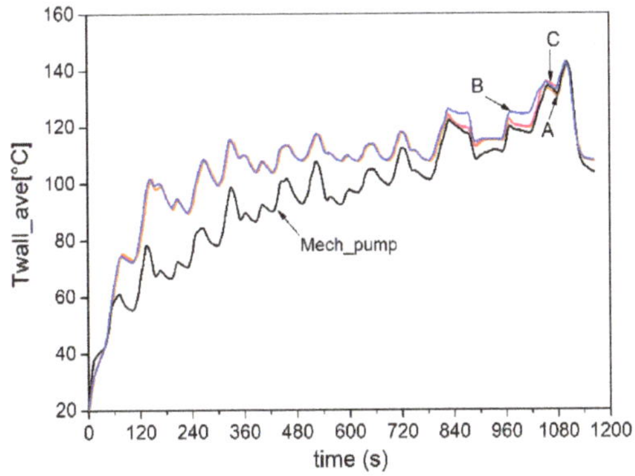

Figure 17. Average metal temperature during the NEDC.

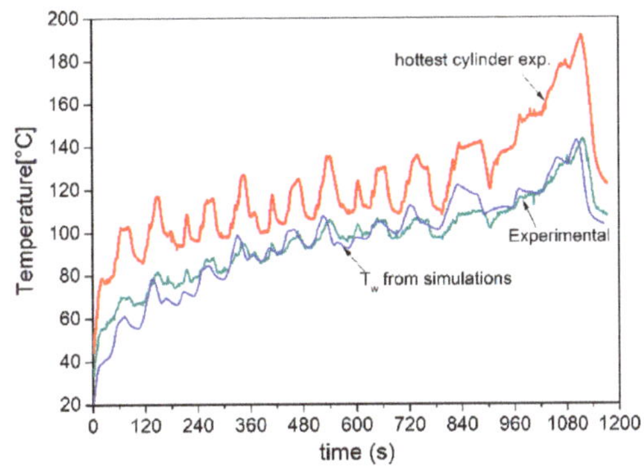

Figure 18. Simulated and measured metal temperatures during the NEDC cycle.

A final comparison regards the fuel economy related to the adoption of the controller. The results are summarized in Figure 19. The amount of fuel consumed to actuate the belt-driven pump is about 9.3% of the fuel consumed over the entire NEDC cycle. The largest amount is consumed during the four urban driving cycles (7.8%) and the remaining 1.5% is consumed during the extra-urban driving cycle. When the controller is adopted, the fuel consumption due to the actuation of the electrically driven pump is considerably lowered. If the controller with low boiling levels is adopted (Controller A), the total fuel consumption for pump actuation is reduced from 9.3% to 2.1%, with a fuel-saving of about 7.2%. In this case, only a very small amount of fuel is consumed during the urban part of the cycle (~0.42%), owing to the low coolant flow rate adopted; the remaining 1.7% is consumed during the extra-urban cycle, where the coolant flow rate rises. For the case of the controller, which allows the same boiling level as the mechanical pump, namely Controller B, the total fuel consumption for pump actuation is reduced to 1.2%, i.e., 8.1% less than the standard belt-driven pump. The fuel consumed during the urban cycle is the same as Controller A (~0.42%), where the coolant flow rate is the minimum, while a significant reduction occurs in the extra-urban cycle, where a fuel consumption of about 0.8% is achieved. Finally, when the controller with the highest boiling level is adopted (Controller C), the fuel consumption for pump actuation is the lowest; in this case, the fuel consumption is reduced at 1.1% of the total fuel consumed in the NEDC. The fuel consumed during the UDC is about 0.4%, which is the same as the previous cases, while it reduces to 0.7% in the EUDC, where the coolant flow rate is the lowest when compared to the other cases.

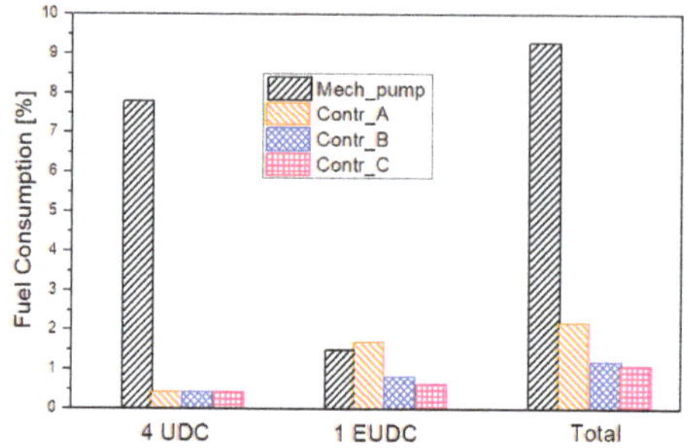

Figure 19. Fuel consumption reduction achieved with the use of the controller during the NEDC cycle.

The results in terms of fuel economy are reported in Table 4. From the above considerations, Controller B is the most effective, as it allows the achievement of the same boiling level as the belt-driven pump, with a significant fuel economy. This controller will be, therefore, taken into consideration for evaluating its performance for the engine operating under different conditions from the NEDC cycle.

Table 4. Fuel consumption for pump actuation during the NEDC.

Fuel Consumption (Total over the NEDC = 513 g)			
	NEDC	**4 UDC**	**1 EUDC**
Belt-driven pump	47.9 g (9.3%)	40.1 g (7.8%)	7.8 g (1.5%)
Controller A	10.7 g (2.1%)	2.1 g (0.4%)	8.5 g (1.7%)
Controller B	6.3 g (1.2%)	2.1 g (0.4%)	4.2 g (0.8%)
Controller C	5.4 g (1.1%)	2.1 g (0.4%)	3.3 g (0.7%)

5. Other Test Cases

The selected controller (Controller B) is tested under various operating conditions, which differ from the NEDC cycle, in order to assess its robustness during real engine operations. The operating conditions include the cold start (wall and coolant temperatures starting from 0 °C and from −20 °C), a prolonged high load operation with a maximum fuel flow rate of 9.0 kg/h and a prolonged high load operation with a maximum fuel flow rate of 18.0 kg/h.

5.1. NEDC with $T_{C0} = 0$ °C and $T_{W0} = 0$ °C

The controller is applied to the engine, which performs the NEDC in a typical winter environment where the starting temperatures are about 0 °C. The controller output is plotted in Figure 20, which shows that, even though the wall and coolant temperatures start from 0 °C, the trajectory falls inside the control regions, and no further ellipsoids, for covering different engine operating points, are needed.

The desired specifications of the controller are satisfied also in this case, as displayed in Figure 21. In particular, the main difference with the previous case, where an ambient temperature of 20 °C was considered (Section 4.2), occurs for the wall temperature and for the NB_Index in the early 160 s of the cycle; after that time interval, the two cases practically overlap.

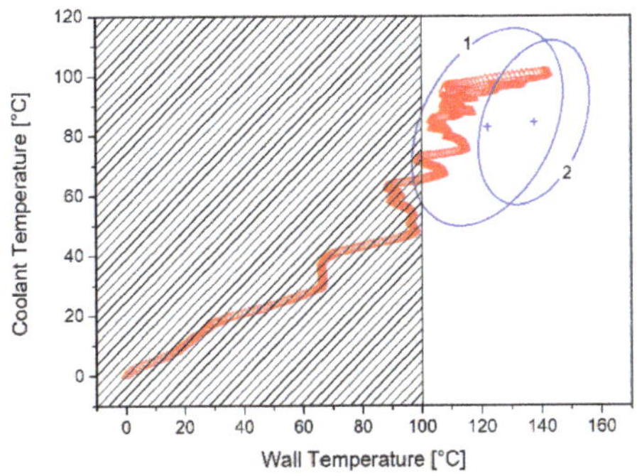

Figure 20. Coolant and wall temperatures during the NEDC starting from 0 °C environment.

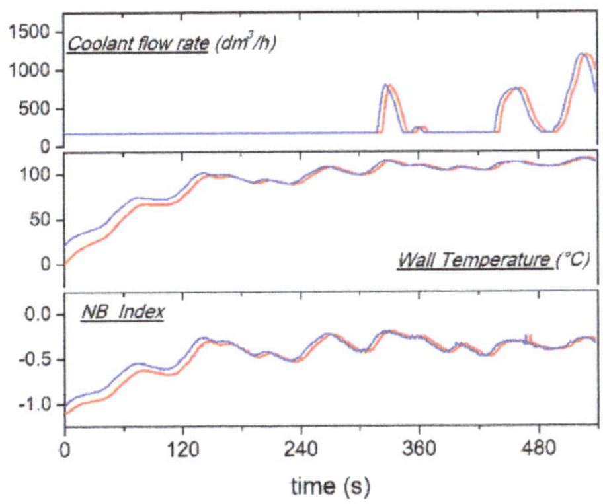

Figure 21. Controlled coolant flow rate (**top**), average wall temperature (**middle**) and NB_Index (**bottom**) during the NEDC cycle starting from 20 °C (red line) and 0 °C (blue line) environment.

5.2. One Urban Driving Cycle and One Extra-Urban Driving Cycle with Prolonged High Load Operation at $\dot{m}_f = 9$ kg/h

The case where the engine performs one urban driving cycle, one extra-urban cycle and is then kept at high load operation (constant fuel flow rate and engine speed) for about 60 s, is simulated (Figure 22). The maximum fuel flow rate during the extra-urban cycle is about 9 kg/h; this value is the upper limit of the constraint on the fuel flow rate during the controller design procedure (Table 2). The engine speed is 3700 rpm.

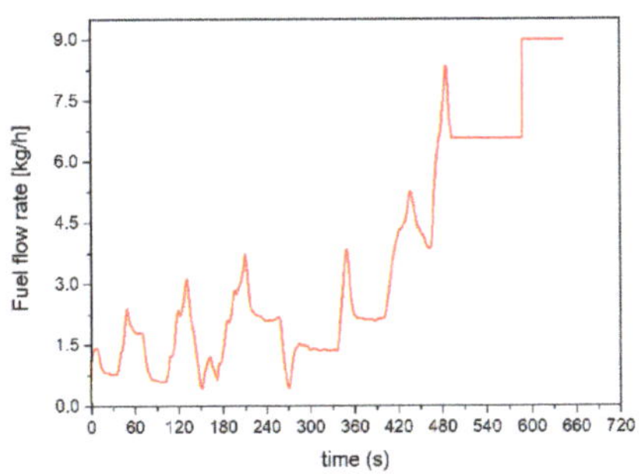

Figure 22. Fuel flow rate variation for prolonged high load operation.

Figure 23 shows that, also for this case, the trajectory falls inside the control regions owing to the fact that the constraints on the input variables, mainly the one on fuel flow rate, are satisfied.

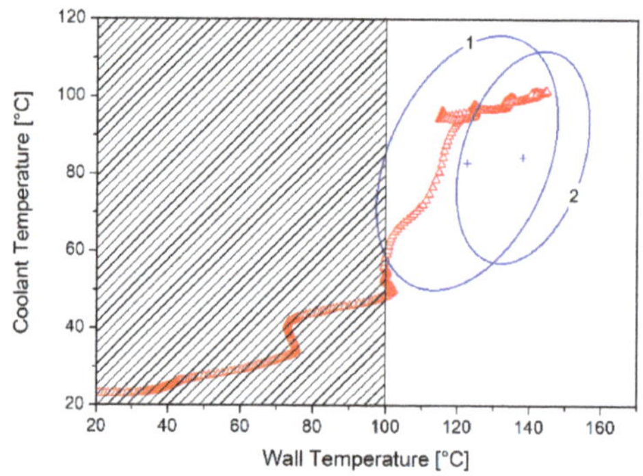

Figure 23. Coolant and wall temperatures for prolonged high load operation ($\dot{m}_f$ = 9 kg/h).

The results in terms of coolant flow rate, wall temperature, and NB_Index are displayed in Figure 24, where a comparison with the crankshaft-driven pump is also included. As expected, the controller allows the adoption of lower coolant flow rates than the mechanically driven pump; however, during the high load operation, the coolant flow rate is blocked at a value of 3300 dm^3/h (at 3700 rpm) for the belt-driven pump, while it increases at 3800 dm^3/h for the controlled pump. Despite the wall temperature levels being almost the same, the NB_Index exceeds the upper limit of 0.6 in the case of the belt-driven pump, while it remains in the prescribed limits with the controller (NB_Index = 0.55).

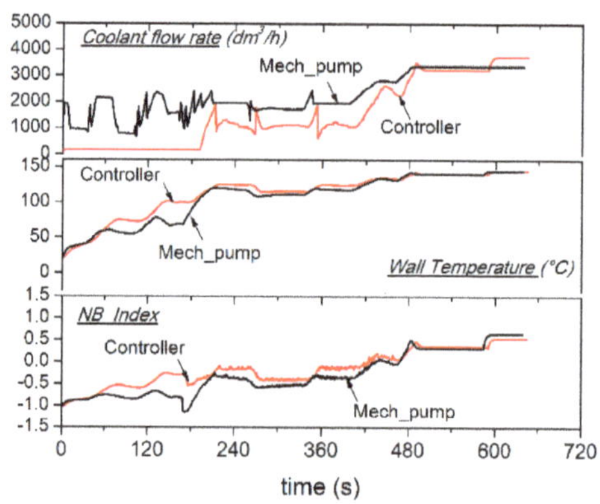

Figure 24. Controlled coolant flow rate (**top**), average wall temperature (**middle**), and NB_Index (**bottom**) for prolonged high load operation ($\dot{m}_f$ = 9 kg/h).

5.3. One Urban Driving Cycle and One Extra-Urban Driving Cycle with Prolonged High Load operation at $\dot{m}_f$ = 18 kg/h

The last simulated case is similar to the previous one, except for the high load operation. In this case, in fact, the fuel flow rate is fixed at about 18 kg/h, far beyond the engine operating range and the controller design constraints; this case is presented only for an illustrative purpose and is not a realistic case.

As Figure 25 shows, at high load operation the trajectory exceeds the ellipsoids borders, indicating therefore that, as the constraints on the input variables are not satisfied, the controller is not able to guarantee that the coolant flow rate is adequately corrected. In fact, Figure 26 shows that, during the high load operation, the controller enforces the highest possible coolant flow rate allowed by the actuator (~7000 dm^3/h); this value is higher by the one fixed by the belt-driven pump, which at 3700 rpm delivers a coolant flow rate of about 3300 dm^3/h. The metal temperatures are similar for both cases, while a relevant difference occurs in the NB_Index. With the belt-driven pump, a severe boiling occurs (NB_Index ~ 1.8); the controller contributes to reducing considerably this value to about 0.7, which, however, slightly overcomes the maximum allowed value (NB_Index = 0.6).

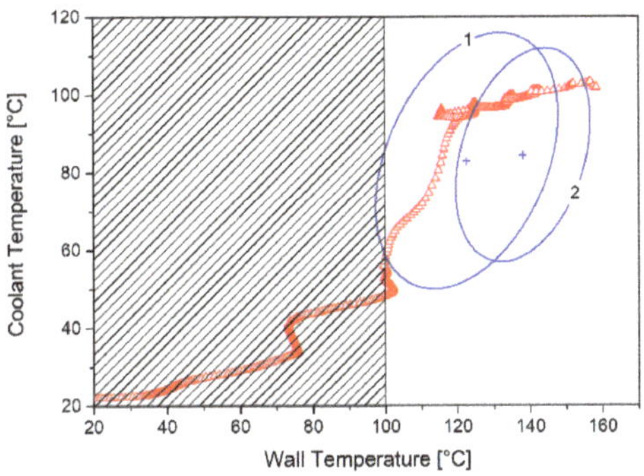

Figure 25. Coolant and wall temperatures for prolonged high load operation ($\dot{m}_f$ = 18 kg/h).

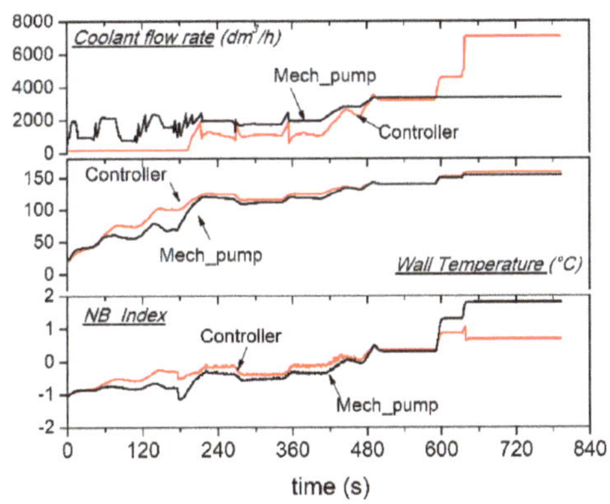

Figure 26. Controlled coolant flow rate (**top**), average wall temperature (**middle**), and NB_Index (**bottom**) for prolonged high load operation ($\dot{m}_f$ = 18 kg/h).

6. Discussion

The novel approach was validated at the experimental test rig during a standard driving homologation cycle and its effectiveness was demonstrated. Several control strategies were simulated, with three different allowable boiling levels. In particular, a controller, which allows a lower boiling level (NB_Index_{max} = 0.3) than the one reached with the belt-driven pump under the same engine operating conditions (NB_Index_{max} = 0.6) was first analyzed. Then, a second controller, which permits the same boiling level as the one achieved with the belt-driven pump (NB_Index_{max} = 0.6) was simulated. Finally, a controller which tolerates a maximum boiling level of about NB_Index_{max} = 0.8 was tested.

The findings can be summarized as follows:

- Increasing the allowed boiling level determines lower coolant flow rates; the metal temperatures are always kept inside the defined reliability limits.

- A reduced coolant flow rate under part-load conditions generally reduces fuel consumption. The 1.2 dm^3 engine considered burns about 500 g during the NEDC and about 9.3% of this fuel is spent for the mechanical pump, while the fuel consumption for the electric pump, in the case of the low boiling level controller (NB_Index_{max} = 0.3), is about 2%; the fuel-saving is therefore in the order of 7%. The benefit increases for the case of the high boiling level controller, when about 1.1% of the fuel is used by the electric pump and a fuel-saving of about 8% is achieved.
- The benefits of the proposed approach also occur under prolonged high-load, moderate-engine speed conditions in terms of engine reliability. In these conditions, in fact, the controlled electric pump delivers considerably higher coolant flow rates than the standard belt-driven pump and keeps the boiling levels within the safety limits (NB_Index = 0.7, in contrast with NB_Index > 1 for the case of the belt-driven pump).

7. Conclusions

In this work, a tool for designing a controller for the optimal thermal management of a spark-ignition engine was presented. The proposed approach aims at improving the engine efficiency for all those cases where the traditional belt-driven pump determines engine overcooling and at improving the engine reliability, with higher coolant flow rates than the traditional system, for all those cases where overheating could occur. This goal was achieved with the adoption of an electrically driven pump, instead of the standard crank-shaft driven one, which is driven by a specific controller. The controller makes use of a lumped-parameters model of the thermal exchange within the engine and of a metrics of the nucleate-boiling level, the NB_Index.

The results presented demonstrate that a controller is a useful tool for appropriately cooling the engine under the whole field of possible operating conditions. This facilitates an increase in engine efficiency without substantial modifications to the engine cooling system layout: just the substitution of the belt-driven pump with an electrically driven one is required.

The possibility of further efficiency improvements will be investigated as a future development by applying the proposed methodology to knock mitigation.

Author Contributions: Conceptualization, T.C. and P.M.; Data curation, L.F.; Formal analysis, L.F.; Investigation, P.M. and D.P.; Methodology, T.C.; Project administration, S.B.; Software, D.P.; Writing—original draft, T.C. All authors have read and agreed to the published version of the manuscript.

Funding: The investigation started in the framework of the project PON01_01517, CUP: B21H11000400005, with financial support of the European Commission and the Italian Ministry of University and Research.

Conflicts of Interest: The authors declare no conflict of interest.

References

1. Edwards, R.; Hass, H.; Larive, J.H.F.; Lonza, L.; Maas, H.; Rickeard, D. Well-to-Wheels Analysis of Future Automotive Fuels and Powertrains in the European Context, WELL-to-WHEELS Report Version 4a 2014, European Commission, Joint Research Centre, Institute for Energy and Transport, Brussel. Available online: https://ec.europa.eu/jrc/en/publication/eur-scientific-and-technical-research-reports/tank-wheels-report-version-4a-well-wheels-analysis-future-automotive-fuels-and-powertrains (accessed on 28 February 2020).
2. Pischinger, S. Antriebsentwicklung der Zukunft. *ATZextra* **2011**, *16*, 136–141. [CrossRef]
3. Regulation 443/2009/EC and amending Regulation (EU) 333/2014 - passenger cars - and Regulation (EU) 510/2011/EC and amending Regulation (EU) 253/2014. Brussel.
4. Boretti, A. The Future of the Internal Combustion Engine after "Diesel-Gate". *SAE Technical Paper 2017-28-1933* **2017**. [CrossRef]
5. Johnson, T. Vehicular Emissions in Review. *SAE Int. J. Engines* **2012**, *5*, 216–234. [CrossRef]
6. Sellnau, M.; Kunz, T.; Sinnamon, J.; Burkhard, J. 2-Step Variable Valve Actuation: System Optimization and Integration on an SI Engine. *SAE Technical Paper 2006-01-0040* **2006**. [CrossRef]

7. Bozza, F.; De Bellis, V.; Teodosio, L. A Numerical Procedure for the Calibration of a Turbocharged Spark-Ignition Variable Valve Actuation Engine at Part Load. *Int. J. Engine Res.* **2017**, *18*, 810–823. [CrossRef]
8. De Bartolo, C.; Algieri, A.; Bova, S. Simulation and Experimental Validation of the Flow Field at the Entrance and within the Filter Housing of a Production Spark-Ignition Engine. *Simul. Model. Pract. Theory* **2014**, *41*, 73–86. [CrossRef]
9. Marchitto, L.; Tornatore, C.; Valentino, G.; Teodosio, L. Impact of Cooled EGR on Performance and Emissions of a Turbocharged Spark-Ignition Engine under Low-Full Load Conditions. *SAE Technical Paper 2019-24-0021* **2019**. [CrossRef]
10. Algieri, A.; Amelio, M.; Bova, S.; Morrone, P. Energy Efficiency Analysis of Monolith and Pellet Emission Control Systems in Unidirectional and Reverse-Flow Designs. *SAE Int. J. Engines* **2010**, *2*, 684–693. [CrossRef]
11. Algieri, A.; Morrone, P.; Settino, J.; Castiglione, T.; Bova, S. A Comparative Analysis of Active and Passive Emission Control Systems Adopting Standard Emission Test Cycles. *SAE Technical Paper 2017-24-0125* **2017**. [CrossRef]
12. Cipollone, R.; Di Battista, D.; Gualtieri, A. A Novel Engine Cooling System with Two Circuits Operating at Different Temperatures. *Energ. Convers. Manag.* **2013**, *75*, 581–592. [CrossRef]
13. Kang, H.; Ahn, H.; Min, K. Smart Cooling System of the Double Loop Coolant Structure with Engine Thermal Management Modeling. *Appl. Therm. Eng.* **2015**, *79*, 124–131. [CrossRef]
14. Support for the Revision of Regulation (EC) No 443/2009 on CO2 Emissions from Cars TNO, AEA, CE Delft, Okopol, TML, Ricardo, HIS Global Insight Service request #1 for Framework Contract on Vehicle Emissions Framework Contract No ENV.C.3./FRA/2009/0043. Final report (November 25th, **2011**), Brussel. Available online: https://www.google.com.hk/url?sa=t&rct=j&q=&esrc=s&source=web&cd=1&cad=rja&uact=8&ved=2ahUKEwiD1pee8_3nAhWQd94KHfqvD80QFjAAegQIBhAB&url=https%3A%2F%2Fec.europa.eu%2Fclima%2Fsites%2Fclima%2Ffiles%2Fdocs%2F0048%2Fsr1_final_report_en.pdf&usg=AOvVaw2_IGLG7pxVdWJZ2ocXyfHm (accessed on 3 March 2020).
15. Oezdemir, O.; Huttinger, K.; Bargende, M.; Rienacker, A. Friction Reduction by Optimization of Local Oil Temperatures. *SAE Technical Paper 2019-24-0177* **2019**. [CrossRef]
16. Perrone, D.; Falbo, L.; Castiglione, T.; Bova, S. Knock Mitigation by Means of Coolant Control. *SAE Technical Paper 2019-24-0183* **2019**. [CrossRef]
17. Brace, C.; Burnham-Slipper, H.; Wijetunge, R.; Vaughan, N. Integrated Cooling Systems for Passenger Vehicles. *SAE Technical Paper 2001-01-1248* **2001**. [CrossRef]
18. Ap, N.; Golm, N. New Concept of Engine Cooling System (Newcool). *SAE Technical Paper 971775* **1997**. [CrossRef]
19. Brace, C.J.; Hawley, G.; Akehurst, S.; Piddock, M.; Pegg, I. Cooling System Improvements - Assessing the Effects on Emissions and Fuel Economy. *Proc. Inst. Mech. Eng. Part. D J. Automob. Eng.* **2008**, *222*, 579–591. [CrossRef]
20. Clough, M. Precision Cooling of a Four Valve per Cylinder Engine. *SAE Technical Paper 931123* **1993**. [CrossRef]
21. Maciejowski, J.M. *Predictive Control.: With Constraints*; Pearson Education: London, UK, 2002.
22. Bakosova, M.; Oravec, J. Robust Model Predictive Control for Heat Exchanger Network. *Appl. Therm. Eng.* **2014**, *73*, 924–930. [CrossRef]
23. Oravec, J.; Bakosova, M.; Meszaros, A.; Mikova, N. Experimental Investigation of Alternative Robust Model Predictive Control of a Heat Exchanger. *Appl. Therm. Eng.* **2016**, *105*, 774–782. [CrossRef]
24. Rausen, D.J.; Stefanopoulou, A.G.; Kang, J.-M.; Eng, J.A.; Kuo, T.-W. A Mean-Value Model for Control of Homogeneous Charge Compression Ignition (HCCI) Engines. *J. Dyn. Syst.-T ASME* **2005**, *127*, 355–362. [CrossRef]
25. Ferreau, H.; Ortner, P.; Langthaaler, P.; del Re, L.; Diehl, M. Predictive Control of a Real-World Diesel Engine Using an Extended Online Active Set Strategy. *Annu. Rev. Control.* **2007**, *31*, 293–301. [CrossRef]
26. Huang, M.; Zaseck, K.; Butts, K.; Kolmanovsky, I. Rate-Based Model Predictive Controller for Diesel Engine Air Path: Design and Experimental Evaluation. *IEEE T Contr. Sist. T* **2016**, *99*, 1–14. [CrossRef]
27. Zhu, Q.; Onori, S.; Prucka, R. Pattern Recognition Technique Based Active Set QP Strategy Applied to MPC for a Driving Cycle Test. In Proceedings of the 2015 American Control Conference, Chicago, IL, USA, 1–3 July 2015; pp. 4935–4940.

28. Zhu, Q.; Onori, S.; Prucka, R. Nonlinear Economic Model Predictive Control for SI Engines Based on Sequential Quadratic Programming. In Proceedings of the 2016 American Control Conference, Boston, MA, USA, 6–8 July 2016.
29. Wiese, A.; Stefanopoulou, A.; Karnik, A.; Buckland, J. Model Predictive Control for Low Pressure Exhaust Gas Recirculation with Scavenging. In Proceedings of the 2017 American Control Conference, Seattle, WA, USA, 24–26 May 2017.
30. Pizzonia, F.; Castiglione, T.; Bova, S. A Robust Model Predictive Control for Efficient Thermal Management of Internal Combustion Engines. *Appl. Energy* **2016**, *169*, 555–566. [CrossRef]
31. Piccione, R.; Bova, S. Engine Rapid Shutdown: Experimental Investigation on the Cooling System Transient Response. *J. Eng. Gas. Turb. Power* **2010**, *132*. [CrossRef]
32. Bova, S.; Piccione, R.; Durante, D.; Perrussio, M. Experimental Analysis of the After-Boiling Phenomenon in a Small I.C.E. SAE Technical Paper 2004-32-0091. *Trans. J. Engines* **2004**. [CrossRef]
33. Castiglione, T.; Morrone, P.; Bova, S. A Model Predictive Approach to Avoid After-Boiling in ICE. *SAE Technical Paper 2018-01-0779* **2018**. [CrossRef]
34. Bova, S.; Castiglione, T.; Piccione, R.; Pizzonia, F. A Dynamic Nucleate-Boiling Model for CO_2 Reduction in Internal Combustion Engines. *Appl. Energy* **2015**, *143*, 271–282. [CrossRef]
35. Castiglione, T.; Perrone, D.; Algieri, A.; Bova, S. A Contribution to Improving the Thermal Management of Powertrain Systems. *SAE Int. J. Engines* **2020**, *13*. [CrossRef]
36. Dittus, F.W.; Boelter, L.M.K. *Heat Transfer in Automobile Radiators of the Tubular Type*; University of California Publications in Engineering: Berkeley, CA, USA, 1930; Volume 2, pp. 443–461.
37. Chen, J.C. Correlation for Boiling Heat Transfer to Saturated Fluids in Convective Flow, Industrial and Engineering Chemistry Process Design and Development. *Ind. Eng. Chem. Process Des. Dev.* **1966**, *5*, 233–329.

energies

MDPI

Article

Effect of the Intake Valve Lift and Closing Angle on Part Load Efficiency of a Spark Ignition Engine

Michelangelo Balmelli *, Norbert Zsiga, Laura Merotto and Patrik Soltic

Automotive Powertrain Technologies Laboratory, Empa Swiss Federal Laboratories for Materials Science and Technology, 8600 Dübendorf, Switzerland; Norbert.Zsiga@empa.ch (N.Z.); laura.merotto@empa.ch (L.M.); Patrik.Soltic@empa.ch (P.S.)

* Correspondence: michelangelo.balmelli@empa.ch; Tel.: +41-58-765-47-00

Received: 11 February 2020; Accepted: 20 March 2020; Published: 3 April 2020

Abstract: This study provides an experimental evaluation of the effectiveness of Miller cycles with various combinations of lift and intake valve closing angle for a passenger car engine with premixed combustion in naturally aspirated operation. A fully variable electro-hydraulic valve train provided different valve lift profiles. Six load points, from 1.5 up to 5 bar brake mean effective pressure at a constant engine speed of 2000 min^{-1}, were tested with 6 different intake valve lift/intake valve closing angle combinations. The intake valve closing angle was always set before bottom dead center to achieve the desired load with unthrottled operations. Experimental comparison with throttled operation outlines an indicated efficiency increase of up to 10% using high intake lift with early valve closing angle. Furthermore, this analysis outlines the influences that early intake valve closing angle has on fuel energy disposition. Longer combustion duration occurs using early intake valve closing angle because of turbulence dissipation effects, leading to slight reductions in the heat-to-work efficiency. However, overall pressure and temperature levels decrease and consequently heat losses and losses due to incomplete combustion decrease as well. Overall, we found that combustion deterioration is compensated/mitigated by the reduction of the heat losses so that reductions of pumping losses using early intake valve closing can be fully exploited to increase the engine's efficiency.

Keywords: Miller cycles; early intake valve closing; electro hydraulic valve train; energy balance; heat losses

1. Introduction

This work outlines an experimental investigation of unthrottled Miller cycles using a self-developed electro hydraulic valve train [1,2]. While valve lift is set for all cylinders on the intake and exhaust side, respectively, the valve timing is set individually for the intake- and exhaust valves for each cylinder.

In conventional stoichiometrically operated spark ignition engines, which control the amount of aspirated gas by throttling the intake airflow, pumping losses make the engine efficiency deteriorate significantly, in particular at low loads. This leads to high fuel consumption during typical operation of non-hybrid powertrains [3]. Controlling the amount of fresh gas by adjusting the intake valve closing (IVC) angle promises to reduce or eliminate pumping losses. The intake valves can be closed either before or after bottom dead center. The resulting cycles are usually called Miller and Atkinson, respectively. Late IVC (Atkinson) as a general strategy for load control is problematic, because late intake closing interferes at low load with the ignition angle. Therefore, early IVC (Miller) was used throughout the work presented here.

Whether avoiding pumping losses by early IVC in a spark-ignited concept increases the engine's brake efficiency or not is, however, not clear a priori because of the following detrimental effects:

1. Miller cycles are normally used to increase the efficiency and are characterized by a larger expansion ratio compared to the compression ratio [4]. To maintain an effective compression ratio, the geometrical compression ratio is increased. However, load control by phasing the IVC while keeping the geometrical compression ratio reduces the effective compression ratio with decreasing load.

2. Although Miller timing reduces pumping losses, it also reduces turbulence intensity. Turbulent kinetic energy is generated during the intake flow process and dissipated later because no momentum addition is present to compensate for the viscous losses [5]. The fast turbulence dissipation for the same engine as used in this research with its standard valve timing settings, but with direct gas injection, is documented in [6]. The earlier the valve closes, the lower the turbulence present during combustion therefore is, which leads to a lower turbulent flame speed. Using early intake closing, the combustion is expected to be slower and less efficient [7].

Miller cycles are often achieved using camshaft-based systems, limiting the possible advantages. A paper by Unger and Schwartz [8] lists the progress and challenges of unthrottled Miller cycles achieved with a serial-production continuously variable valve train where valve timing it strongly coupled to the valve lift. As a result, at low load operating points the maximum valve lift is even lower than 1 mm [9]. This leads unavoidably to strong throttling effects for short valve openings, mainly because the valve's opening and closing velocities become very low. This drawback does not exist for the electro hydraulic valve train used here, as the valve movement (i.e., the time from fully closed to fully open and vice versa) is independent from valve lift or valve timing.

Therefore, the work presented here allows experimental quantification of the net effects of early intake valve closing and valve lift to the engine's efficiency independently. The analysis also focuses on the identification and quantification of the individual losses arising from Miller cycles.

2. Experimental Setup

Experiments were performed on a Volkswagen engine with four cylinders and a displacement of 1.4 L, which was operated with port-fuel-injected natural gas and controlled under stoichiometric conditions. The complete camshaft system—composed of the camshaft, gears, chain, and timing adjustment—was removed and the engine was fitted with the electro hydraulic valve train (shown in Figure 1, left).

Figure 1. Electro hydraulic valve train on the VW EA111 short block (**left**), engine on the test bench (**right**).

The cylinder head was modified to hold the variable valve system, while the entire short block remained unchanged from factory specifications. The entire engine control and valve train control were based on a single dSPACE Microautobox rapid prototyping system and the control functionalities were all fully developed in house. Besides the control of injection, ignition, and valve train, the engine control also recorded the cylinder pressure signals of all cylinders and provided closed-loop control of the center of combustion. The center of combustion, estimated online with the Rassweiler and Withrow method [10], was set to 368 °CA by closed-loop-control of the ignition angle. This value was chosen for all the experiments as it gives the highest efficiency (or maximum brake torque, respectively) and it ensured stable operation [11]. Figure 1 (right) provides a picture of the test-bench setup (engine and entire control unit) and Table 1 lists the main characteristics of the experimental setup.

Table 1. Main characteristics of the experimental setup.

Engine basis	Volkswagen 1.4 L, production code EA111
Bore/stroke/number of cylinders	76.5 mm/75.6 mm/4
Compression ratio	10
Fueling	Port fuel injection of natural gas using Bosch NGI2 injectors
Valve train	Self-developed, electrohydraulic, water-glycol (engine coolant) used as the hydraulic working fluid
Engine control	Rapid prototyping system (dSPACE)
In-cylinder pressure transducers	M10 Piezoelectric, water-cooled (Kistler 6061B)
Stoichiometry sensor	Wide band (Bosch LSU 4.9)
Valve lift measurement	Linear encoders (Elgo Electronic), using custom-developed converters
Test bench	Horiba Dynas LI250
Test bench automation	Horiba STARS Engine
Torque measurement	HBM T40 torque transducer
Fuel flow measurement	Coriolis sensor (Rheonik RHM015)
Combustion air flow measurement	ABB Sensyflow P
Emission measurement system	Horiba Mexa-One-D1-EGR

The working principle of the valve train is basically a hydraulically excited spring–mass-system [12] which is able to recover energy. Because of the nature of a spring-mass–system, its oscillation frequency (hence the opening or closing duration in the time domain) is independent in the amplitude (hence the valve lift). The oscillation frequency is set, for a given mass of the moving parts, by choosing the stiffness of the springs. This is done in such a way that an adequate valve opening duration is achieved at the engine's maximum speed. In the current setup, a duration for valve opening of around 3 ms was chosen. Constant opening and closing durations in the time domain translate to the faster movements in the crank angle domain as the engine speed was reduced. This lead to a very fast release or blocking of the gas flow at lower engine speed.

On the intake side, the valve lifts were set by the hydraulic pressure level. The opening of the valves was initiated by energizing a magnetic coil [13]. The valves remained open until the current to the coil was cut. The valve lift profiles are divided into three main phases: opening, holding and closing. All three phases were mechanically automated to allow opening, keeping open, closing, recuperation of hydraulic fluid and soft seating without the need of electronical feedback control of the valve lift and with just one magnetic coil per actuator [1]. Because of the efficient recuperation, the energy demand of the electro hydraulic valve train was lower than that of cam-driven valve systems.

Figure 2 shows a scheme of the test-bench and depicts the relevant signals (Data acquisition, control units and main mechanical parts).

Figure 3 shows intake valve lift measurements at an engine speed of 2000 min^{-1} recorded during the fired engine experiments explained in detail in the following sections. These experiments were performed in order to characterize different Miller cycles, as well as the influence that different combinations of IVC time and lift have on pumping losses variation and efficiency. For the sake of consistency, all the valve timings except IVC angle and intake lift were kept constant for all experiments. As Figure 3 shows, the time to open or close the valves remained constant, independent of the valve lift, which ensures a fast release of the flow area even at low valve lifts. This fast opening, even for small valve lifts, is a major advantage of the system described here.

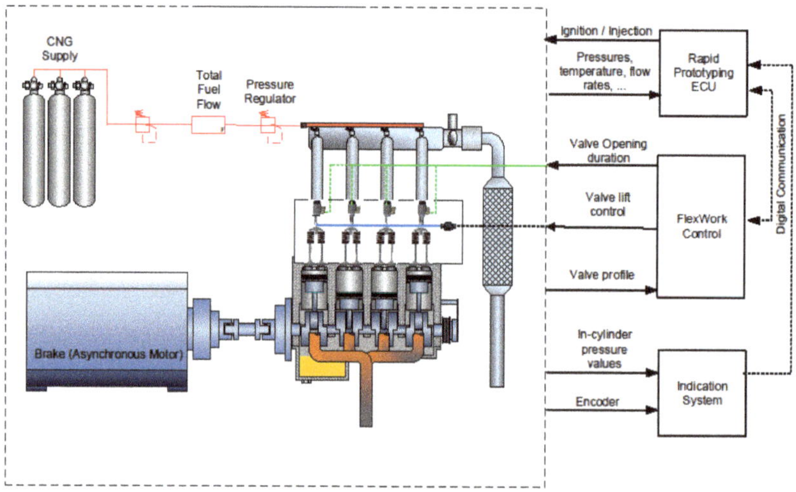

Figure 2. Scheme of the engine's and valve train's setup.

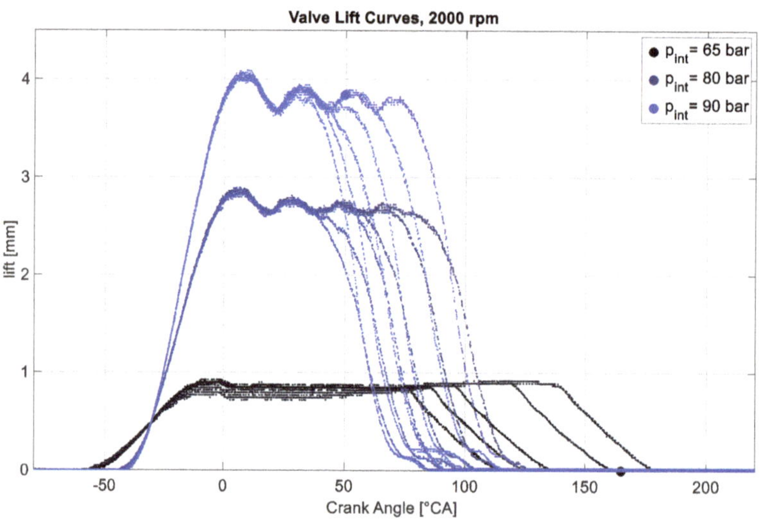

Figure 3. Measured valve lift profiles for different lift- and duration setting of the intake valves.

The intake valve lift was measured for each cylinder using the linear encoder (reported in Table 1) and it depended only on the hydraulic pressure level, which is the control input. For the sake of simplicity, only the maximum achieved lift for each pressure level was used throughout this analysis.

3. Theoretical Considerations and Description of Procedures

The mass flow ($\dot{m}$) through the intake valve can be described using an approximated form of the isenthalpic orifice equation for compressible fluids, as reported in Appendix A. The approximation was used to qualitatively describe the flow through the intake valve. When the effective area is small ($c_d A$), namely at low lifts, the mass flow decreases, hence lowering the in cylinder pressure during the aspiration phase and increasing the pumping losses [14]. The valve throttling effect is less significant at lower piston speeds, which occur around top dead center and bottom dead center, and generally at lower engine speeds.

Intake valve opening time, exhaust valve opening and closing instants, and exhaust valve lift were kept constant during all experiments described here. To compensate for cylinder-to-cylinder distribution and for the small difference between valve profiles during each test, the indicated mean effective pressure (IMEP) was kept constant in all the cylinders by closed-loop-control of IVC timing for each cylinder individually. Since the IMEP was the same in all cylinders, the mass per cylinder was approximated as one fourth of the total mass flow measured. This approximation holds by assuming that the indicated efficiency is the same in all the four cylinders.

A constant load level can be achieved with different combinations of IVC time and intake lift. For similar loads, four concurring influences driving the fuel efficiency were identified:

1. Variation in pumping losses due to different valve lift.
2. Variation in combustion efficiency due to earlier IVC time resulting in slower combustion due to a lower turbulence.
3. Influence due to lower overall pressures and temperatures during the high-pressure loop IMEP using earlier IVC timing.
4. Influence of lowering the compression ratio while keeping the expansion ratio constant.

The burned rate variation was analyzed following the procedure outlined in [15]. With this methodology, the combustion characteristics were evaluated through the mass fraction burned (x_B). The calculation procedure is reported in Appendix B. The positive and negative IMEP were calculated from cylinder pressure and volume. The area enclosed by the part of the cycle that runs clockwise (counterclockwise) represents the positive (negative) work. The negative work represents the pumping mean effective pressure (PMEP). In contrast, the 360 degree integration between top dead center and top dead center enables a correct reading in terms of total work, but it provides a false division between positive and negative IMEP [16].

To compare the Miller unthrottled experiments with normal throttled operations, the valve timings were not changed, with exception of IVC, which was set to 200 °CA after top dead center and lift, which was kept at 3.5 mm. The lift was not increased any further because, for the engine speed analyzed, no differences were observed when using higher lifts. To achieve the same load as with IVC variation, the intake manifold pressure was controlled using a throttle.

To evaluate the experimental uncertainty, the measurement uncertainty of the individual sensors has to be taken into account. Table 2 lists the uncertainties for each used measurement (m).

Table 2. Measurements uncertainty.

Measurement	Uncertainty
Fuel mass flow	$\begin{cases} 0.2\%\ mass\ flow > 1.8kg/h \\ 0.5\%\ mass\ flow < 1.8kg/h \end{cases}$
Cylinder pressure	0.6%
Air to fuel ratio	0.7%
Lower Heating Value	0.1%
Exhaust Gas Temperature	0.4%

The experimental uncertainty (ΔV) was calculated according to Equation (1), where m indicates the measurement. This calculation assumes the worst-case scenario of linear error propagation.

$$\Delta V_i = \sum \left| \frac{\partial V_i}{\partial m_i} \right| \Delta m_i \quad (1)$$

4. Results and Discussion

In the first step, described in Section 4.1, the experiments were analyzed as a function of IMEP. This was to gain a basic insight about the different influences of early intake valve closing angle and valve lift on overall efficiency, pumping losses, and combustion deterioration. Afterwards, in Section 4.2, a brief comparison between unthrottled and throttled operation is presented. In the third step, in Section 4.3, a Willans analysis was used, as explained in [17], to distinguish between influences on thermodynamic efficiencies and gas exchange losses. Finally, in Section 4.4, the first law energy balance [18] is described for a constant fuel mass flow, different valve timings, and manifold pressures. This analysis provides insights about the disposition of the initial fuel energy [19].

All experiments and analyses were performed at an engine speed of 2000 min^{-1} for different engine loads. This engine speed is chosen because it represents a typical maximal engine speed during urban driving cycles [20] where pumping losses have a big impact on the efficiency.

4.1. Intake Valve Closing Angle and Lift Combinations (IMEP Analysis)

The same loads were achieved at different valve lifts using different IVC angles. The goal was to reconstruct the influences driving the efficiency variation for different IVC and lift combinations. Thanks to the fast opening of the valves, lifts as low as 0.9 mm were relevant for analysis. Lifts higher than 4.1 mm were not analyzed because, for the chosen engine speed and loads, no effects were seen using higher lifts. Figure 4 illustrates: (a) the pumping losses; (b) the indicated efficiencies; and (c) the experimental uncertainty of the indicated efficiency as a function of IMEP and intake valve lift. Plot (d) shows the pressure volume diagram of three experiments in double logarithmic scale. The engine output of all three experiments was the same, 3 bar brake mean effective pressure (BMEP), but it was achieved with different IVC and lift combinations. These three experiments are highlighted in the (a), (b) and (c) plots in Figure 4 with blue red and green points. It has to be pointed out that the energy demand for valve actuation depends, among other parameters, on the valve lift. Smaller valve lifts mean less losses for valve actuation, especially because a large part of the actuation energy for exhaust valves is dissipative and cannot be recovered with any valve actuation system. Therefore, choosing the minimum valve lift for best indicated efficiency is generally beneficial.

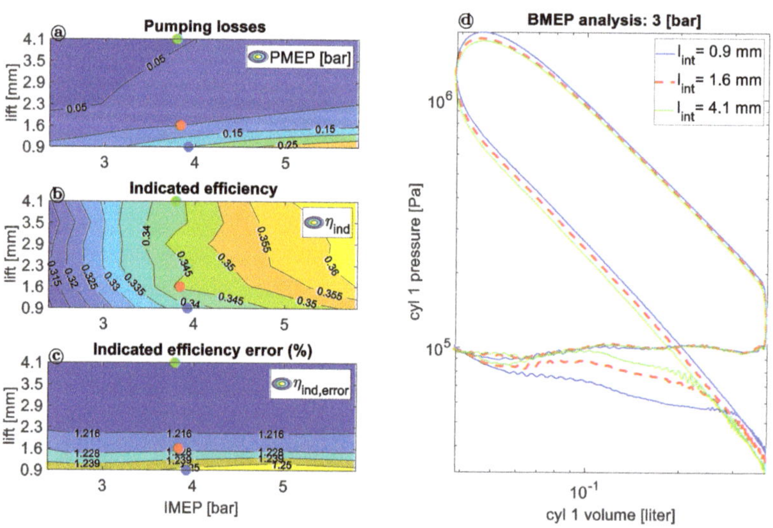

Figure 4. Analysis of efficiency and pumping losses at an engine speed of 2000 min^{-1}.

Pumping losses, plot (a) in Figure 4, considerably decreased at constant IMEP for higher lifts and they slightly increased at constant lift for higher IMEP. For higher lifts, the pressure difference between in cylinder and intake manifold (which is close to ambient pressure with throttle fully open) decreased (by same mass flow), reducing the pumping losses. To achieve the same in cylinder mass, the IVC was therefore advanced. The increase of PMEP at higher loads and constant lift can be explained by the cylinder motion. In fact, the cylinder accelerated from top dead center to maximal speed at half stroke and decelerated in the second half stroke. For this reason, the in-cylinder pressure during intake in the middle of the stroke decreased. For constant valve lift, higher pumping losses occurred at higher loads. Two factors influenced the variation of indicated efficiency, plot (b) in Figure 5. Firstly, the efficiency increased at higher loads. This is a normal trend in internal combustion engines, as some losses (e.g., wall heat losses) do not proportionally scale with load. Additionally, at higher loads, the thermodynamic efficiency increases due to the more efficient combustion. Secondly, the efficiency increased at higher lifts, thanks to the reduction in PMEP. No clear decrease in indicated efficiency was visible at high lifts. In fact, early IVC time could not only be responsible for reduced PMEP, but also for combustion deterioration, and hence lower efficiency due to lack of turbulence. Nevertheless gas exchange improvement at the engine speed of 2000 min^{-1} investigated here was found to always overcome the possible degradation of the combustion process. The valve train was able to achieve valve lifts up to 9 mm, but for the engine speed of 2000 min^{-1} discussed here lifts above 4 mm did not show an additional gain. As higher valve lifts need more energy for the valve train itself, which would negatively affect the engine's effective work, an adaption of the valve lift to the engine speed is beneficial.

The pV diagram, plot (c) in Figure 4, depicts the same loads, achieved with different combinations of IVC angle and lift. The three analyzed experiments are also shown as points in plots (a), (b) and (c) in Figure 5. The important parameters for these experiments are listed in Table 3.

Table 3. pV diagram experiments details. IVC – intake valve closing; PMEP – pumping mean effective pressure.

	Blue	Red	Green
IVC	111 °CA	97 °CA	84 °CA
Lift	0.9 mm	1.6 mm	4.1 mm
PMEP	0.190 bar	0.099 bar	0.047 bar
η_{ind}	33.6%	34.6%	34.4%

At higher lifts, the IVC timing was advanced to keep the load constant. Earlier IVC corresponds to lower effective compression ratio, lowering the overall pressure during the high pressure loop (shown in Figure 4 plot d). As expected, the pumping losses increased at lower lifts (up to four times). During the gas exchange process (where pumping losses occur), the lines diverged most between 0.1 and 0.3 L (visible in Figure 4 plot d). The pumping losses were mainly present in the middle of the stroke, where the piston moves fastest. Comparing these unthrottled operating points, the fuel consumption was reduced up to 3% (relativ) at 3 bar BMEP with higher lifts and earlier valve closing.

To understand the magnitude of the combustion deterioration, the burned rate variation was analyzed. Figure 5 depicts the flame development phase (0–5% of mass fraction burned), the main combustion duration (5–95% of mass fraction burned), and the IVC angle as a function of IMEP and lift.

The flame development phase and combustion duration show inverse correlation with IVC time. This can be explained taking into account the turbulence dissipation inside the cylinder. Once the intake valves were closed, the turbulent kinetic energy dissipated, and the lack of turbulence slowed down the flame propagation. At constant load, a higher lift corresponded to earlier IVC and lead, therefore, to a longer combustion and flame development phase. In spite of the fact that the combustion duration increased for higher lift and earlier IVC, the efficiency increased as a combined effect of combustion deterioration, changes in heat losses (see also Section 4.4) and a decrease in pumping loss.

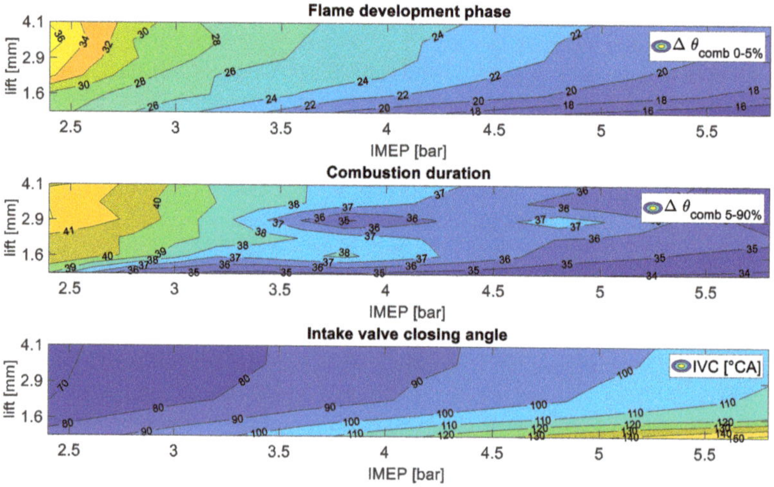

Figure 5. Combustion duration and IVC angle.

Cycle-to-cycle-variation is shown in Figure 6. It depicts the coefficient of variation (COV) of IMEP as a function of lift and IMEP. The COV_{IMEP} was calculated according to Equation (3) and was evaluated for 167 consequently recorded cycles.

$$COV_{IMEP} = \frac{std(IMEP)}{mean(IMEP)} \times 100 \tag{2}$$

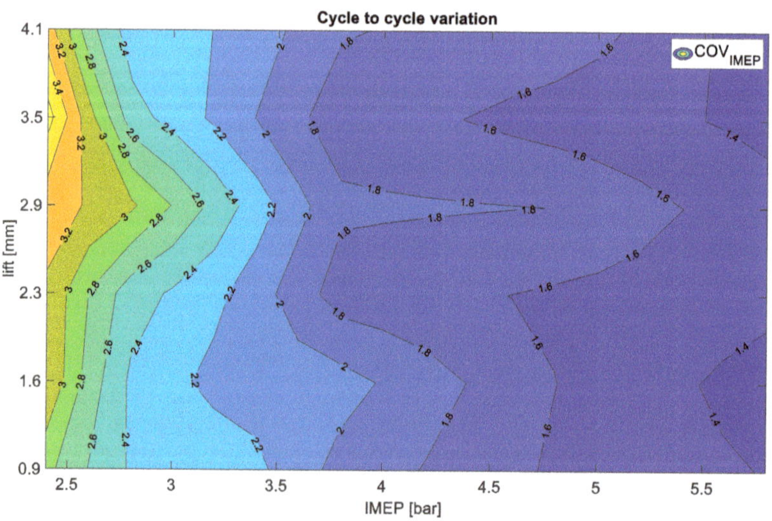

Figure 6. Cycle to cycle variation.

No significant influence on lift/IVC on cycle-to-cycle-variation was observed; COV_{IMEP} depends mainly on the in cylinder mass. All levels were unproblematic as it is usually accepted that COV_{IMEP} values below 5% lead to very smooth engine operation. It must be mentioned that cyclic variation can be more prominently influenced by changing the valve overlap, which was not done in this analysis.

4.2. Throttled versus Unthrottled Operation

Measurements were made to compare the standard throttled Otto cycle, where IVC is set to 200 °CA and load is controlled by adjusting the pressure in the intake manifold, with the Miller cycle where the intake valve timing is adjusted while the throttle upstream the intake manifold remains open. Table 4 lists, for three IMEP values, the indicated efficiencies and the pumping losses for the throttled Otto cycles, as well as for the Miller cycles using best efficiency IVC lift/timing combinations.

Table 4. Comparison between throttled and unthrottled operation.

	IMEP [bar]	Indicated Efficiency	Pumping Losses [bar]	Fuel Saving Compared to Throttled Operation
Throttled Otto cycles	2.5	28.8%	0.389	
	3.5	32.1%	0.365	
	4.5	34.0%	0.324	
Unthrottled Miller cycles	2.5	31.8%	0.032	9.4%
	3.5	34.1%	0.053	6.1%
	4.5	35.4%	0.057	3.9%

For all three IMEP values (2.5, 3.5 and 4.5 bar) significant fuel savings were achieved using Miller cycles instead of throttled operation.

Figure 7 depicts the pressure volume diagram of the throttled and unthrottled case for 4 bar BMEP.

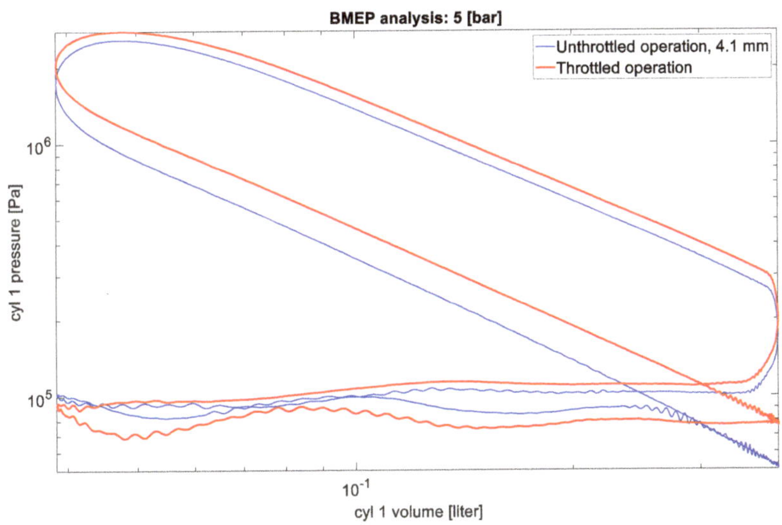

Figure 7. Pressure volume diagramm.

Even at medium load, where the differences in pumping losses are low (0.3 and 0.06 bar), an improvement in efficiency of 3.9% was recorded. This suggests that the combustion deterioration losses due to lack of turbulence are often overestimated or mitigated by other factors.

4.3. Willans Approximation to Decouple Thermal Efficiency and Pumping Losses

Typically, the energetic behavior of an energy conversion or transmission device is described by its efficiency, which is the ratio of the useful to the invested energy. Energy conversion devices always have intrinsic dissipative losses, which means that they need a certain input to cover dissipation, even if no output is delivered. This leads to nonlinear efficiency versus load curves, especially towards low output. In order to systematically analyze the energetic input-to-output behavior of an energy conversion device, the direct representation of output versus input-energy turns out to be much more meaningful than the study of the efficiency, or the specific fuel consumption, versus load. This kind of representation is usually called a "Willans plot" [21], referring to observations of Peter Willans [22], who saw in the late 19th century that input versus output power on high-speed steam engines can be represented by an affine relationship.

Equation (3) outlines the affine Willans approximation for internal combustion engines, which returns the IMEP (output) as a function of fuel mean effective pressure (FuelMEP, input). PMEP represents the gas exchange losses (calculated as stated in previous chapter) and the Willans efficiency (e_W) represents the thermodynamic properties of the engine [17].

$$\underbrace{IMEP}_{-\frac{1}{V_d}\oint pdV} = e_w \underbrace{FuelMEP}_{\frac{2\cdot\dot{m}_{fuel}\cdot LHV}{n\cdot V_d}} - \underbrace{PMEP}_{\substack{negative\ area \\ pv\ diagram}} \tag{3}$$

FuelMEP, which is proportional to the fuel quantity, can be interpreted as the IMEP level an engine would deliver without pumping losses (PMEP = 0) and with a hypothetical efficiency of 100% ($e_W = 1$, i.e., perfect conversion of the fuel's thermal energy into work) [17]. The Willans efficiency e_W can be interpreted as the engine's "inner efficiency" driven by the quality of the thermodynamic cycle.

Equation (4) outlines the energy conservation equation for a control volume that surrounds the engine [19].

$$\dot{m}_{fuel}Q_{LHV} = P_p + \dot{Q}_w + \dot{H}_e + \dot{H}_{e,ic} \tag{4}$$

The fuel energy power calculated with the fuel mass flow ($\dot{m}_{fuel}$) and fuel lower heating value (Q_{LHV}) is divided between power delivered to the pistons (P_p), wall heat losses ($\dot{Q}_w$), exhaust enthalpy ($\dot{H}_e$), and exhaust enthalpy loss due to incomplete combustion ($\dot{H}_{e,ic}$). By integrating Equation (4) over one engine cycle and dividing it by the displacement volume (V_d), the specific per cycle mean effective pressures were derived. Equation (5) outlines the energy equation in mean effective pressure terms.

$$\begin{gathered} FuelMEP = IMEP + Q_wMEP + H_eMEP + H_{e,ic}MEP \\ FuelMEP = IMEP^+ + PMEP + Q_wMEP + H_eMEP + H_{e,ic}MEP \end{gathered} \tag{5}$$

In Equation (5), the IMEP is further divided into positive IMEP ($IMEP^+$) and PMEP. The remaining terms are wall heat losses mean effective pressure (MEP), incomplete combustion enthalpy MEP, and exhaust enthalpy MEP. The latter was calculated according to Equation (6) [17]. The temperature used in Equation (6) is the one measured at exhaust manifold entrance and the mass (m) is the total mass flowing through the engine in one engine cycle.

$$\begin{gathered} H_eMEP = \frac{c_p m T}{V_d} \\ c_p = \begin{cases} 573\ T^{0.097} & T < 700K \\ 392\ T^{0.155} & T > 700K \end{cases} \end{gathered} \tag{6}$$

Since precise calculation of wall heat losses and incomplete combustion enthalpy need complex modeling, these two terms were combined and named rest heat losses mean effective pressure ($Q_{rest}MEP$). They were calculated as the remaining part of the FuelMEP according to Equation (7).

$$Q_{rest}MEP = FuelMEP - Q_EMEP - IMEP \tag{7}$$

This approximation is valid for two reasons. First, the remaining term is mostly dependent on the wall heat losses, which normally have a minimum share of 20% of fuel energy compared to a maximal share of incomplete combustion of 5% [19]. Second, both wall heat losses and incomplete combustion (quenching and blowby) are expected to decrease for lower pressures and temperatures. For the same load, shorter IVC time with higher lift results in lower pressures and therefore lower temperatures, reducing the wall heat losses. For lower pressure, the amount of fuel in the crevice volume and the amount of leakage decreased, lowering the incomplete combustion enthalpy.

As the Willans approximation was used to split the influences into pumping losses and fuel conversion efficiency at constant IMEP, it allowed us to separate the effect that early IVC has on gas exchange and thermodynamic properties. Figure 8 depicts, from left to right, the pumping losses PMEP, the Willans efficiency e_W, and the fuel mean effective pressure FuelMEP as a function of lift for three different IMEP.

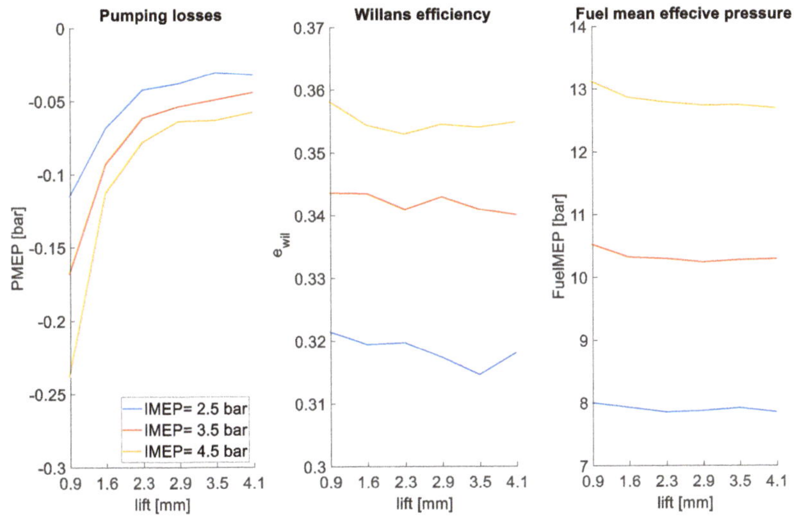

Figure 8. Willans pararameters.

Higher Willans efficiency was reached at higher IMEP because higher in cylinder mass generally results in better Willans conversion efficiency. The Willans efficiency at constant IMEP slightly decreased with higher valve lifts (earlier IVC), which was a combined effect of slower combustion, lower heat losses (detailed discussion see below), and lower in cylinder mass. As expected, when using early IVC to control the load a clear reduction in PMEP was visible at higher lifts, and therefore earlier IVC for all three IMEP. FuelMEP decreased with increasing valve lift (earlier IVC) and, since FuelMEP is proportional to the fuel use, this suggest that increasing valve lift (earlier IVC) has a net beneficial effect.

The variations in Willans efficiency are driven by various influences. Figure 9 illustrates some factors affecting these variations. It depicts, from left to right, the combustion duration, the rest heat losses mean effective pressure (composed from wall heat losses and incomplete combustion), and the exhaust heat losses mean effective pressure as a function of intake valve lift for the three IMEP levels discussed.

An increase in combustion duration was observed at constant IMEP for higher lifts because of the aforementioned turbulence decrease with earlier IVC, which reduce the Willans efficiency. At the same time, the wall heat losses for higher lifts decreased and the exhaust heat losses remained approximately constant. The combined effect was, as shown in Figure 8, a slight reduction of the Willans efficiency.

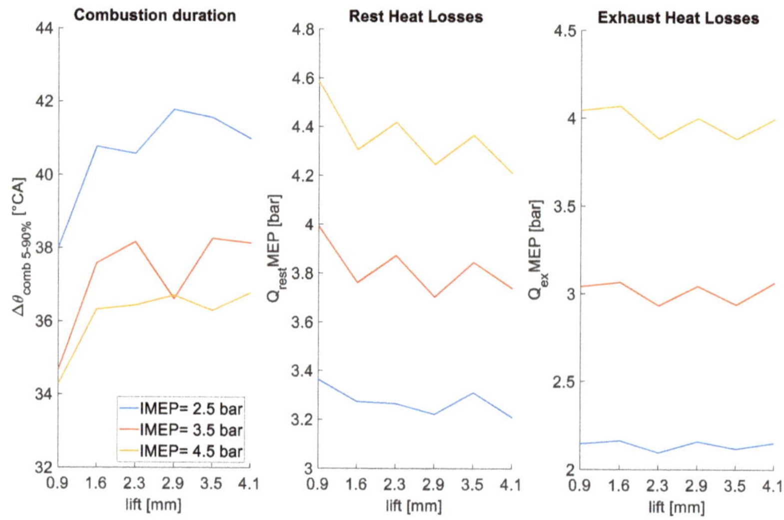

Figure 9. Influence of intake valve lift on Willans efficiency.

Based on the Willans analysis for the engine considered, it can be concluded that the positive effect on engine efficiency of reducing pumping losses with early IVC is stronger than the losses attributed to early IVC. A tradeoff, or at least an attenuation of the disadvantage of the slightly less efficient combustion, can be recognized in the variation of the heat losses and in the incomplete combustion due to the lower pressures and temperatures achieved with early IVC. This experimental observation is in opposition to some results from the literature [7,23].

4.4. Engine Energy Balance

Variation in the Willans efficiency at constant IMEP and different IVC lift combination depend on the variation of in cylinder mass, combustion deterioration, exhaust heat losses, wall heat losses, and incomplete combustion heat losses. In order to interpret the influences driving the efficiency variation, an energy balance analysis was carried out for a constant FuelMEP level. Since FuelMEP is directly proportional to the fuel mass and the engine is strictly run at $\lambda = 1$, this resulted in a constant in cylinder air- and fuel mass. For this analysis, the losses arising from wall heat losses and incomplete combustion were grouped together into rest heat MEP. In this way, the FuelMEP (i.e., the hypothetical IMEP a 100% efficient engine would produce) was divided between IMEP, exhaust heat MEP, and rest heat MEP.

Figure 10 shows this analysis for 9 bar and 13 bar FuelMEP. On the left sides, the FuelMEP was divided into exhaust heat MEP, IMEP, and the rest heat MEP. On the right side, the IMEP was further split into positive IMEP and PMEP. Four points with the same FuelMEP were compared. The first point represents throttled operation (TB), the remaining ones represents three Miller cycles achieved with increasing valve lift (therefore earlier IVC). The right plot in Figure 10 shows, beside the division between the different terms of the energy balance, their relative experimental uncertainties as a percentage of FuelMEP, which was calculated according to Equation (1). IMEP and PMEP uncertainty arises only from pressure measurement, whereas the exhaust MEP uncertainty is derived from the air to fuel ratio, fuel mass flow, and exhaust temperature measurements. The remaining MEP uncertainty is derived from the sum of the previous uncertainties and from the FuelMEP uncertainty, which depends on air to fuel ratio, fuel mass flow, and lower heating value measurements.

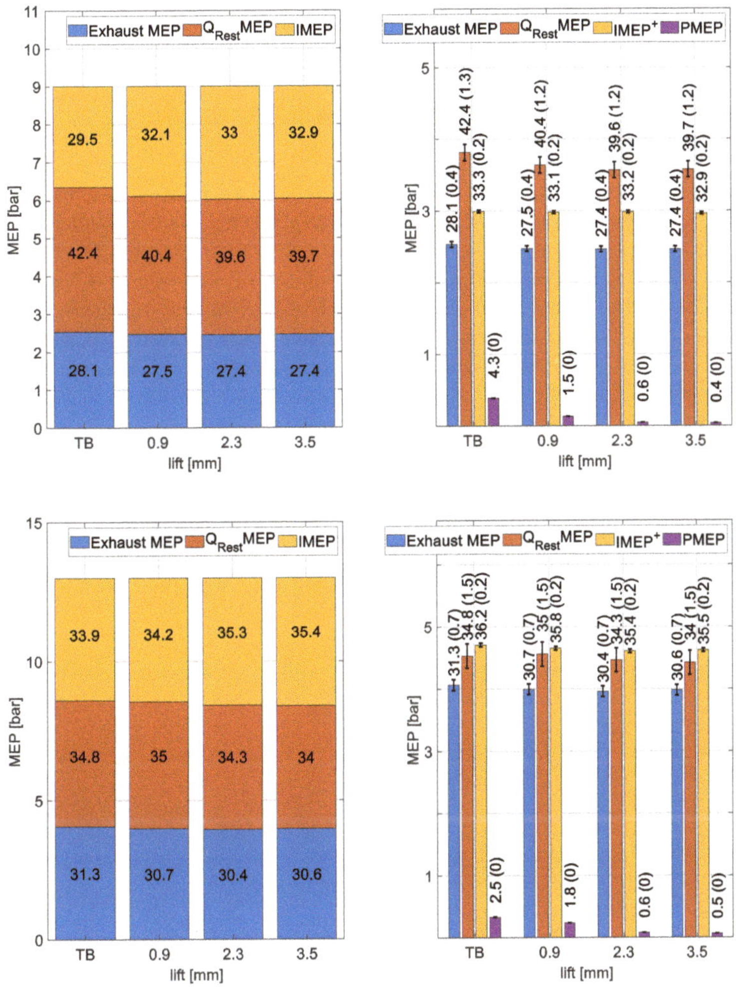

Figure 10. Energy balance at 9 bar (above) and 13 bar (below) FuelMEP for throttled (TB) operation and three levels of valve lift (three values of IVC).

On the left side in Figure 10, the increase in IMEP (from left to right columns) depicts the increase in indicated efficiency with higher valve lift (earlier IVC). To compensate for this effect, both wall heat losses MEP and exhaust heat MEP decreased. As the right side shows, the increase in efficiency was dominated by the reduction in PMEP. In spite of the fact that the wall heat losses and the exhaust heat losses decreased, the positive IMEP slightly decreased with higher valve lift (earlier IVC) because of the turbulence effect discussed.

5. Conclusions

This paper presents the experimental results achieved by applying early IVC (Miller cycle) to a four-cylinder engine operated with natural gas and fitted with a fully variable electro-hydraulic valve train. The focus of the work presented is the effect of IVC at low loads, and IVC effects on boosted operation were not considered. Loads ranging from 1.5 up to 5 bar BMEP were achieved in unthrottled operation with different combinations of IVC and intake lift, and the results were compared to throttled operation. Pumping losses reduction achieved with early IVC explains the efficiency gains achievable using early IVC. An indicated efficiency increase of up to 10% was observed when compared to throttled operations using high lift with an early IVC angle. Efficiency increases of up to 3% were observed by comparing unthrottled operations with higher lifts. As expected, a longer combustion duration resulted from earlier IVC angles. The earlier the valve closes, the more the lack of turbulence during combustion slows down the flame propagation. Therefore, the heat supply is less efficient. To understand the net effects, a Willans approximation was used to split the influences into pumping losses and fuel conversion efficiency at constant IMEP for various IVC and lift combinations. The Willans efficiency decrease for earlier IVC angles was weaker than expected, the reason being that multiple factors affect the Willans efficiency variation (in cylinder mass, heat losses, and combustion efficiency). The simplified energy balance for constant FuelMEP outlines that the earlier IVC angles lower heat losses (wall, incomplete combustion and exhaust ones), since early IVC strategies reduce in-cylinder temperature levels. For example, at low loads (ca. 3 bar IMEP), by advancing IVC the combustion deterioration was almost completely compensated by the reduction in heat losses so that the effect of reduced pumping losses with early IVC is fully exploited.

Studies in the literature typically report a tradeoff between pumping losses and combustion deterioration for premixed combustion [4,7,23]. The present experimental research shows that the behavior of the heat losses is an additional factor that also has to be considered, as it plays an important role in the overall energy balance. It can therefore be concluded that early IVC does not only reduce turbulence (therefore increasing combustion duration), but it also lowers pressure and temperature, thus leading to a significant reduction in wall heat losses, blowby, incomplete combustion, and also exhaust gas temperature.

Author Contributions: Conceptualization, M.B., N.Z. and P.S.; formal analysis, M.B, L.M.; writing—original draft preparation, M.B.; writing—review and editing, L.M. and P.S.; visualization, M.B.; supervision, N.Z. and P.S.; funding acquisition, P.S. All authors have read and agreed to the published version of the manuscript.

Funding: This research is part of the project "FlexWork" which is funded by the Swiss Federal Office of Energy, grant number SI/501619-01.

Acknowledgments: The authors express their gratitude to Wolfgang Schneider (Wolfgang Schneider Ingenieurbüro, Thun) and Andyn Omanovic (Empa) for their contributions in inventing and realizing the electro hydraulic valve train. Special thanks go also to the technical staff of Empa (H. Ehrensperger, R. Spühler, R. Graf) who set-up the experiments and modified the engine with great care.

Conflicts of Interest: The authors declare no conflict of interest.

Appendix A

The flow through a valve can be described through the isenthalpic orifice equation for compressible fluids. The approximation is valid for constant isentropic coefficient of 1.4 and a pressure ratio over the valve bigger then 0.5. Furthermore, the approximation assumes no losses in the accelerating part and fully turbulent flow after the narrowest point. The mass flow depends on the upstream (p_{in}) and downstream (p_{out}) pressure of the valve, as well as on the area of the valve (A), the discharge coefficient (c_d), and the fluid density (ρ) according to Equation (A1) [17].

$$\dot{m}(t) \approx c_d A(t) \sqrt{2\rho\, p_{out}\left[1 - \frac{p_{out}}{p_{in}}\right]} \tag{A1}$$

Appendix B

The burned rate variation was analyzed following the procedure outlined in [15]. With this methodology, the combustion characteristic was evaluated through the mass fraction burned (x_B), which was estimated from the cylinder pressure trace (p), the volume (V), and from the per cylinder per cycle mass flow (m). By neglecting quenching, the fuel energy converted (ΔQ_f) is transferred to the gas (ΔQ_n) and partially to the walls (ΔQ_w). The heat transferred to the gas (Equation (A2)) was approximated using Rassweiler and Withrow's assumption [10].

$$\Delta Q_n = mc_v(T_{i+1} - T_i) = \frac{c_v}{R} V_{i+1}\left[p_{i+1} - p_i\left(\frac{V_i}{V_{i+1}}\right)^k\right] \tag{A2}$$

The heat transferred to the wall was approximated with Hohenberg's assumption (Equation (A3)) where n is the engine speed and $\Delta\alpha$ is the angular increment. The heat transfer coefficient (α_w) was calculated as a function of in cylinder pressure (p), temperature (T), volume (V), and mean piston speed (c_m).

$$\begin{gathered} \Delta Q_w = [\alpha_w A_w (T - T_w)] \frac{\Delta\alpha}{6n} \\ \alpha_w = 0.0013 V^{-0.06} p^{0.8} T^{0.4} (c_m + 1.4)^{0.8} \; J/m^2K \end{gathered} \tag{A3}$$

The temperature (T) isentropic coefficient (k) and the specific heat at constant volume/pressure (c_v/c_p) were calculated according to Equation (A4).

$$\begin{gathered} T_i = \frac{p_i V_i}{mR} \\ k = \frac{c_p}{c_v} \\ c_v = 700 + 0.255T \; [J/kgK] \end{gathered} \tag{A4}$$

This methodology returns the mass fraction burned as a function of the crank angle, as shown in Equation (A5).

$$x_B(\theta_i) = \frac{\sum_0^i \Delta Q_f}{\sum_0^N \Delta Q_f} = \frac{\sum_0^i \Delta Q_n + \Delta Q_w}{\sum_0^N \Delta Q_n + \Delta Q_w} \tag{A5}$$

The heat release rate was calculated, from spark to bottom dead center, on the expansion stroke. The calculations were discreetly done in 0.2 °CA steps using measurements at the current angle (subscript i) and on the next discrete value (subscript i+1). The end of combustion wass set at the location of the first negative chemical fuel heat release transferred to the gas (subscript n) [24].

The two blue lines in Figure A1 represent the heat release rate (full line) and the wall heat losses (dashed line). The red points depict the mass fraction burned evolution calculated according to Equation (A5) for a 5 bar BMEP achieved with a 0.9 mm lift and an IVC angle of 155 °CA.

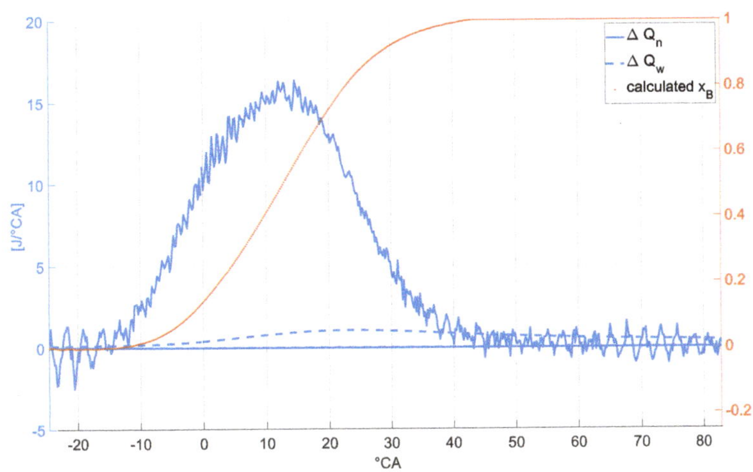

Figure A1. Mass fraction burned calculation.

References

1. Zsiga, N.; Omanovic, A.; Soltic, P.; Schneider, W. Functionality and Potential of a New Electrohydraulic Valve Train. *MTZ Worldw.* **2019**, *80*, 18–27. [CrossRef]
2. Zsiga, N.; Soltic, P.; Schneider, W. FlexWork—Lastregelung bei einem Motor mit elektrohydraulischem, vollvariablen Ventiltrieb. Presented at the Ladungswechsel und Emissionierung, 11, Stuttgart, Germany, 24 October 2018.
3. Shiao, Y.; Dat, L.V. Efficiency Improvement for an Unthrottled SI Engine at Part Load. *Int. J. Automot. Technol.* **2012**, *13*, 885–893. [CrossRef]
4. Li, T.; Gao, Y.; Wang, J.; Chen, Z. The Miller cycle effects on improvement of fuel economy in a highly boosted, high compression ratio, direct-injection gasoline engine: EIVC vs. LIVC. *Energy Convers. Manag.* **2014**, *79*, 59–65. [CrossRef]
5. Lanzanova, T.; Nora, M.D.; Zhao, H. Investigation of Early and Late Intake Valve Closure Strategies for Load Control in a Spark Ignition Ethanol Engine. *SAE Int. J. Engines* **2017**, *10*, 858–872. [CrossRef]
6. Schmitt, M.; Hu, R.; Wright, Y.M.; Soltic, P.; Boulouchos, K. Multiple Cycle LES Simulations of a Direct Injection Natural Gas Engine. *Flow Turbul. Combust.* **2015**, *95*, 645–668. [CrossRef]
7. Knop, V.; Mattioli, L. An analysis of limits for part load efficiency improvement with VVA devices. *Energy Convers. Manag.* **2015**, *105*, 1006–1016. [CrossRef]
8. Unger, H.; Schwarz, C.; Schneider, J.; Koch, K.-F. The Valvetronic: Experience from Seven Years of Mass Production and a Discussion of Future Prospects. *MTZ Worldw.* **2008**, *69*, 30–37. [CrossRef]
9. Luttermann, C.; Schünemann, E.; Klauer, N. Enhanced VALVETRONIC Technology for Meeting SULEV Emission Requirements. *SAE Technol. Pap.* **2010**, *1*, 7. [CrossRef]
10. Mendera, M.S.K.Z.; Spyra, A. Mass fraction burned analysis. *J. KONES Intern. Combust. Engines* **2002**, *3*, 193–201.
11. Soltic, P.; Hilfiker, T.; Hänggi, S. Efficient light-duty engine using turbulent jet ignition of lean methane mixtures. *Int. J. Engine Res.* **2019**, *70*, 123–134. [CrossRef]
12. Schechter, M.M.; Levin, M.B. Camless Engine. *SAE Tech. Pap.* **1996**. [CrossRef]
13. Zsiga, N.; Omanovic, A.; Soltic, P.; Schneider, W. Wirkungsgradvorteile beim Ottomotor unter Verwendung einer nockenwellenlosen, vollvariablen Ventilsteuerung gegenüber gedrosseltem Betrieb. Presented at the 8, VDI-Tagung Ventiltrieb und Zylinderkopf, Würzburg, Germany, 25 June 2019.
14. Chiang, C.J.; Yang, J.L.; Lan, S.Y.; Shei, T.W.; Chiang, W.S.; Chen, B.L. Dynamic modeling of a SI/HCCI free-piston engine generator with electric mechanical valves. *Appl. Energy* **2013**, *102*, 336–346. [CrossRef]
15. Tomić, M.V. A quick, simplified approach to the evaluation of combustion rate from an internal combustion engine indicator diagram. *Therm. Sci.* **2008**, *12*, 85–102. [CrossRef]
16. Shelby, M.H.; Stein, R.A.; Warren, C.C. A New Analysis Method for Accurate Accounting of IC Engine Pumping Work and Indicated Work. Presented at the SAE 2004 World Congress & Exhibition, Detroit, MI, USA, 8–11 March 2004. [CrossRef]
17. Guzzella, L.; Onder, C.H. *Introduction to Modeling and Control of Internal Combustion Engine Systems*; Springer Science & Business Media: Berlin, Germany, 2010.
18. Weberbauer, F.; Rauscher, M.; Kulzer, A.; Knopf, M.; Bargende, M. Generally Applicate Split of Losses for New Combustion Concepts. *MTZ Worldw.* **2005**, *66*, 17–19. [CrossRef]
19. Heywood, J.B. *Internal Combustion Engine Fundamentals*; McGraw-Hill: New York, NY, USA, 1988.
20. Zhao, J.; Xi, Q.; Wang, S.; Wang, S. Improving the partial-load fuel economy of 4-cylinder SI engines by combining variable valve timing and cylinder-deactivation through double intake manifolds. *Appl. Therm. Eng.* **2018**, *141*, 245–256. [CrossRef]
21. Greene, A.B.; Lucas, G.G. *The Testing on Internal Combustion Engines*; English University Publisher: London, UK, 1969.
22. Hills, R.L. *Power from Steam: A History of the Stationary Steam Engine*; Cambridge University Press: Cambridge, UK, 1989.
23. Sun, T. Experimental research on pumping losses and combustion performance in an unthrottled spark ignition engine. *Proc. Inst. Mech. Eng. Part A J. Power Energy* **2018**, *232*, 888–897. [CrossRef]
24. Central, P. Combustion analysis and cycle-by-cycle variations in spark ignition engine combustion Part I: An evaluation of combustion analysis routines by reference to model data. *Proc. Inst. Mech. Eng. Part D J. Automob. Eng.* **1998**, *212*, 381–399.

energies

MDPI

Article

New Design of Copper–Inconel 601 Ground Electrode Spark Plug Based on a Thermo-Mechanical Model

Chawki Tahri [1,2,3,*], Helmut Klocker [3], Bernard Beaugiraud [4], Christophe Bertoni [2], Eric Feulvarch [1] and Jean-Michel Bergheau [1]

1 LTDS, UMR 5513 CNRS, Ecole Nationale d'Ingénieurs de Saint-Etienne, Université de Lyon, 42023 Saint-Etienne, France; eric.feulvarch@enise.fr (E.F.); jean-michel.bergheau@enise.fr (J.-M.B.)
2 Federal Mogul Powertrain Ignition Products SAS, 42140 Chazelles sur Lyon, France; Christophe.bertoni@tenneco.com
3 LGF, UMR 5307 CNRS, Mines Saint-Etienne, Université de Lyon, 42023 Saint-Etienne, France; klocker@emse.fr
4 LTDS, UMR 5513 CNRS, Ecole Centrale de Lyon, Université de Lyon, 69134 Ecully, Lyon, France; bernard.beaugiraud@ec-lyon.fr
* Correspondence: chawki.tahri@enise.fr; Tel.: +33-7-69-24-56-90

Received: 25 June 2020; Accepted: 22 July 2020; Published: 24 July 2020

Abstract: Inconel 601 is one material of choice for intermediate- to high-temperature protective coatings for spark plugs' ground electrodes. Production of ground electrodes of spark plugs implies the following operations: the tamping of the copper core in an Inconel 601 cup, cold-forming of the assembly, annealing, welding, and bending of the final spark plug. On the production line, the use of Inconel 601 as a protective coating for ground electrodes leads to possible cracking in the welded area after bending. In the present paper, possible causes of cracking are analyzed. It is clearly shown that a combination of Copper –Inconel interface oxidation, Inconel yielding during the heat treatment, and micro-movements during bending lead to cracks in the welded area of the ground electrode. First, the detrimental effect of gaps, between Copper and Inconel 601, is shown experimentally. Second, a thermo-mechanical analysis combined with SEM (Scanning Electron Microscopy) observations identified the annealing treatment and interface oxidation as the main cause of gaps. Third, bending simulations show the relation between these gaps and cracking. Finally, a new ground electrode design, preventing cracks, is suggested.

Keywords: spark plug; new design; experimental analysis; thermo-mechanical model; Inconel 601

1. Introduction

Considering the increased cost of oil and the stringent emission regulations, engine design should satisfy three important expectations: lower fuel consumption, lower emission, and correct function of engine. The combustion of air/fuel mixture is the main task of spark plugs in spark ignition engines. Spark plugs produce about a thousand sparks per minute in a combustion chamber under high pressures and temperatures [1]. Engine modifications to decrease emission and lower fuel consumption cause intensified thermal loads on the spark plugs [2]. To resist high-temperature wear and corrosion, ground electrodes of spark plugs formed by a copper core and a high-temperature corrosion-resistant Nickel-based alloy mantle have been developed. Inconel 601 is the material of choice for high-temperature corrosive environments [3,4].

Coating with Inconel 601 will be done by *lubricated cold working* (Figure 1a,b). Then, the ground electrodes are submitted to an annealing (in a very large furnace) at a temperature of 1040 °C during 20 min in a protected atmosphere containing hydrogen and nitrogen (Figure 1c). Cooling is performed using liquid nitrogen. Finally, the assembly is assembled to a steel shell by resistance braze welding

(Figure 1d), and bent (Figure 1e,f). Bending is done in a specially designed tool. Several operations are used to guarantee perfect alignment of the tool and the ground electrode. Resistance braze welding is commonly used for assembling different kind of materials (copper, nickel-based alloy, and steel). The effect of different welding parameters was analyzed in detail in Reference [5].

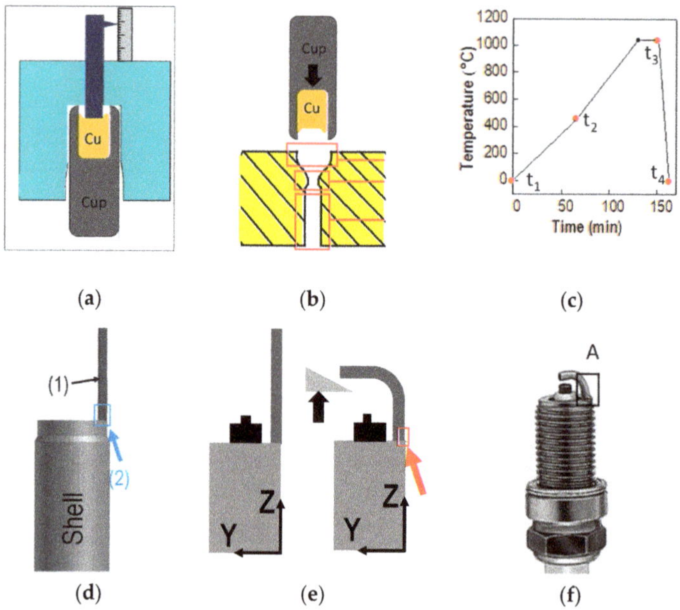

Figure 1. Manufacturing processes of a ground electrode. (**a**) Tamping of the copper in Inconel 601 cup, (**b**) cold working of the assembly, (**c**) annealing: temperature variation, (**d**) welding of the ground electrode to the steel shell [5], (1) and (2) indicate respectively the ground electrode and the welded zone, (**e**) bending inducing cracks (red rectangle), (**f**) final spark plug with crack in A.

The use of Inconel 601 led to some unanticipated difficulties during the different production phases. After bending, *a small proportion (ppm)* of spark plugs revealed unexpected cracking in the welded volume. The first investigations of spark plugs with cracks systematically revealed a gap between the copper core and the Inconel 601 mantle (Figure 2). Experimental evidence thus pointed to a strong correlation between cracking in the welded zone and the presence of gaps. High-temperature corrosion of Inconel 601 has been analyzed extensively in the literature [3,4]. However, the present paper analyzes the synergetic effects between Cu–Inconel 601 interface pollution, heat treatment-induced Inconel yielding, and cracking due to micro-movement during bending. Only the combination of experimental characterizations with thermo-mechanical analysis of the heat treatment and the bending procedure allow to understand these synergetic effects. To the best of our knowledge, this is completely novel and leads to an innovative ground electrode design.

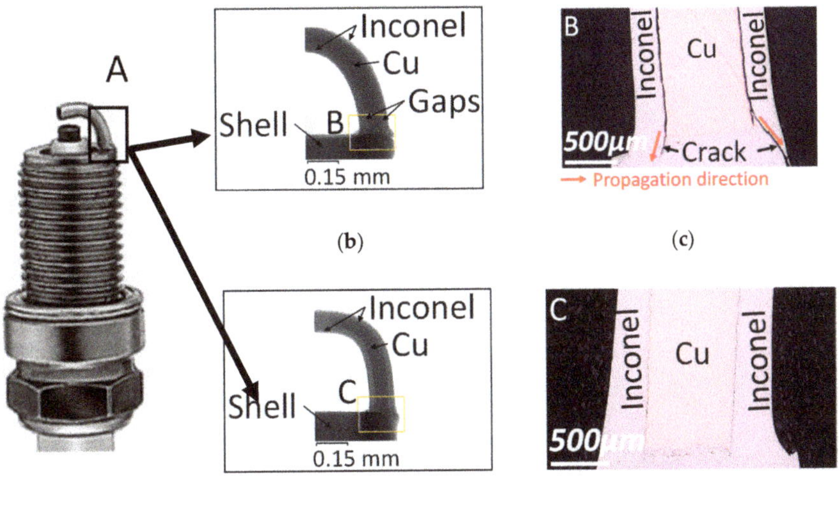

(**a**) (**b**) (**c**) (**d**) (**e**)

Figure 2. (**a**) Spark plug, Tomographic (**b**,**d**) and light microscopic (**c**,**e**) observations of the welded zone after bending (zone A in (**a**)). After cold forming, the ground electrode (**d**,**e**) was cleaned prior to the annealing treatment under lab conditions (Section 2.1).

Section 2 is devoted to the experimental analysis. First, samples were prepared under laboratory conditions, permitted to generate two series of ground electrodes with and without gaps, respectively. A systematic analysis confirmed that gaps cause cracking during bending. At this point, the detrimental effect of gaps is thus confirmed. However, the possible cause of gaps is still to be elucidated. Therefore, Section 2 ends with a microscopic characterization of samples exhibiting gaps. These analyses clearly reveal an oxide layer at the copper–Inconel interface. Thus, bad interface adhesion may be suspected as the main cause of gaps. This led to the thermo-mechanical analysis of the annealing treatment offered in Section 3. This model confirms the tendency of gap formation during heat treatment for small adhesion between the copper core and Inconel mantle. The welding was analyzed in detail in Reference [5]. Gaps induced by the heat treatment will not disappear during welding. Sections 2 and 3 confirm the causes of gap formation: bad adhesion and yielding of the Inconel mantle during the heat treatment. However, at this stage, the detrimental effect of possible gaps during the bending operation still has to be clarified. Possible micro-movements during bending were suspected to cause cracking. The bending model in Section 4 compares the stress distribution in electrodes with and without gaps. Effectively, gaps lead to stress concentrations in the welded volume prone to cracking. At the end of Section 4, all the possible causes of cracks induced by residual gaps are clearly understood. Clearly, gaps have to be avoided at any cost. As mentioned above, only a few parts per million of ground electrodes exhibit heat treatment-induced gaps and consequent cracking during bending. The microscopic observations of Section 2 suggest bad adhesion due to oil pollution at the interface. This pollution corresponds to oil residues after the cold forming of the Inconel mantle. Consequently, two solutions may be considered. The first solution consists of very costly cleaning and control procedures to guarantee oil-free interfaces and consequently, good adhesion between the copper core and Inconel mantle. Based on the mechanical analysis of the heat treatment, we opted for a less expensive solution. Section 5 presents a new ground electrode design preventing large plastic deformations of the Inconel mantle and, hence, avoiding the gap formation.

2. Experimental

2.1. Materials and Process

Two materials were used in this study: Cu-OF (Free-Oxygen Copper)high-purity copper wires and Inconel 601 [6–9] coatings. The chemical compositions are given in Table 1. Oil residues after cold forming were suspected to cause bad adhesion, and consequently, gaps after the heat treatment. Hence, samples were produced under laboratory conditions. To ensure cleanliness, the Inconel buckets were thoroughly washed, and cleaned in an ultrasonic bath (96° ethyl alcohol). Then, 18 specimens without oil and 18 specimens with *intentionally added oil* were prepared. For all samples, the copper wires were tamped in the Inconel 601 mantle. Then, the assembly was cold-formed under the presence of CONDAMAX ALS oil (commercial brand). Finally, all samples were annealed at 1040 °C during 20 min in a hydrogen- and nitrogen-protected atmosphere. Cooling was achieved in liquid nitrogen.

Table 1. Chemical compositions of materials.

Wt (%)	Nickel	Copper	Chromium	Iron	Aluminum	Carbon	Magnesium	Bismuth	Sulfur	Silicon
Inconel 601	58–63	1	21–25	18	1–1.7	0.1	1	-	0.015	0.5
Copper	-	99.95	-	-	-	-	-	$<510^{-3}$	-	-

2.2. Interface Gaps after Heat Treatment

Gaps between the copper core and Inconel 601 were determined after heat treatment by tomographic observations on a North Star Imaging® topographer (with two-dimensional (2D)/three-dimensional (3D) radiology, a voltage range of 10 to 160 kV) under a magnification of 2000 on 36 samples at different locations on each sample. Figure 3 shows a typical cross-section obtained by tomography with indication of the measurement zones. In the following, Z_i designates the value (in μm) of the considered gap. The corresponding measurement locations (in mm) are given in Figure 3b. On each sample, the lateral gaps were measured in the center (Z_3). Gaps, Z_1, between the upper end of the copper core and the Inconel bucket were also recorded. Thus, for each series of samples (with or without oil), there were 18 tomographic measurements for Z_1 and 36 for Z_3. The gaps at locations Z_1 to Z_4 were checked by optical light microscopy on ten samples. The correlation between tomographic and optical observations was satisfactory. The largest difference between the two measurement techniques was smaller than 4%.

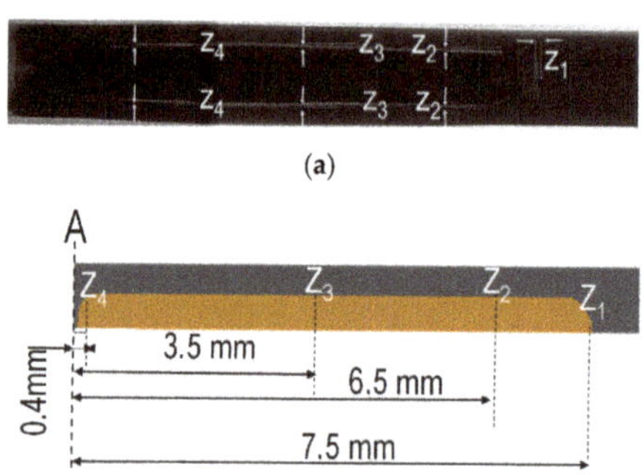

(b)

Figure 3. Measuring gaps: (**a**) typical cross-section after tomographic observation, (**b**) schematic representation of half of the ground electrode with exact locations of the gap measurements. Z_1, Z_2, Z_3, and Z_4 are microscopic gaps between Copper and Inconel, measured respectively at 7.5, 6.5, 3.5, and 0.4 mm from AA'.

Table 2 summarizes the gaps at Z_1 and Z_3, determined by tomography on 18 oiled and 18 non-oiled samples. At Z_3, two measurements (each side) were made on each electrode. Only the maximum value of both was recorded. Both gaps Z_1 and Z_3 are much larger on oiled samples than on the non-oiled ones. For both series, the standard deviation is very large. Let us consider an interval of 1 standard deviation on each side of the mean value. Gap Z_1 leads to overlapping intervals for both series (oiled and non-oiled). However, these intervals do not intersect for gap Z_3. In the following section, the relation between the measured gaps and the tendency to crack during bending is analyzed.

Table 2. Gaps measured at room temperature after annealing by tomographic observations at locations Z_1 and Z_3 on 18 specimens with and 18 specimens without oil. Mean and standard deviation (SD) are shown.

	With Oil		Without Oil	
Gap Values	***Mean***	***SD***	***Mean***	***SD***
Z_1 (μm)	164	128	84	52
Z_3 (μm)	23	11	6.1	2.9

2.3. *Cracking during Bending*

All electrodes of both series were welded and bent. During the welding operation, the maximum gap, Z_3, was on the outer side of the electrode (red rectangle in Figure 4). After welding and bending, all electrodes were observed, once more, by tomography to detect eventual cracks. Figure 4 summarizes the observations on the 18 oiled and 18 non-oiled electrodes. In the (Z_1, Z_3)-space, clearly, two domains appear. For Z_3 smaller than 10 μm, no cracks were detected. All electrodes with gaps (after heat treatment) larger than 12 μm exhibited cracks after bending. Determining the exact boundary shared by both domains would need a larger number of tests, however, this is not the purpose of the present work.

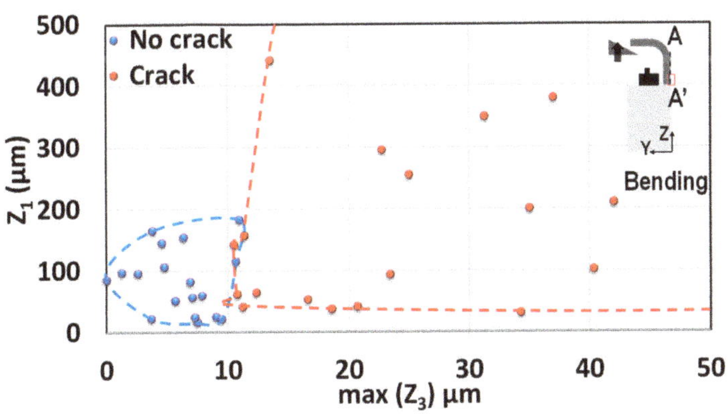

Figure 4. Relation between observed gaps and cracking after bending. The admissible gaps are shown in blue. Lateral gaps larger than 10 μm systematically lead to cracking.

All "oiled" electrodes exhibited cracks after bending, while all "non-oiled" electrodes underwent bending without leading to cracks. Thus, large gaps (Z_3 > 14 μm) systematically led to cracking. Cleaning the samples with the lab procedure leads to heat treatment-induced gaps smaller than this threshold (samples represented in blue in Figure 4). This corresponds to the main results on the production line. However, all oiled samples exhibited large gaps and cracks after bending. These samples represent a small percentage on the production line. *To show the detrimental effect of oil "residues", the detrimental effect of intentional added oil on otherwise clean samples has been proven.* The copper–Inconel interface of "oiled" samples has been characterized thoroughly. Section 2.4 presents the main results of these characterizations needed for the mechanical analyses.

2.4. Characterization of the Inconel–Cu Interface

Oxidation of Inconel 601 was thoroughly analyzed in the scientific literature [8,9]. Even at temperatures above 1000 °C, a protective Cr_2O_3 layer exists (see References [1,2,10–13]), and the oxidation rate seems to be controlled by transport through this layer [1,8].

In the present work, the oxidation kinetics is not addressed. The focus is on the formation of oxides at the Inconel–Cu interface. First, the chemical compositions of the different components, present at the Inconel–Cu interface after annealing of "oiled" samples, were determined by Raman Microscopy (XPloRA ONE Raman, laser with a wavelength of 532 nm, and a range between 150 and 4000). The spectra exhibit typical peaks [14] for NiO, Cr_2O_3, and Al_2O_3 at the Inconel–Cu interface.

The presence of oxides in oiled samples was confirmed by EDX (energy dispersion X ray)analyses (SEM (Scanning Electron Microscopy): ZEISS SUPRA 55 at 20 keV and WD (Working distance) = 10 mm, 10^{-3} Pa with EDX detector: Oxford XMAX 80 mm^2). Figure 5 offers detailed EDS maps of the Cu–Inconel interface. All the maps were done in the white rectangle (continuous line on the SEM observation). Close to the interface with Cu, the Inconel 601 presents increased Cr and O contents corresponding to Cr_2O formation, whereas the Inconel core is Cr-depleted and enriched in aluminum and oxygen. At a distance larger than 10 µm inside the Inconel 601, Al was detected in grain boundaries. The latter suggests Al_2O_3 formation in the Inconel grain boundaries. Thus, EDS (energy dispersion spectrometry) maps indicate the formation of a Cr_2O_3 layer at the Inconel surface and the formation of an underlying (inside the Inconel) layer with high Al_2O_3 content, preferentially at the Inconel grain boundaries. Several authors [15–17] have mentioned oxide formation at Inconel grain boundaries (Cr_2O_3 and Al_2O_3).

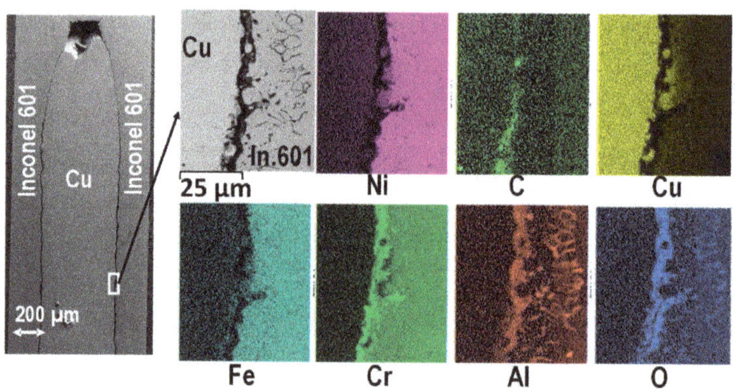

Figure 5. EDS mapping of the interface between copper and Inconel 601 after heat treatment of a sample prepared with oil. Mapping was done in the white rectangle.

Both oxides (Al_2O_3 and Cr_2O_3) are formed at high temperature (during the heat treatment). Both oxides exhibit much smaller coefficients of thermal expansion than the Inconel matrix. During cooling, the Cr_2O_3 layer will be submitted to compressive stresses. However, the Al_2O_3, present at the grain boundaries, will undergo large tensile stresses. This may lead to localized grain boundary cracks. Figure 6a–g depicts optical observations of the Inconel 601–copper interface prior to welding and bending. These observations show the damaged interface after heat treatment. Figure 6h shows cracks after welding and bending. Figure 6a–e corresponds to the upper half of one sample and Figure 6h corresponds to the lower half of another sample. The optical observations essentially highlight a damaged interface. Figure 6f shows a SEM observation in zone B. On this SEM observation, grain boundary damage in the Inconel 601 is visible. The same zone (same sample) was used for the

EDS mappings in Figure 5. Moreover, crack propagation starting from this interface (during bending) is shown in Figure 6h. Prior to welding and bending, the microstructure of region C corresponds to Figure 6f. However, the micro-cracks are smaller. Crack formation in region C (Figure 6h), during bending, is controlled by the presence of micro-cracks and the particular boundary conditions during the bending operation. This is addressed in Section 4.

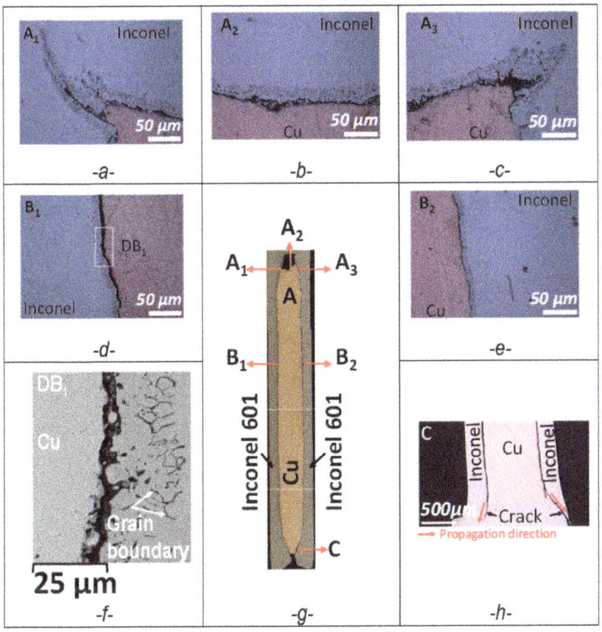

Figure 6. Samples prepared with oil, interface damage (after heat treatment), and consequent cracking (after welding and bending): optical and SEM (Scanning Electron Microscopy) observations of damage close to the Cu–Inconel 601 interface. (**g**) Overall view of the sample. (**a**–**f**) Correspond to a sample after heat treatment (prior to welding and bending) and (**h**) corresponds to another sample (after bending). (**a**–**c**) Copper–Inconel 601 interface in regions (A_1, A_2, A_3), (**d**–**e**) copper–Inconel 601 interface in regions (B_1, B_2). (**f**) SEM observation of the interface in region B_1. (**h**) Optical observation of the interface in region C after welding and bending with crack propagation through the Inconel 601.

Thus, all the results point to oxide formation at the Inconel grain boundaries. A similar mechanism was reported by Garat et al [17]. These oxides grow at high temperature during the annealing treatment. They initiate grain boundary damage in the Inconel 601 during cooling to room temperature. Hence, in the following section, the annealing treatment is modeled assuming zero adherence between Copper and Inconel 601.

3. Thermo-Mechanical Analysis of Current Ground Electrode Design

The thermal strains and stresses induced by the annealing and cooling were modelled using Sysweld™ [18]. The effect of plastic dissipation on heat transfer is neglected. Therefore, the simulation is based on a thermal followed by a mechanical analysis [19–23]. The metallurgical phenomena are not taken into account. The heat transfer analysis is based on the solution of the classical heat equation [23]. The mechanical analysis is based on the usual equations describing the static equilibrium and additive small strain decomposition $\varepsilon = \varepsilon_e + \varepsilon_p + \varepsilon_{th}$, where ε, ε_e, ε_p, and ε_{th} are the total, elastic, plastic, and thermal strains, respectively. All temperature-dependent material properties are given in Tables 3 and 4. The coefficients of thermal expansion of the two materials were determined by dilatometry tests.

Table 3. Physical properties of Inconel 601 [9].

T (°C)	20	100	200	300	400	500	600	700	800	900	1000
λ (W/m·k)	11.2	12.7	14.3	16.0	17.7	19.5	21.0	22.8	24.4	26.1	27.8
C (J/Kg·k)	448	469	498	523	548	578	603	632	657	686	712
α ($10^{-6}K^{-1}$)	9.8	10.98	11	12.25	12.19	11.1	13.12	13.88	13.36	12	13.32
E (GPa)	206	202	197	191.2	184.8	178.2	170.8	161.3	150.2	137.9	124.7
Rp_{02} (MPa)	240	230	200	170	150	150	160	160	160	120	90

Table 4. Physical properties of copper [6].

T (°C)	20	100	200	300	400	500	600	700	800	900	1000
λ (W/m·k)	410	398	386	380	360	353	348	340	335	330	320
C (J/Kg·k)	386	396	400	410	415	420	440	450	460	470	500
α ($10^{-6}K^{-1}$)	13.5	13.96	14.25	14.62	15.47	15.09	14.01	15.12	14.84	15.35	19.17
E (GPa)	117	119	114	98	68	50	-	-	-	-	-
Rp_{02} (MPa)	170	160	150	120	90	50	-	-	-	-	-

Different geometries will be studied depending on the presence or absence of initial gaps (prior to annealing) between copper and Inconel 601. Two-dimensional (2D) generalized plane strain models will be used. The generalized plane strain assumption in Sysweld™ [18], allows 2D models with non-zero (but constant) strain σ_{zz}. The latter adds a degree of freedom to the problem. It is well-known that such thermo-mechanical models overestimate the σ_{zz} stresses due to the constraint imposed by the generalized plane strain condition. In order to check the ability of such models to predict gap formation in the (xy) plane, a comparison with a full 3D model will be done in one case (without initial gap). First-order finite elements with selectively reduced integration will be used both in 2D and 3D [18], due to their ability to simulate complex elastoplastic problems.

3.1. Geometry and Finite Element Mesh

Symmetry allows modelling ½ of the electrode in 2D, and ¼ in 3D. Figure 7a depicts the dimensions of the ground electrode. X designates an axis of symmetry. Two contact geometries were simulated: without initial gap (Figure 7b) and with an initial gap of 5 µm (Figure 7c).

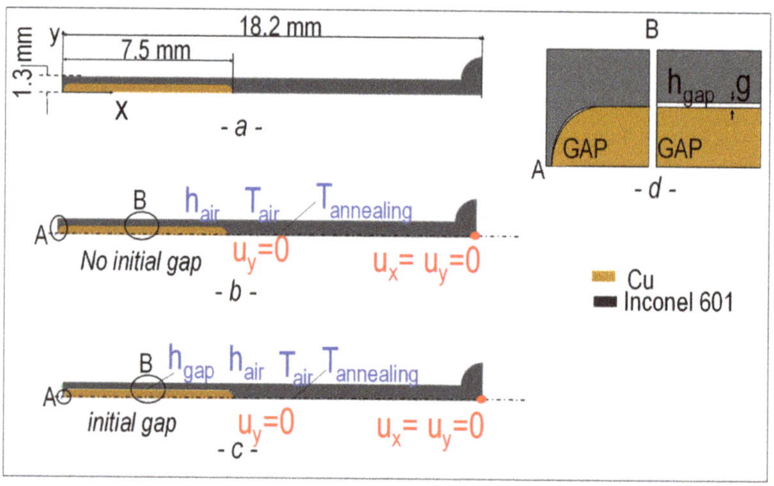

Figure 7. Considered geometries (**a**) overall dimensions, (**b**) geometry without initial gap between the copper and Inconel, (**c**) with an initial gap of 5 µm, (**d**) zoom on the initial gap.

Several mesh densities were tested. Figure 8 depicts the optimum mesh finally used in all further analyses. With 10 elements in the y direction, corresponding to an element size of 0.06 mm, mesh refinement would not lead to significantly different results.

On interfaces between solid bodies, Sysweld™ uses a master-slave algorithm. Mechanical contact is defined between interacting faces using an algorithm which strictly enforces a non-penetration condition. Perfectly sliding conditions are assumed at the interface.

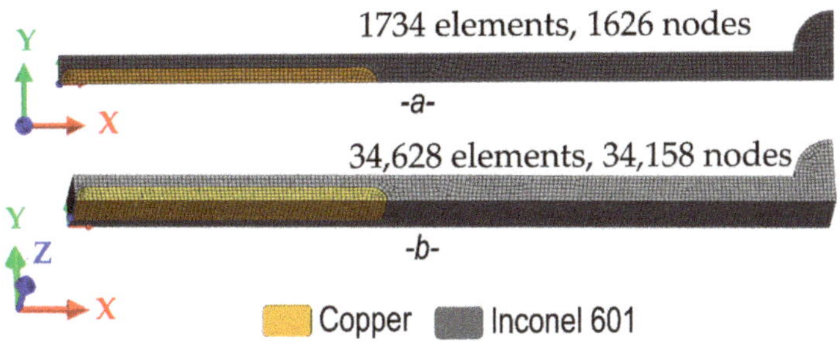

Figure 8. Finite element meshes for (**a**) generalized plane strain model and (**b**) three-dimensional (3D) model.

3.2. Thermo-Mechanical Loading

The ground electrode was placed in a furnace with homogeneous temperature. The annealing temperature as a function of time is depicted in Figure 1c. There was an initial increase from room temperature to 1040 °C during 20 min, followed by a plateau, then rapid liquid nitrogen cooling, and finally, slow air cooling to room temperature. As the heating is long, a homogeneous temperature was applied to all the nodes of the ground electrode. During cooling, a heat exchange coefficient between the external surface of the Inconel 601 and the nitrogen atmosphere, $h_{nitrogen}$ (3000 W K/m^2), was considered (Figure 7). The external temperature was set to −150 °C. The heat exchange coefficient between air existing in the gap and both Inconel 601 and copper was set to h_{gap} (Figure 7). For thermal contact, a geometric search was performed to find the element on the master surface onto which each element on the slave surface is subjected. The equivalent gap conductibility was defined as the inverse of the gap width (g) multiplied by the thermal conductivity of the air (k_g) in the gap:

$$h_{gap} = k_g / g \quad (1)$$

Displacement components are blocked so as to respect symmetry conditions for the 2D and the 3D models. At point A, the copper is linked to Inconel so that its body motion will be blocked (Figure 7). A sliding contact between copper and Inconel 601 (Figure 7) was assumed. Time steps of 10 s and 1 s were used for the heating and the cooling, respectively. For the mechanical calculations, a time step of 0.5 s was used.

3.3. Stress Distribution in the Electrode during Annealing

Figure 9 depicts the distribution of axial stress, σ_{xx}, at different times during the annealing treatment (Figure 1c). The results correspond to an initial gap (prior to annealing) of 5 μm. The 3D and 2D generalized plane strain models predict very close axial stress distributions in the Inconel. After cooling to room temperature, a difference in the copper core is observed. The sole difference between 2D and 3D models concerns the compressive stresses (σ_{zz}) predicted in the copper core, which are higher with the generalized plane strain model. As damage is observed in the Inconel 601, we focus essentially on the stress field in this material. Concerning the Inconel 601 coating, only both ends of the electrode are submitted to significant stresses at high temperature (t_2 and t_3). At room temperature

(t_4), no noteworthy stress subsists in the Inconel. At high temperature, the stress in the Inconel 601 varies continuously from less than −80 MPa to more than +80 MPa through the thickness at both ends of the electrode. This stress distribution corresponds to bending of the Inconel 601 coating. The same stress levels are achieved without an initial gap of 5 μm.

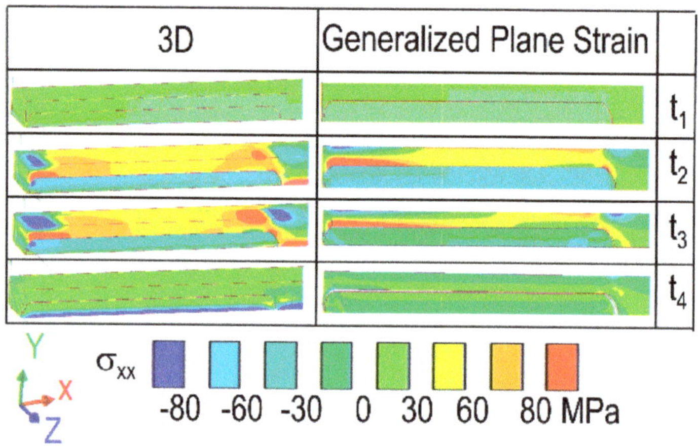

Figure 9. Axial stress, σ_{xx}, predicted by the three-dimensional (3D) and the two-dimensional (2D) generalized plane strain models with an initial gap of 5 μm.

Figure 10b depicts the axial stress, σ_{xx}, at time t_2 (beginning of temperature plateau) in the Inconel coating. At the very top of the electrode (x = 0), the stress is close to zero. This corresponds to the Inconel 601 closing of the electrode. For x > 0, the outer Inconel 601 (lines L_3, L_4) is submitted to compressive stresses, whereas the inner Inconel 601 (lines L_1, L_2) is submitted to positive axial stresses. The coating is submitted to bending around the z-axis.

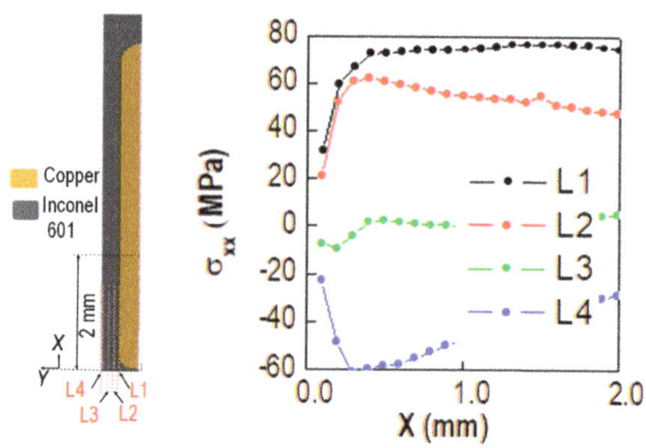

(**a**) (**b**)

Figure 10. Stress distribution in the sample. (**a**) Schematic electrode geometry with 4 lines (L_i) defined in the Inconel coating. (**b**) Axial stress (σ_{xx}) in the Inconel along lines L_i.

During heating, copper expansion induces compressive loading along the copper–Inconel interface. At the electrode top (upper part in Figure 10a), thermal expansion of copper is hindered most. At the "free" bottom, thermal expansion is hindered less. Effectively, the induced bending stresses (σ_{xx}) decrease close to the free end. This point will be of prime significance to suggest a new electrode design.

The consequence of residual plastic deformation in the Inconel (the gaps at room temperature) is discussed thoroughly in the Section 3.4.

3.4. Gaps versus Time

Predictions of the 2D plane strain model were checked in the preceding section on the stress maps. Gaps *after* annealing and cooling have been simulated with this model for both assumptions: an initial gap *prior* to heat treatment of 5 μm and no initial gap.

Figure 11 shows the gaps as a function of time predicted by the 2D model with an initial gap of 5 μm and without an initial gap. For the gap at Z_1 (Figure 11a), a small time delay as well as a difference in the maximum value are observed. In both cases, with or without initial gap, no contact at the top of the ground electrode is observed. Gaps Z_2 to Z_4 (along the electrode axis) are only slightly affected by an initial gap of 5 μm (Figure 11b). The final gap in the middle of the electrode (Z_3) depends on the presence or absence of an initial gap. If no initial gap is present, contact loads between Copper and Inconel 601 all along the x axis prevent the free bending movement. This leads to smaller gaps at Z_3 than in the case without an initial gap.

Positive gaps are predicted during the heating. Copper exhibits a larger coefficient of thermal expansion than Inconel 601. Hence, at first glance, a decrease of the gaps would be expected during heating. But, closing the bottom of the electrode prevents the latter from "free" expansion. Hence, locally, at the very bottom of the electrode, the Inconel 601 is submitted to severe contact loads and bending.

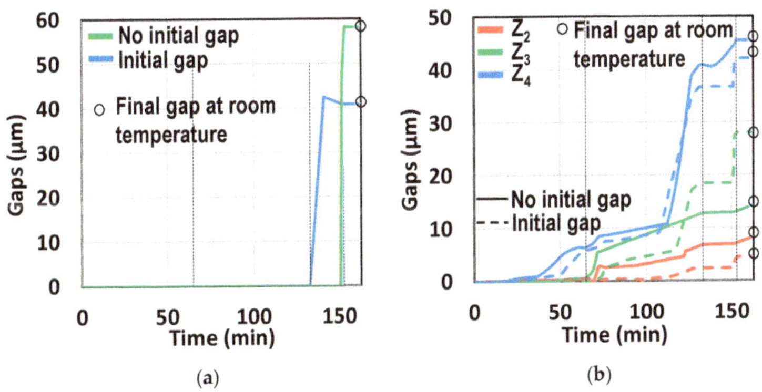

Figure 11. Evolution of the gaps during annealing. (**a**) Z_1 at the top of the electrode, (**b**) Z_2, Z_3, and Z_4 along the electrode's axis.

As the presence of gaps seems to play a key role, the model results were checked at the 4 locations defined in Figure 3 by light optical measurements on 10 samples. To the best of our knowledge, no such measurements have been made on comparable geometries previously. Ishikawa [24] analyzed thermal stresses in a plane cylinder. Bengeri et al [25] modeled shrink fit stresses in circular cylinders. However, thermal stresses were not analyzed in hollow cylinders with a rectangular cross-section. Thus, the only verification available consists in comparing new simulation and new experimental measurements. Table 5 shows simulation results corresponding to a zero initial gap and the experimental data. Considering the complexity of the phenomena involved, a good agreement between experimental observations and mechanical predictions of gaps after annealing is achieved. The presence of gaps after annealing is confirmed. *Persistence of the gaps after cooling clearly confirms irreversible (plastic) deformation of the Inconel 601.*

At this stage, the presence effect of gaps after annealing treatment has been observed experimentally. The existence was attributed to small (no) adhesion between the Copper core and Inconel. This assumption was justified by present thermomechanical analysis. The detrimental effect of the gaps during annealing has been observed experimentally. In the next section, the effect of gaps on the bending behavior is analyzed by a mechanical model.

Table 5. Heat treatment-induced gaps at room temperature, simulation results, and optical measurements. Mean and standard deviation (SD) are shown

Model versus Experiment		Location Z_1	Z_2	Z_3	Z_4
2D plane strain model (μm)		41	1.5	15	40
Light optical measurements (μm)	Mean	29	2.1	17	30.1
	SD	2.2	0.9	2.0	2.5

4. Cracking during Bending Due to Micro-Movements

4.1. Model Description

As explained in Section 3, the generalized plane strain assumption leads to an overestimation of σ_{zz} stress. The experimentally observed cracks appear in the (xy) plane, and (xy) stress maps obtained during forming are simulated accurately by 2D generalized plane strain [26,27]. In order to simplify the problem, and to reduce the calculation time, the bending of the spark plug is modeled in 2D using Sysweld™ [18]. Figure 12 summarizes the geometry, boundary conditions, and finite element mesh. The bending tool and core were modelled as non-deformable solids. The electrode was meshed using linear rectangular elements of size of 0.1 (Figure 12c). The model features 1079 elements and 1094 nodes. Displacements of the shell and the bending core nodes are blocked (Figure 12b). The bending tool is moving along Ox at a speed Vx = 0.01 m/s (Figure 12b). A time step of 10^{-4} is used for the analysis. For contact conditions, the bending tool is declared as master, and the ground electrode is declared as slave. In the presence of gaps, the Inconel 601 presents the master surface, and the Copper is the slave surface. Isotropic von Mises plasticity with Johnson–Cook behavior was considered (Table 6). The general expression of the stress (σ) is given by:

$$\sigma = [A + B\varepsilon^n] \times \left[1 + Cln\frac{\dot{\varepsilon}}{\dot{\varepsilon}_0}\right] \times \left[1 - \left(\frac{\theta - \theta_t}{\theta_{melting} - \theta_t}\right)^m\right] \quad (2)$$

where ε , $\dot{\varepsilon}$, and θ correspond respectively to the strain, strain rate, and temperature, $\dot{\varepsilon}_0$, $\theta_{melting}$, and θ_t designate a reference strain rate, the melting temperature, and the transition temperature, and A, B, and C are material-specific parameters, given in Table 6. Stationary bending simulations were done at room temperature ($\theta = 25$ °C).

Bending has been simulated with no initial gap and with a gap of 40 μm.

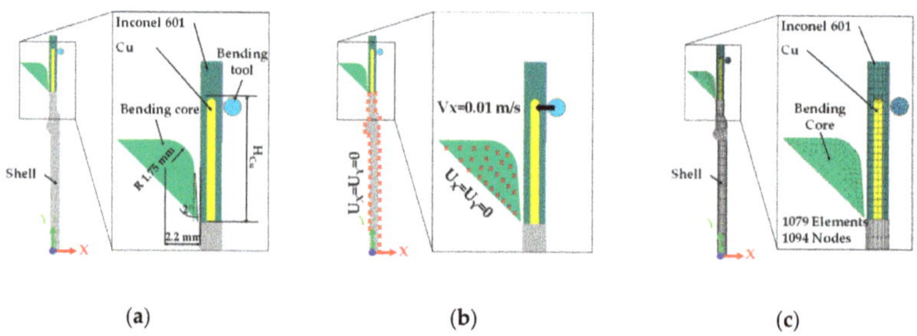

Figure 12. Schematic representation of (**a**) the geometry of the workpiece and the bending tools, (**b**) boundary conditions, and (**c**) finite element meshes.

Table 6. Johnson–Cook's law parameters for different materials [6,9].

	A (MPa)	B (MPa)	n	C	m
Copper Cu-OF	170	405	0.22	0.02	0.72
Inconel 601	450	1700	0.65	0.017	1.3
Steel	244	554	0.217	0.0088	0.047

4.2. Stress Distribution in the Welded Area after Bending

Figure 13 shows the distribution of the maximum principal stress in the ground electrode at the end of the bending operation. Figure 13a,b corresponds, respectively, to a ground electrode with and without gap prior to bending. Of particular interest is area A. Effectively, cracks were observed experimentally in this area after bending. In the presence of gaps (Figure 13a), a significant maximum principle stress was predicted in the area where cracks were observed. However, the ground electrode without gap leads to a much smaller maximum principle stress in the same zone. As expected, the micro-movements allowed by the presence of gaps led to severe stress concentration at the outer border of the ground electrode.

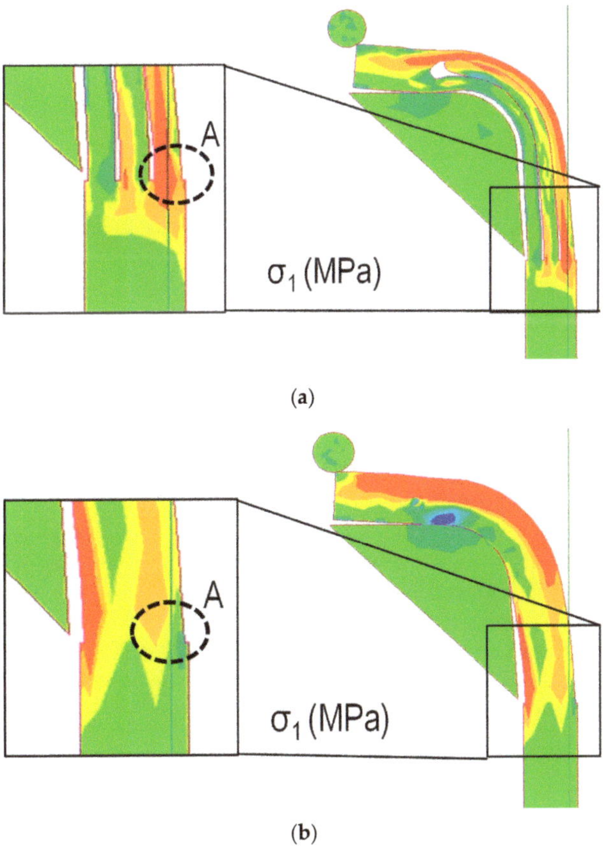

Figure 13. Maximum principal stress distribution (σ_1) in the case (**a**) with gaps, and (**b**) without gaps at the interface between copper and Inconel 601.

The results of Section 2.2 revealed a strong relation between the presence of gaps after annealing and cracks after welding and bending. The results of Section 2.3 revealed the presence of a Cr_2O_3 layer leading to bad adherence between Cu and Inconel 601. Section 3.3 confirmed the presence of gaps after annealing and cooling, and *the present section confirms the high stresses induced during bending under the presence of gaps.* Thus, during annealing and cooling, the Inconel 601 undergoes irreversible (plastic) deformation. In the next section, a new ground electrode design avoiding the yielding of the Inconel 601 during annealing is recommended.

5. New Design of the Ground Electrode Avoiding Micro-Movements

5.1. Geometry and Finite Element Mesh

The thermo-mechanical model of the annealing treatment clearly showed bending of the Inconel due to larger thermal expansion of the copper core. Bending was allowed due to the open bottom of the ground electrode. Based on this finding, 1 mm of material was added to the bottom of the ground electrode (Figure 14a). The same approach as in Section 3 was adopted for annealing modelling. A 2D generalized plane strain model with an initial gap of 5 μm before annealing was used. The same mesh size was chosen (Figure 14b) and the same thermal loading was applied. The additional material was used to prevent gap formation during annealing. After annealing, it has to be cut.

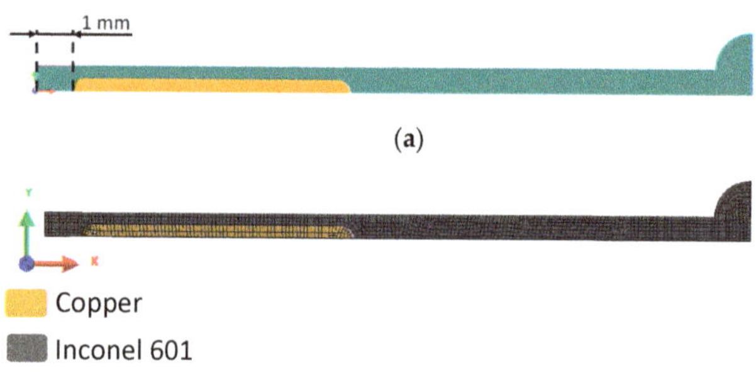

Figure 14. (**a**) New geometry design, and (**b**) finite element meshes for generalized plane strain model.

5.2. Stress Distribution in the Electrode during Annealing

Figure 15a shows the distribution of axial stress at various heat treatment times (Figure 1c). The results correspond to an addition of 1 mm of material to the bottom of the ground electrode. In the Inconel 601 coating, only the two ends of the electrode are subjected to high stresses (at t_2). The stress concentration at intermediate temperature t_2 (red) is due to the copper core pushing the Inconel mantle. This corresponds to the thermal elongation of the copper in the x direction. At high temperature (t_3), softening of the Inconel leads to a smaller stress in the same region. At room temperature (t_4), compressive stresses in a small part of the additional material are found. Otherwise, the additional material undergoes small tensile stresses.

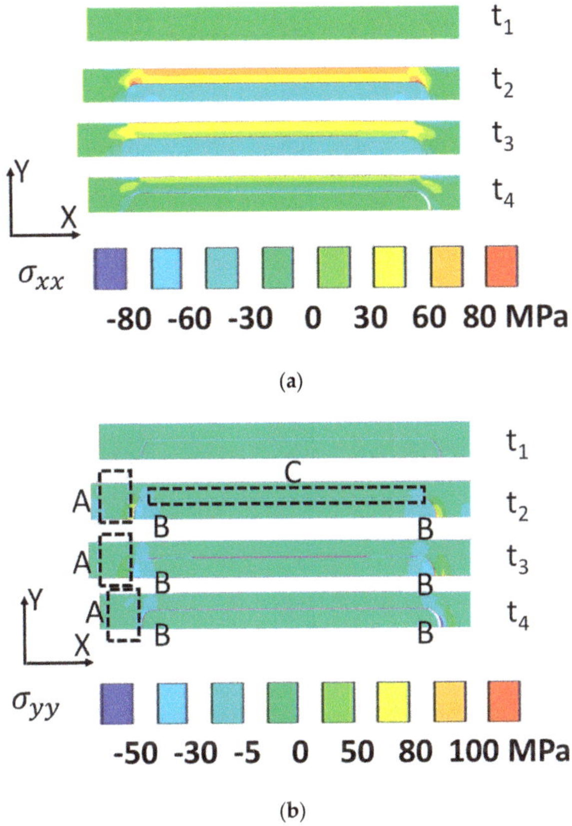

Figure 15. 2D generalized plane strain model with 1 mm of additional material and with an initial gap of 5 μm. (**a**) Axial stress σ_{xx}, (**b**) stress σ_{yy}.

The additional material prevents bending and hence movement of the Inconel in the y direction. To prevent this movement, small tensile stresses, σ_{yy}, appear in the additional material (Figure 15b). At temperature (t_2), the copper core pushes the Inconel. But, this load is equilibrated by tensile stresses in region A. However, a very small outward movement of the Inconel occurs close to the electrode bottom with yielding of the Inconel. But, the additional material (dashed area A) remains essentially elastic. Thus, when cooled to room temperature, the bottom part of the Inconel undergoes compressive stresses. For both materials, the stresses in region C are extremely small. Thus, the additional material leads to a satisfactory new stress distribution. No significant tensile stresses are found at room temperature.

5.3. Gaps versus Time

The aim of adding material was to decrease the final lateral gaps (Z_2 to Z_4). Z_1 is not critical for the bending operation. Figure 16 shows a comparison between predicted gaps without (current geometry) and with additional material (new geometry). All gaps (Z_2 to Z_4) decreased significantly with the additional material. Moreover, all gaps at room temperature were smaller than the threshold of 10 μm. The new geometry leads to better stress distributions in the electrode and acceptable room temperature gaps.

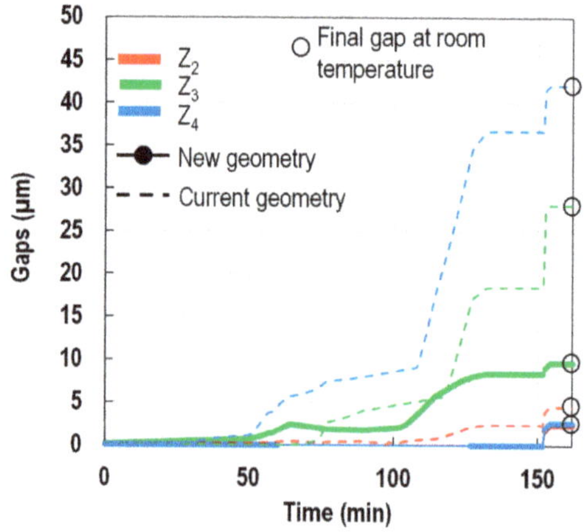

Figure 16. Influence of the additional material on final gap after annealing (Figure 1c) along the electrode's axis.

6. Conclusions

Ground electrodes with Inconel 601 coating were analyzed. On the production line, cracking was observed on a small proportion (ppm) of electrodes with this new Inconel 601 coating. In the present paper, possible causes of cracking were analyzed experimentally and by mechanical models.

Two series of ground electrodes (with and without oil) were prepared and underwent the same production process. This original approach combined with advanced characterization techniques allowed us to identify gaps between Copper and Inconel 601 as the main cause of cracks. A thermo-mechanical model confirmed high stress concentrations during bending due to micro-movements allowed by gaps. A second thermo-mechanical model explained the presence of gaps by Inconel yielding during heat treatment. At this stage, the genuine combination of astute experimental procedure and well-chosen thermo-mechanical models allowed for the following simple conclusion: oil residues may lead to interface gaps, and the latter cause cracking during bending. Based on this knowledge, a completely new ground electrode design could be suggested. This new design is more cost-effective than advanced cleaning.

Author Contributions: Conceptualization, J.-M.B., H.K. and E.F.; methodology, C.T., J.-M.B., E.F., H.K. and C.B.; software, J.-M.B. and E.F.; validation, C.T., H.K., J.-M.B., E.F., C.B. and B.B.; formal analysis, C.T., J.-M.B., E.F. and H.K.; investigation, C.T., H.K., C.B. and B.B.; resources, J.-M.B., C.B. and B.B.; data curation, C.T., J.-M.B. and E.F.; writing—original draft preparation, C.T. and H.K.; writing—review and editing, C.T., H.K., E.F., B.B. and C.B.; visualization, C.T.; supervision, H.K., J.-M.B., E.F. and C.B.; project administration, J.-M.B. and C.B.; funding acquisition, J.-M.B. All authors have read and agreed to the published version of the manuscript.

Funding: This research received no external funding.

Acknowledgments: The authors are thankful to the "Association Nationale Recherche Technologie" and Federal Mogul Powertrain Ignition Products SAS for allowing this research project. Chawki Tahri prepared his PhD under the supervision of the following people; J.-M.B. and E.F. from "Ecole Nationale d'Ingénieurs de Saint-Etienne" and H.K. from "Mines Saint-Etienne". B.B. (from "Ecole Centrale de Lyon") helped with microscopic observations. We thank all the staff members from these institutions, which are all parts of "Université de Lyon", for fruitful discussions.

Conflicts of Interest: The authors declare no conflict of interest.

References

1. Javant, S.; Hosseini, S.V.; Alaviyoun, S.S. An Experimental Investigation of Spark Plug Temperature in Bi-fuel Engine and Its Effect on Electrode Erosion. *Int. J. Automot. Eng.* **2012**, *2*, 21–29.
2. Lee, M.J.; Hall, M.; Ezekoye, O.A.; Matthews, R. Voltage, and Energy Deposition Characteristics of Spark Ignition Systems. *J. Soc. Automot. Engine (SAE)* **2006**, *01*, 231–243.
3. Singh, H.; Sidhu, T.S.; Kalsi, S.S. Microstructure Study of Cold Sprayed 50%Ni–50%Cr Coating on Inconel-601. *Surf. Eng.* **2015**, *31*, 825–831. [CrossRef]
4. Upadhyay, C.; Saurav, D.; Manoj, M.; Siba, S.M. An Experimental Investigation Emphasizing Surface Characteristics of Electro-Discharge-Machined Inconel 601. *J. Braz. Soc. Mech. Sci. Eng.* **2017**, *39*, 3051–3066. [CrossRef]
5. Tahri, C.; Bertoni, C.; Klocker, H.; Feulvarch, E.; Bergheau, J.M. Numerical modeling of the resistance braze welded assembly of a Copper Inconel 601 ground electrode, and a Steel shell. *Numer. Heat Transf. Part A Appl.* **2020**, *78*, 1–20. [CrossRef]
6. Lamineries Matthey. A Unit of Notz Metal SA, Montagu 38, Case Postale CH-2520, LA Neuveville. Available online: https://www.matthey.ch/en/alliages/cuivre/ (accessed on 8 July 2020).
7. Lim, Y.Y.; Chaudhri, M.M. The effect of the indenter load on the nano-hardness of ductile metals: An experimental study on polycrystalline work hardened and annealed oxygen-free copper. *Philos. Mag. A* **1999**, 2979–3000. [CrossRef]
8. Buscail, H.; Perrier, S.; Josse, C. Oxidation Mechanism of the Inconel 601 Alloy at High Temperatures. *Mater. Corros.* **2011**, *62*, 416–422. [CrossRef]
9. Special Metals, 3200 Riverside Drive Huntington, WV 25705-1771. Available online: www.specialmetals.com (accessed on 8 July 2020).
10. Hewidy, M.S.; El-Taweel, T.A.; El-Safty, M.F. Modelling the Machining Parameters of Wire Electrical Discharge Machining of Inconel 601 Using RSM. *J. Mater. Process. Technol.* **2005**, *169*, 328–336. [CrossRef]
11. Gonzalez-Rodriguez, J.G.; Salinas-Bravo, V.M.; Martinez-Villafane, A. Low-temperature stress corrosion cracking of alloy 601 in thiosulfate and chloride solutions. *Corrosion* **1999**, *55*, 1–38. [CrossRef]
12. Nishikata, A.; Numata, H.; Tsuru, T. Electrochemistry of molten-salt corrosion. *Mater. Sci. Eng. A Struct.* **1991**, *146*, 15–31. [CrossRef]
13. Vossen, J.P.T.; Plomp, L.; Dewit, J.H.W.; Rietveld, G. Corrosion behavior of stainless-steel and nickel-base alloys in molten-carbonate. *J. Electrochem. Soc.* **1995**, *142*, 3327–3335. [CrossRef]
14. *RRUFF Project, Department of Geosciences*; University of Arizona: Tucson, AZ, USA; Available online: https://rruff.info/chem=cr%20/display=default/ (accessed on 8 July 2020).
15. Bohua, Y.; Li, Y.; Nie, Y.; Mei, H. High Temperature Oxidation Behavior of a Novel Cobalt-Nickel-Base Superalloy. *J. Alloys Compd.* **2018**, *765*, 1148–1157. [CrossRef]
16. Wallwork, G.R. The oxidation of alloys. *Rep. Prog. Phys.* **1976**, *39*, 401–485. [CrossRef]
17. Garat, V.; Cloue, J.M.; Poquillon, D.; Andrieu, E. Influence of Portevin–Le Chatelier effect on rupture mode of alloy 718 specimens. *J. Nucl. Mater.* **2008**, *375*, 95–101. [CrossRef]
18. SYSWELD™. *User's Manual*; ESI Group: Paris, France, 2019.
19. Khan, J.A.; Xu, L.; Chao, Y.-J.; Broach, K. Numerical Simulation of Resistance Spot Welding Process. *Numer. Heat Transf. A* **2000**, *37*, 425–446.
20. Feulvarch, E.; Robin, V.; Bergheau, J.M. Thermometallurgical and Mechanical Modelling of Welding—Application to Multipass Dissimilar Metal Girth Welds. *Sci. Technol. Weld. Join.* **2011**, *16*, 221–231. [CrossRef]
21. Feulvarch, E.; Bergheau, J.M. Modeling and Numerical Simulation of Resistance Spot Welding Process. In *Encyclopedia of Thermal Stresses*; Richard Hetnarski, B., Ed.; Springer: Dordrecht, The Netherlands, 2014; pp. 3112–3123. [CrossRef]
22. Feulvarch, E.; Rogeon, P.; Carré, P.; Robin, V.; Sibilia, G.; Bergheau, J.M. Resistance Spot Welding Process: Experimental and Numerical Modeling of the Weld Growth Mechanisms with Consideration of Contact Conditions. *Numer. Heat Transf. Part A Appl.* **2006**, *49*, 345–367. [CrossRef]
23. Bergheau, J.M.; Fortunier, R. *Finite Element Simulation of Heat Transfer*; ISTE & Wiley: London, UK, 2008; pp. 1–279. [CrossRef]

24. Ishikawa, H. A thermoplastic solution for a circular solid cylinder subjected to heating and cooling. *J. Therm. Stresses* **1978**, *1–2*, 211–222. [CrossRef]
25. Bengeri, M.; Mack, W. The Influence of the Temperature Dependence of the Yield Stress on the Stress Distribution in a Thermally Assembled Elastic-Plastic Shrink Fit. *Acta Mech.* **1994**, *103*, 243–257. [CrossRef]
26. Pascon, F.; Cescotto, S.; Habraken, A.M. A 2.5D Finite Element Model for Bending and Straightening in Continuous Casting of Steel Slabs. *Int. J. Numer. Methods Eng.* **2006**, *68*, 125–149. [CrossRef]
27. Yabo, G.; Pourboghrat, F. Fourier Series Based Finite Element Analysis of Tube Hydroforming—Generalized Plane Strain Model. *J. Mater. Process. Technol.* **2008**, *197*, 379–392. [CrossRef]

energies

MDPI

Article

Optimal Calibration Strategy of a Hybrid Electric Vehicle Equipped with an Ultra-Lean Pre-Chamber SI Engine for the Minimization of CO_2 and Pollutant Emissions

Fabio Bozza, Vincenzo De Bellis, Enrica Malfi *, Luigi Teodosio and Daniela Tufano

Dipartimento di Ingegneria Industriale, Università di Napoli Federico II, 80125 Napoli, Italy; fabio.bozza@unina.it (F.B.); vincenzo.debellis@unina.it (V.D.B.); luigi.teodosio@unina.it (L.T.); daniela.tufano@unina.it (D.T.)

* Correspondence: enrica.malfi@unina.it; Tel.: +39-081-7683264

Received: 15 June 2020; Accepted: 28 July 2020; Published: 3 August 2020

Abstract: The complexity of modern hybrid powertrains poses new challenges for the optimal control concerning, on one hand, the thermal engine to maximize its efficiency, and, on the other hand, the vehicle to minimize the noxious emissions and CO_2. In this context, the engine calibration has to be conducted by considering simultaneously the powertrain management, the vehicle characteristics, and the driving mission. In this work, a calibration methodology for a two-stage boosted ultra-lean pre-chamber spark ignition (SI) engine is proposed, aiming at minimizing its CO_2 and pollutant emissions. The engine features a flexible variable valve timing (VVT) control of the valves and an E-compressor, coupled in series to a turbocharger, to guarantee an adequate boost level needed for ultra-lean operation. The engine is simulated in a refined 1D model. A simplified methodology, based on a network of proportional integral derivative (PID) controllers, is presented for the calibration over the whole operating domain. Two calibration variants are proposed and compared, characterized by different fuel and electric consumptions: the first one aims to exclusively maximize the brake thermal efficiency, and the second one additionally considers the electric energy absorbed by the E-compressor and drained from the battery. After a verification against the outcomes of an automatic optimizer, the calibration strategies are assessed based on pollutant and CO_2 emissions along representative driving cycles by vehicle simulations. The results highlight slightly lower CO_2 emissions with the calibration approach that minimizes the E-compressor consumption, thus revealing the importance of considering the engine calibration phase, the powertrain management, the vehicle characteristics, and its mission.

Keywords: virtual engine calibration; ultra-lean combustion; hybrid vehicle; 0D-1D engine modelling

1. Introduction

In the 2013, the Diesel-gate scandal erupted in the USA, flooding into Europe and worldwide, involving more than 11 million of vehicles produced by various car manufacturers [1]. The related rumors hit the entire automotive sector, leading to the idea that the traditional internal combustion engine (ICE), fed by fossil fuels, would disappear within a few years. A full-electric mobility, based on battery electric vehicles (BEVs), has been deeply promoted by politicians and public opinion, as the unique available solution to fulfill more and more stringent regulations [2].

At the same time, however, alternative propulsion architectures, such as hybrid electric vehicles (HEVs) and plug-in HEV (PHEVs), due to their lower costs and greater autonomy, continuously gained new market shares. Today, they represent much more than a temporary solution, filling the gap between standard ICE-powered vehicles and full electric mobility.

As known, the choice for the optimal electrification level of a vehicle requires to consider its entire life cycle, including the CO_2 formation during the vehicle manufacturing and the one associated to electrical energy production and distribution. Under this point of view, the overall CO_2 emissions from the two extreme antagonist vehicles (ICEs and BEVs) become comparable and are strongly dependent on the so called "energy-mix" utilized for electrical energy production [3,4]. In particular, BEV or PHEV have consistent benefits only when renewable energy sources are used, leading to the conclusion that an electric mobility cannot be considered as the only practical alternative due to the limited and country-dependent availability of renewable sources. Additionally, as already remarked, the higher ownership cost of the BEV, combined with the high battery charging time and the low autonomy, strongly limits their market acceptance and diffusion [5]. In the light of those concerns, rather than a pure-electric future, a scenario should be expected in the years to come characterized by variegated technologies that are best suited to the contest in which they are employed. This means that ICE-based vehicles, HEVs, PHEVs, BEVs and fuel cell-based vehicles will coexist in the market for a long time, pushing car manufactures to improve the propulsion system efficiency [6] with the aim to respect the stringent regulation in terms of CO_2 and pollutant emissions.

Among the various pathways to improve the thermal efficiency of ICEs, various knock mitigation measures were attempted, such as the variable compression ratio [7], the cooled exhaust gas recirculation (EGR) [8], and the water injection [9]. Benefits of fuel consumption highly depend on the considered operating condition [10]. An increasing research work is nowadays devoted to ultra-lean combustion systems since they have the potential of simultaneously reducing the cylinder-out NOx emissions and the fuel consumption in the whole engine operating domain. Among the various concepts analyzed in the literature to sustain a lean combustion [11,12], such as fuel stratification [13,14], HCCI [15] and SACI [16], the employment of a pre-chamber (PC) ignition system is nowadays considered the most practical solution. In this architecture, a small PC volume is connected to the main chamber (MC) through some orifices. The combustion process starts in the PC and propagates in the MC one, in the form of hot burned jets ejected from the pre-chamber. The high turbulence intensity of these jets ignites the MC mixture and ensures a stable flame propagation even under extremely lean mixtures [17,18].

The aim of this activity is to numerically analyze the calibration and the energy management strategy of a HEV equipped with an ultra-lean PC engine. The overall propulsion system under investigation is developed within the framework of the EAGLE H2020 project (https://h2020-eagle.eu/). The hybrid architecture and the components' sizing has been completely defined by the manufacturer and includes two electric motor/generator units, a battery, and two gearboxes. The system is designed to provide the maximum flexibility and a versatile series/parallel operation. A hybrid series-parallel architecture was chosen to combine the advantages of series powertrain in urban driving (including smooth takeoff) and of the parallel mode over rural/highway routes. Close-to-optimal operations for the ICE are guaranteed by a four-speed gearbox placed between the ICE and differential. A gearbox is preferred to a CVT power-split device to get a higher mechanical efficiency. The switch between series and parallel modes is handled by three clutches, which also help to decrease the mechanical losses when one of the motors is not used.

The thermal unit is equipped with an active pre-chamber ignition system and a flexible intake and exhaust VVT device. The very high air-flow demand of the ultra-lean architecture requires the adoption of a two-stage boosting system composed of a variable geometry turbocharger, coupled in series to an E-compressor (E-Comp). The presence of an electrically driven compressor implies that the energy input to the engine is not just the chemical energy given by the fuel. The energy input in the E-Comp also plays a role on the effective engine efficiency. The overall boost level can be then shared between the turbocharger compressor and the E-Comp, depending on the electrical energy input given to the second one. The latter is provided by the same battery pack of the hybrid propulsion system. In other words, the possibility to modify the overall boost level sharing between the LP compressor and the E-comp opens the way to two calibration variants. From one side, an intensive use of the E-Comp leads to a reduced backpressure and an improved brake thermal efficiency (BTE). A more frequent battery discharge/recharge is, however, to be expected along a driving mission. The opposite situation occurs when the boost level is mainly provided by the standard turbocharger. The main objective of this paper is hence to provide some insights on the calibration and energy management strategy, which provides the minimum CO_2 and pollutant emission along representative vehicle missions.

To this aim, the pre-chamber engine under investigation is schematized in a 1D modeling framework. An "in-house" quasi-dimensional model is employed to reproduce the main in-cylinder phenomena, such as mixture formation, combustion, turbulence, and emissions. Its reliability has been validated in a prototype single-cylinder unit in previous activities [19,20]. Here, the same model, with few enhancements, is applied to foresee the behavior of the multi-cylinder engine as well.

The engine control parameters are set by coupling the 1D model to an external optimizer. The latter is instructed to maximize the engine BTE or, alternatively, a properly defined overall BTE, which also considers the electrical energy absorbed by the E-Comp (strategies #1 and #2). Optimization results are obtained at two engine speeds over a load sweep. They are then utilized to define a number of heuristic rules implemented within several PID controllers in the 1D model, mimicking the mathematical optimization procedure. This rule-based (RB) calibration strategy is preliminary validated through a direct comparison with the results of the external optimizer. Then it is utilized to compute two sets of complete engine performance maps, summarizing the calibration strategies following the maximum BTE and maximum overall BTE (OBTE), respectively. Finally, these two sets are embedded in a vehicle simulation able to quantify the CO_2 and pollutant emissions over two different driving cycles, namely the worldwide harmonized light vehicles test cycle (WLTC) and a real driving emission (RDE) cycle. This framework is schematized and summarized in the flowchart in Figure 1.

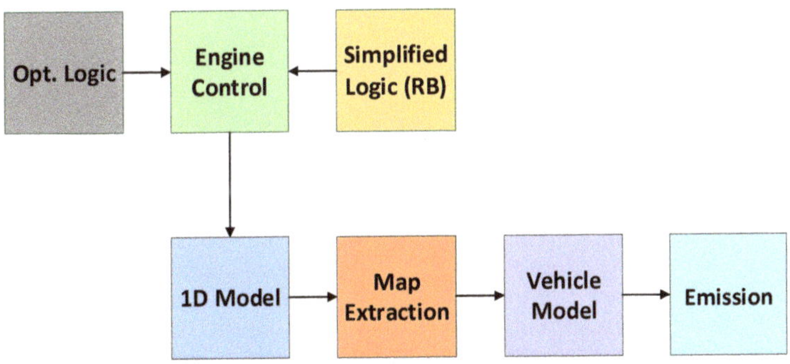

Figure 1. Flowchart of the methodology followed in the work.

The paper is organized as follows: firstly, the ultra-lean engine layout is described and the employed combustion model suitable for a pre-chamber engine is briefly recalled. Then, the numerical RB calibration procedure is illustrated and verified against the results of a multi-purpose optimizer. Finally, the energy management strategy developed by the authors is summarized [21] and the vehicle simulation outcomes are shown. In order to quantify the potential benefit of this HEV combined with this high-efficiency engine, the vehicle emissions are also compared against the current regulation limits (Euro 6d) and the CO_2 European targets.

2. Engine Description

The SI engine under study, outlined in Figure 2, is a prototype power unit constituted by four cylinders and equipped with active pre-chambers. Main features of the investigated engine are summarized in Table 1. The PC features four orifices of 1mm [22], and each cylinder has two intake and exhaust valves. The engine presents a cam phaser for both the intake and the exhaust camshaft. Moreover, two different cams profiles can be selected on the intake camshaft, allowing for an early valve closure (Standard Miller Lift-SML), or an extremely advanced valve closure (Extreme Miller Lift-EML) [22–24]. The latter is particularly suitable to limit the knock occurrence at high load.

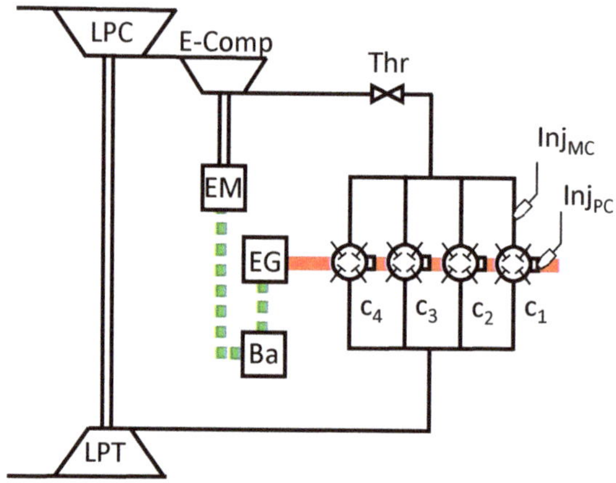

Figure 2. Schematic engine layout.

Table 1. Engine main features.

Multi-Cylinder Pre-Chamber Engine			
Bore	76mm	$V_{pre\text{-}chamber}/V_{TDC}$	~3%
Stroke	90mm	PC hole number	4
Displacement	1633 cm^3	PC hole diameter	1 mm
Compression ratio	15.8	$A_{jet\ holes}/V_{pre\text{-}chamber}$	~0.03 cm^{-1}
Fuel in main chamber	PFI, RON 95	Intake Valve Opening	344–370 CAD (@0.7 mm lift)
Fuel in pre-chamber	DI RON 95	Exhaust Valve Closure	350–380 CAD (@0.7 mm lift)
Pre-chamber volume	~1000 mm^3	Start of PFI	140 CAD BTDC
Minimum PC injection duration	300 µs	Start of in-PC injection	300 CAD BTDC

The engine is fed with liquid gasoline through four port fuel injectors located just upstream the intake valves (Inj_{MC}), and four direct injectors located into each PC (Inj_{PC}). A two-stage boosting system provides the needed boosting level to match the prescribed load target. It is composed by a variable geometry turbocharger (LPC and LPT) and a high-pressure (HP) compressor (E-Comp), driven by an electric motor (EM). An electric generator (EG) mounted on the engine shaft provides the electrical energy to recharge the battery (Ba) and to power the EM. The electric fluxes are shown as a dashed line in Figure 2.

As experimentally and numerically demonstrated [18–20], the introduction of a pre-chamber allows to strongly extend the lean burn limit of a conventional SI engine. This engine is designed to operate with very lean air/fuel mixtures to improve the engine efficiency and reduce the NO_X emission in a zone as wide as possible for the operating domain.

The presence of a high number of degrees of freedom makes the calibration of such a complex engine very challenging. In particular, the engine has nine control parameters, namely the relative air/fuel ratio (λ) in the chambers (labelled as λ_{PC} and λ_{MC}), the LPT rack position, the power input to the E-Comp, the intake and exhaust cam phases, the spark timing (or the combustion phasing, which can be represented by the angular position when half of the in-cylinder fresh charge is burned—50% of mass fraction burned, MFB_{50}), the throttle valve position, and the SML/EML cam profile.

3. D Engine Model Description and Validation

Consistently with the engine geometry, a detailed 0D/1D scheme is developed in a commercial modeling framework. The description of the flow inside the intake and exhaust pipes is based on a

1D approach, whereas "in-house" 0D sub-models are used to reproduce the in-cylinder phenomena typical of this novel architecture, such as air/fuel mixture preparation, combustion, turbulence and heat transfer. Regarding PFI and DI injections, it is assumed that the 30% of the injected fuel instantaneously evaporates, without considering spray evolution and liquid wall film formation. An empirical correlation is used for the estimation of the mechanical friction losses, dependent on engine speed, combustion phasing, and in-cylinder pressure peak. The steady-state flow coefficients for both the intake and the exhaust valves have been measured on a similar single-cylinder research engine and are here assigned to predict the flow permeability through the cylinder head [19,20,22]. A standard map-based approach is employed to reproduce the boosting system.

Each cylinder is schematized by two 0D volumes, a constant one (PC) and a variable one (MC), connected by means of an equivalent orifice. In particular, the overall cross-sectional area of the four holes is assigned in the above orifice. The orifice model also takes into account a discharge coefficient, fixed to a constant value of 0.65, which gives the best agreement with the PC-MC pressure difference along the compression stroke. According to a classical filling-emptying technique, mass and energy equations are solved in both volumes to estimate the mass exchange between them, based on pressure difference and overall effective cross-sectional area. The combustion process is described by a rearranged quasi-dimensional fractal model, developed by the authors in the last years [25]. It is based on a two-zone (burned and unburned gases) approach, and it is applied in a similar way in both MC and PC. In this context, the overall burn rate can be expressed as the sum of "classical fractal flame propagation" and a "turbulent jet combustion," expressed by the Equation (1).

$$\left(\frac{dm_b}{dt}\right)_{overall} = \left(\frac{dm_b}{dt}\right)_{fractal} + \left(\frac{dm_b}{dt}\right)_{jet} \tag{1}$$

The first term of Equation (1), applied in both PC and MC, describes the burning rate through the fractal approach under the hypothesis that the flame front locally propagates at a laminar speed, S_L, and that the combustion process is promoted by the turbulence-induced flame wrinkling, Σ, leading to the following expression:

$$\left(\frac{dm_b}{dt}\right)_{fractal} = \rho_u A_L S_L \Sigma \tag{2}$$

where ρ_u is the unburned gas density and A_L is the laminar flame front area. The second term of Equation (1) models the burning rate contribution in the MC due to the turbulent jets, under the hypothesis that the jets entrain fresh charge (air and fuel) and rapidly burn it, in turn releasing heat. This is described by the equations below:

$$\left(\frac{dm_b}{dt}\right)_{jet} = \frac{dm_{b,entr}}{dt} = \frac{m_{entr} - m_{b,entr}}{\tau}; \quad \tau = \frac{\Lambda_T}{S_L} \tag{3}$$

$$\frac{dm_{entr}}{dt} = c_{jet}\dot{m}_{jet}\sqrt{\frac{\rho_{PC}}{\rho_{MC}}} \tag{4}$$

The jet-induced burning rate is assumed proportional to the difference between the current entrained mass (m_{entr}) and its burned portion ($m_{b,entr}$). The Equation (3) resembles the well-known eddy burn-up approach [26,27], for which τ is a characteristic time scale. This last is calculated as the ratio between the Taylor length scale, Λ_T, and the laminar flame speed, S_L. Finally, the total entrained mass, m_{entr} is computed by the integration of its time derivative, according to Equation (4), which in turn is estimated by the semiempirical correlation reported in [28]. This entrained mass is a function of the mass flow rate from the PC, $\dot{m}_{jet}$, on a tuning constant c_{jet} ad on the density ratio between PC and MC. Equally, the burned entrained mass, ($m_{b,entr}$), is estimated by the integration of Equation (3).

The summarized combustion model requires the evaluation of a few parameters—namely the flame wrinkling Σ and the Taylor scale Λ_T—which depend on the turbulent field established in the combustion chamber. An "in-house" developed phenomenological K-k-T turbulence sub-model is employed to estimate those quantities [29]. The turbulence model takes into account the energy flux transferred from the mean flow kinetic energy, K, to the turbulent one, k. The model includes proper balance equations for tumble and swirl angular momentum, T and S. The model is applied to both PC and MC, and also considers the turbulence production induced by the incoming/outcoming flow through the orifices [20].

The description of the knock phenomenon is based on the calculation of the auto-ignition (AI) delay time of the air/fuel mixture in the unburned zone. In particular, AI is carried out with a tabulated approach [30], employing preliminary auto-ignition chemical kinetic simulations performed in a homogeneous reactor at a constant pressure. The kinetic scheme adopted for the estimation of the table includes five elements, 201 species, and 1548 reactions [31] and an additional skeletal sub-mechanism for toluene oxidation to handle a toluene reference fuel. The AI time, τ_{AI}, is stored in the table as a function of pressure, temperature, equivalence ratio, and residual content. As known, due to the presence of the cyclic dispersion, faster-than-average pressure cycles may induce the presence of a stochastic soft-knock, even when the average cycle is working in a safe, non-knocking combustion. Since, in the adopted approach, the AI integral, Equation (5), is computed on the average cycle, a small safety margin is applied, and the knock event is considered to occur when the AI integral exceeds a tunable lower-than-unity threshold level.

$$\int \frac{dt}{\tau_{AI}} \tag{5}$$

The model allows to estimate the regulated cylinder-out emissions, namely CO, HC, and NO. For the evaluation of CO and NO, a multizone-zone approach is applied in both PC and MC to estimate the burned gas stratification. Each burned parcel is compressed/expanded adiabatically according to the in-cylinder pressure. The estimated local zone temperatures are employed within the well-known extended Zeldovich mechanism for the estimation of the NO [32], whereas the CO is computed with a two-step reaction scheme [33]. The experimental findings show that in ultra-lean engines the noxious emissions are mainly composed of NO_2, therefore in the adopted approach the NO production derived by the Zeldovich mechanism is assumed to completely oxidize the NO_2 when the burned gases evolve along the exhaust pipes. Concerning the HC, the crevice contribution and the post-oxidation is considered only for the MC, neglecting the PC contribution. In the employed model, the HC filling and emptying within the crevice regions is followed [34]. The temperature in this volume is considered to be the same as the piston wall, while the pressure equals the one in the cylinder. When unburnt fuel is released from the crevices, a proper boundary layer temperature is computed, at which the released HC partly oxidize according to the kinetic rate proposed in [35].

The quasi-dimensional combustion model was tuned with reference to a research engine [19,20], having geometrical and architectural characteristics similar to the four-cylinder engine under study for which experimental data are not yet available. The model was tuned by a trial-and-error procedure aiming to obtain the best agreement with pressure signals in both PC and MC. A unique set of tuning constants was selected for all the operating conditions, characterized by very different speeds, loads, and air/fuel ratios (λ up to 2.4 in the main chamber) [19,20].

4. Optimization Approach for Engine Calibration

With the aim of calibrating the engine, a value for each of the nine control parameters listed in the "Engine Description" section must be specified in each operating condition. The purpose of this task is the optimization of the engine performance over the entire operating domain.

The calibration procedure here targets the maximum possible efficiency, fulfilling several constraints—i.e., maximum pressure inside MC and PC, knock intensity, turbo speed, boost level, E-Comp power, internal EGR amount, etc.—prescribed to confine thermal and mechanical stresses and operating reliability for the engine and for its components.

To find the optimal set of the control parameters at variable speeds and loads, an automatic optimization is performed, realized by using an external multi-purpose optimizer. The optimization is performed considering two different control strategies with differentiated fuel and electric consumptions. Particularly, the first control strategy (strategy #1) aims to maximize the conventional BTE, namely:

$$\text{BTE} = \frac{P_{\text{ICE}}}{\dot{m}_f \text{LHV}} \tag{6}$$

where P_{ICE} is the brake power at the engine shaft, $\dot{m}_f$ is the total injected fuel flow rate, and LHV is the lower heating value of the fuel.

The second one (strategy #2) is oriented to the optimization of the overall BTE:

$$\text{OBTE} = \frac{P_{\text{ICE}} - P_{\text{HPC}}/\eta_{\text{HPC}}}{\dot{m}_f \text{LHV}} \tag{7}$$

where P_{HPC} is the mechanical power at the E-Comp shaft and η_{HPC} is the electro-mechanical efficiency of the E-Comp. Based on its definition, the OBTE also considers the electric energy required by the E-Comp and drained from the battery.

For both examined control strategies, the optimization is realized along a brake mean effective pressure (BMEP) sweep at two different engine speeds, namely 2000 and 3000 rpm. Low/medium engine rotational speeds are selected because of their relevance regarding the vehicle homologation cycles. The load sweep is automatically realized by alternatively maximizing or minimizing the BMEP, together with the BTE/OBTE, hence performing four different multi-objective problems for each engine speed. Each optimization problem hence identifies a single branch of the BTE/BMEP or OBTE/BMEP Pareto frontiers.

The selected optimization tool is the genetic MOGA-II algorithm, which represents the best choice for the case of a multi-variable/multi-objective problem. The nine control parameters previously listed, apart from the relative air-fuel ratio in the main chamber which is always metered to fix a value of two, represent the independent variables of the optimization problem. Each variable is changed within a prefixed allowable range. The selection of the intake valve lift profile is indeed handled by introducing a fictitious discrete variable, associating values of 0/1 to the Standard/Extreme Miller valve lifts, respectively.

The logical scheme of the optimization process was already reported in a previous author's work [36], and here it is partially modified by the addition of SML/EML and EVC (exhaust valve closure) variables. The optimizer algorithm, at each step with an iterative process, selects the eight engine control variables, passing them to the 1D engine model to evaluate the variables of interest. At the end of each simulation, the estimated values of the objective function are returned to the optimizer to start the next iteration. This process pursues as long as the optimal levels, being part of the Pareto frontier, are achieved.

As said, the optimizer evaluations are post-processed and filtered to check that some monitored variables do not exceed prescribed threshold levels. As an example, the maximum in-cylinder pressure is limited to 180 bar, the maximum plenum pressure to 4.5 bar, the maximum E-Comp power to 10 kW, the maximum AI integral to 0.8, etc.

5. Assessment between RB and Optimizer Calibrations

The previously described optimization procedure is too time consuming to be directly employed to compute the complete engine performance map. To speed up the numerical calibration, a set of heuristic rules were defined to reproduce as best as possible the behavior of the mathematical optimization (rule-based calibration). The RB was applied by including a network of logical switches, PID controllers, and "math functions" in the 0D/1D environment.

Preliminarily, a prefixed running line was chosen on the LPC map as a tradeoff between an appropriate surge margin and a suitably high-pressure ratio to avoid an excessive E-Comp shaft power absorption (below 10 kW). Under these operating conditions, the E-Comp running line was directly identified to reach the HP boost level necessary to fulfill a prescribed full load (FL) target, depicted in red in Figure 3. By changing the LPC running line, a different sharing of the overall boost level between the LPC and the E-Comp would have been accordingly identified. However, whatever the choice for the LPC running line, at very high engine speeds (usually over 4500 rpm) for the selected boosting devices, the constraints on the maximum E-Comp power and/or maximum boost level could not be fulfilled. For this reason, the only way to achieve the FL target was to inject more fuel in the main chamber, resulting in a $\lambda_{MC} < 2$. The L2 line in Figure 3 identifies the BMEP region below which a $\lambda_{MC} = 2$ operation can be sustained (light yellow domain), while the area above the L2 line defines the operating domain with $\lambda_{MC} < 2$ (highlighted in orange).

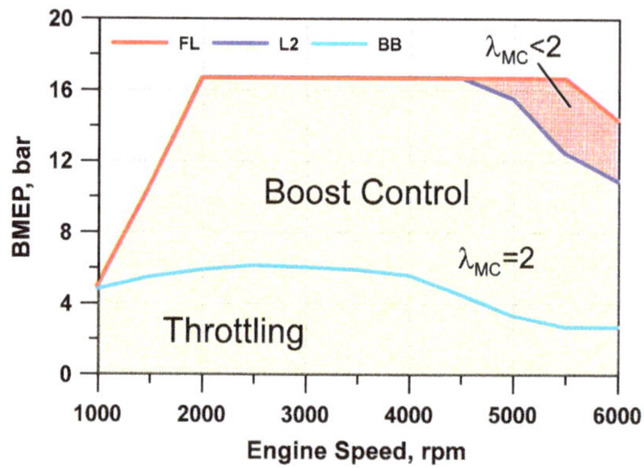

Figure 3. BMEP–rpm map containing the BMEP target (FL), and the L2 and BB lines.

In the low-load region, an additional line is defined corresponding to the BMEP levels reached with the LPT fully open and no power given to the E-Comp, named Base Boost Line (BB in cyan in Figure 3). In these conditions, the E-Comp is bypassed, acting on dedicated control valves. Above the BB line, the engine load is regulated by a progressive closing of the LPT and a proportional increase of the E-Comp power. Below the BB line, the LP and HP compressor settings continue to be fixed, while the load is only regulated by the progressive closure of the throttle valve.

Concerning the VVT position of intake and exhaust camshafts, simple rules are defined to select an advanced IVC (intake valve closure) at low load to have a pumping losses reduction. The intake valve advancing is, however, limited at very low load to avoid an internal EGR rate above 20%. In this way, a stable combustion process can be guaranteed in the real engine at the expense of a certain pumping losses increase.

In the mid-load range, the intake VVT is moved towards a later IVC to allow for increased air flow rates, while, at a very high load, an early IVC is once again preferred to reduce the effective compression ratio for knock control. The exhaust valve is accordingly controlled to limit the valve overlap and/or to increase the expansion work at a low load.

The actual positions of the BB and L2 lines also depend on the selection of the intake lift profile, namely the SML or the EML. To reduce pumping losses, the EML is always specified when the computed airflow rate is below 100 kg/h. By analyzing the results of the previous mathematical optimization, it was indeed found that for strategy #1 (max BTE), the SML is only selected below a specific BMEP limit (namely 6.5 bar), while it is always selected above 100 kg/h airflow rate in the strategy #2 (max OBTE). This can be easily explained by observing that the SML requires a lower boost level for the same load. For this reason, a lower E-Comp power is required, which improves the

OBTE. Moreover, to better reproduce the optimization results, a different running line is selected for the strategy #2, moving the boost sharing mainly on the LPC and reducing the E-Comp contribution.

The proposed rule-based methodology is quite refined to provide engine operations very close to the optimal BTE and OBTE, as confirmed in the following. In addition, the advantage of this RB strategy is related not only to its easy implementation in the used 1D software, but also to its much faster execution compared to a set of multi-objective optimizations.

To quantify the reliability of the two implemented calibration strategies, their outcomes were compared to the optimization results. The optimization outcomes are shown in the next figures with open symbols and label Opt and refer to the outcomes belonging to the Pareto frontier.

Figure 4 reports the comparisons of BTE and OBTE at 2000 rpm for the considered calibration strategies (#1 and #2). In both cases, a considerable agreement was realized along the BMEP sweep between Opt and RB procedures. Figure 4 also shows the potentiality of the analyzed engine architecture to reach a very high BTE level of about 50% at the higher load. In the latter conditions, lower OBTE is reached due to the relevant power request by E-Comp (see OBTE definition in Equation (5)). Indeed, OBTE assumes a maximum value around 42% at medium loads, with a slight reduction at the highest BMEP levels.

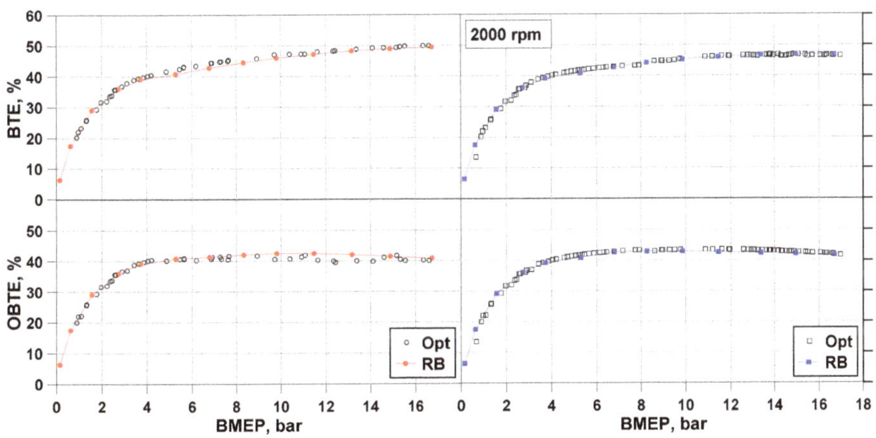

(**a**) strategy #1 (**b**) strategy #2

Figure 4. Brake thermal efficiency (BTE) and overall brake thermal efficiency (OBTE) comparison in a BMEP sweep at 2000 rpm for strategies #1 (**a**) and #2 (**b**).

Quite satisfactorily, RB/Opt agreements were found for the most relevant engine calibration variables depicted in Figures 5 and 6. The first assessment regards the intake valve strategy and the intake/exhaust valve timings, which are plotted in Figure 5. Considering the strategy #1, the Extreme Miller was mainly preferred all along the BMEP sweep, while in the strategy #2 at medium/high loads, the Standard Miller strategy was chosen.

On a general viewpoint at both low and high loads, IVC is selected near to the most advanced timing setting. This option, as said, is preferred for minimizing the pumping losses at low loads and mitigating the knock occurrence at high loads [37]. In the medium BMEP range, especially for strategy #2, IVC is slightly delayed, allowing for higher air flow rates and resulting in a better agreement with the Opt outcomes. EVC indeed exhibits an approximately constant trend with load.

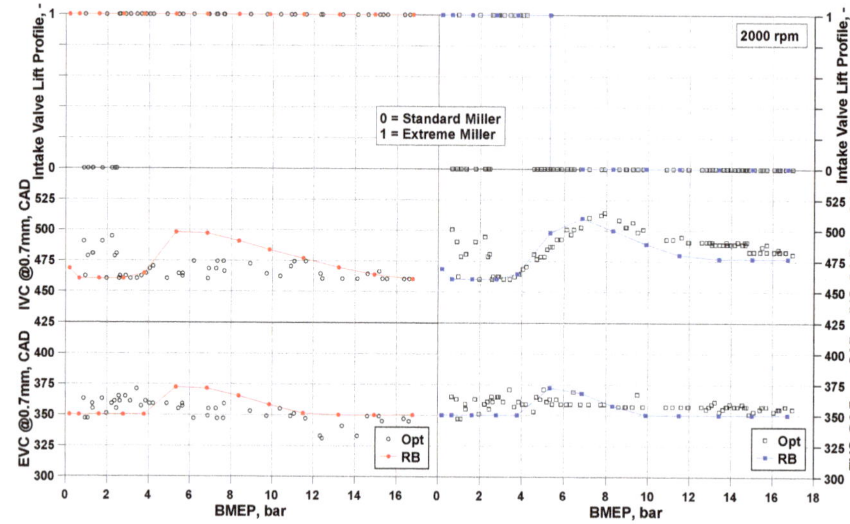

(**a**) strategy #1 (**b**) strategy #2

Figure 5. Selected intake valve lift profile, IVC and EVC, comparison in a BMEP sweep at 2000 rpm for strategies #1 (**a**) and #2 (**b**).

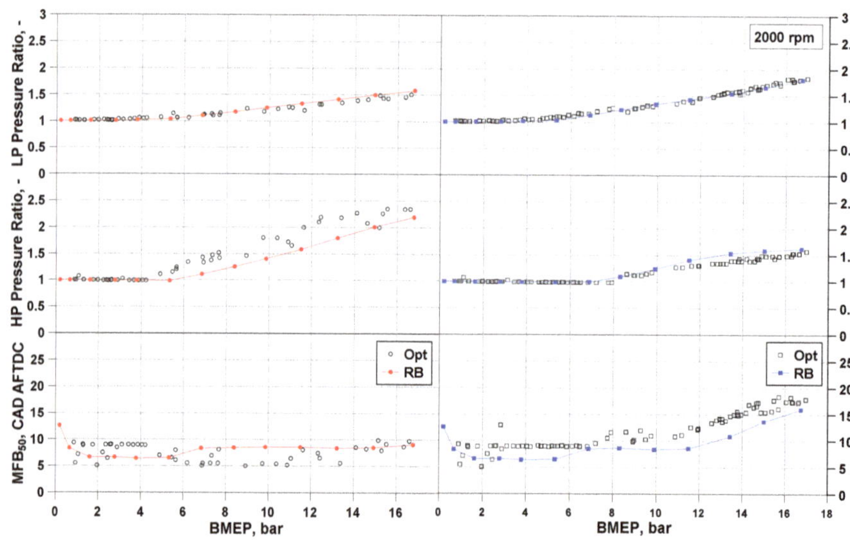

(**a**) strategy #1 (**b**) strategy #2

Figure 6. Low/High pressure ratios and MFB_{50} comparison in a BMEP sweep at 2000 rpm for strategies #1 (**a**) and #2 (**b**).

Figure 6 shows the sharing of the overall engine boosting between the two compressors, depicting the HP and the LP pressure ratios. As expected, strategy #1 favored a higher boosting by the E-Comp instead of the LPC. Conversely, for strategy #2 the boosting was principally realized by the LPC; instead, the E-Comp worked by having a pressure ratio as low as possible. This choice involved an increased engine exhaust backpressure, but it resulted in a lower power absorbed by the E-Comp.

The results in Figure 6 also show that HP and LP pressure ratios assume unit values below 5 bar BMEP, suggesting a throttle-based load control. Above 5 bar, both LP and HP boosting are progressively regulated by a partial closing of the LPT rack and by simultaneously increasing the E-Comp power.

It is worth underlining that the RB strategy #1 between 4–10 bar shows some discrepancies if compared to the optimal outcomes. More specifically, the RB approach determined a delayed IVC, which is compensated by a lower HP boost pressure. Nevertheless, this reflects a slight BTE lowering compared to the optimizer output (less than 1%).

As a further assessment between RB and Opt methodologies, in Figure 6 the BMEP trends of MFB_{50} are reported, highlighting a satisfactory RB/Opt agreement. For this combustion parameter, the optimal value is usually around 6–10 CAD AFTDC. Indeed, in the case of later phasing, the combustion develops during the expansion stroke, leading to a less effective work exchange between the piston and the gas, while, in the case of too advanced MFB_{50}, a relevant portion of the combustion process takes place during the compression stroke, increasing the compression work exerted by the piston on the fluid. An almost flat trend was obtained for strategy #1 in a wide load range (2–15 bar BMEP) with most of the points in the band 5–10 CAD AFTDC. A slightly delayed MFB_{50} at both higher and lower BMEP values was detected. At high loads, this is due to knock mitigation, while at low loads, it is because of the combustion lengthening and of the limitation on the maximum allowable spark advance. For the strategy #2, MFB_{50} showed a different trend with a more pronounced delay of the MFB_{50} at increasing loads (above 10 bar BMEP). This, of course, is the only possibility for knock control once the Standard Miller profile for the intake valves has been selected. This quite different behavior was very well captured by the RB strategies.

Although not discussed here for sake of brevity, another outcome of the optimization is that a close-to-stoichiometric mixture is preferred in the pre-chamber for whatever is the BMEP. A certain mixture leaning is required at high loads to limit the pressure peak in the PC.

To further prove the consistency of the RB calibration, Figures 7 and 8 present additional comparisons between RB and Opt strategies, referring to the engine speed of 3000 rpm. Once again, a satisfactory RB/Opt agreement for all the considered variables was reached all along the BMEP sweep. Figure 7 shows that, for this higher speed, the optimal control involved more relevant modifications of intake valve lift profiles and IVC/EVC timings at changing loads.

For strategy #1, different from the previous speed, the Standard Miller was selected in a restricted low/medium load range (4–5.5 bar BMEP) to enhance the effective compression ratio and hence the efficiency in the light of a minor knock tendency at this speed. However, the Extreme Miller was once again preferred as soon as the load increases (Figure 7). For strategy #2, an earlier transition from the Extreme Miller to the Standard Miller by increasing the BMEP was detected if compared to the results at 2000 rpm. Once again, at medium loads the IVC was delayed, allowing for increased airflow rates. EVC trends appear almost load insensitive regardless of the calibration strategy, and a modest discrepancy between RB and Opt procedures was detected especially at very low BMEPs (0–2 bar in Figure 7). This is probably due to a more refined calibration given by the optimizer, which tends to reduce the valve overlap with the aim to limit the internal EGR even more than the logics of the RB approach. Referring to the results plotted in Figure 8, similar considerations to the ones already carried out for the case at 2000 rpm still hold.

Concerning the combustion phasing differences between 2000 and 3000 rpm, a greater delayed MFB_{50} was observed at low loads for the engine speed of 3000 rpm, mainly ascribed to a pronounced combustion lengthening of its initial stage due to an excessive rich mixture in the pre-chamber at the spark event. This is due to the lower bound of the PC injection duration of 300 μs (see Table 1). At high loads, a less pronounced MFB_{50} delay occurred for the strategy #2 due to the lower knock tendency at 3000 rpm. In summary, the above discussed results highlight that the RB calibration demonstrates to be extremely trustworthy for various load levels and for two different speeds, allowing to reach efficiencies and control variables very close to the ones obtained by the optimizer-based methodology. In the light of the latter consideration, the developed RB calibration was broadened to the full engine speed/load range in order to investigate the close-to-optimal engine performance in the whole operating domain, as reported hereinafter.

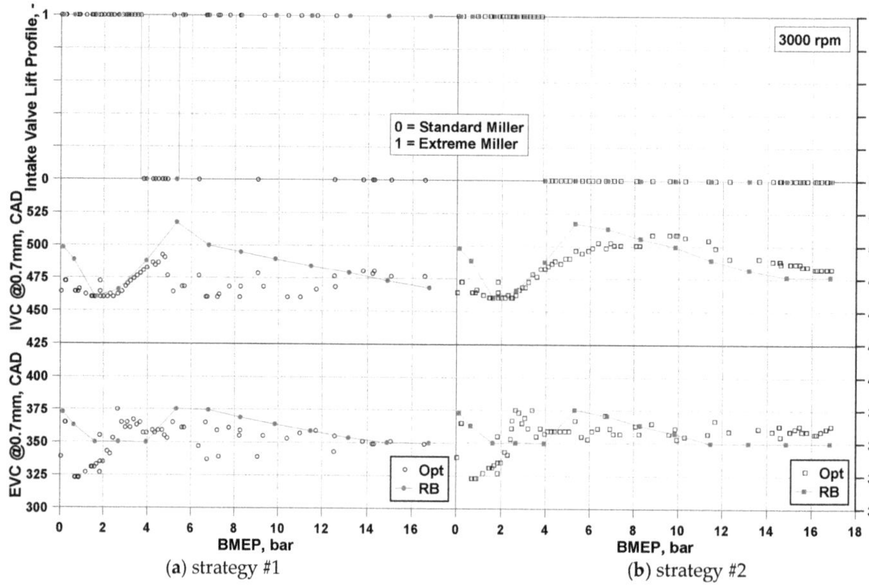

Figure 7. Intake valve lift profile, IVC, and EVC comparison in a BMEP sweep at 3000 rpm for strategies #1 (**a**) and #2 (**b**).

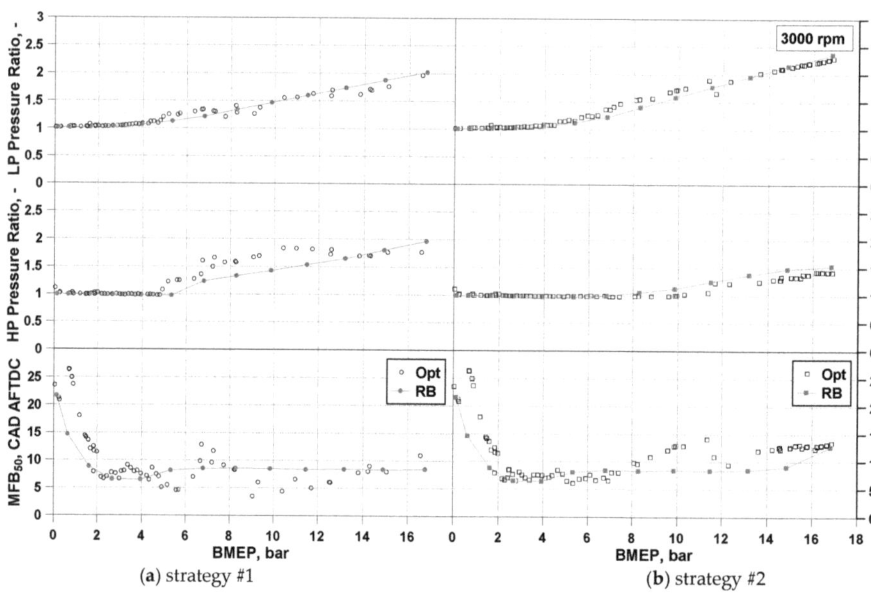

Figure 8. Low/High pressure ratios and MFB_{50} comparison in a BMEP sweep at 3000 rpm for strategies #1 (**a**) and #2 (**b**).

6. Map of Engine Performance

According to the RB control strategy, the entire engine operating map was computed (consisting in 143 operating points—11 rpm × 13 BMEP) for the two different calibration strategies above discussed. In Figure 9, the BTE maps for calibration strategies #1 (a) and #2 (b) are shown. The maximum levels for both strategies occur at 2000 rpm, close to the full load region. The OBTE maps for the different strategies are shown in Figure 10. For these, the maxima occur at medium speeds (2000–2500 rpm) and medium–high loads (8–14 bar BMEP). According to the selected optimization targets, a 3% higher peak BTE was obtained in the strategy #1 (Figure 9a), while a 1% higher peak OBTE was obtained in the strategy #2 (Figure 10b).

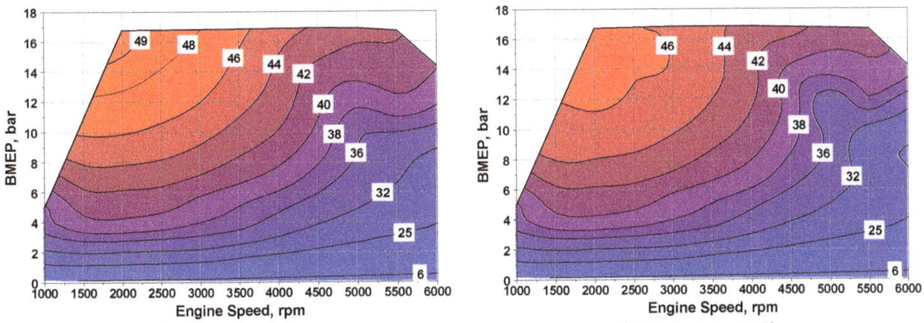

(a) BTE strategy #1 (b) BTE strategy #2

Figure 9. Map of BTE (%) for strategies #1 (a) and #2 (b).

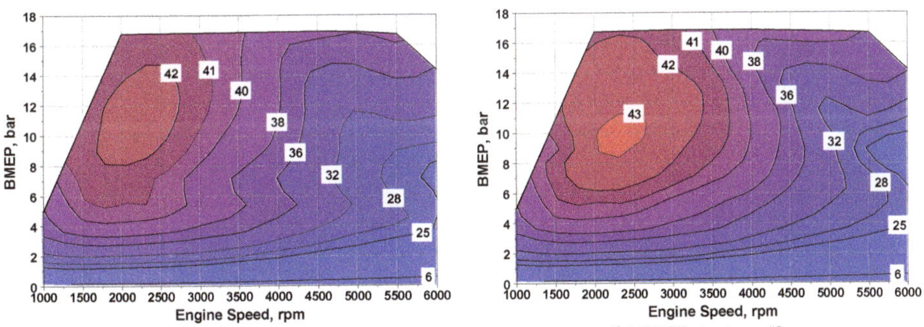

(a) OBTE strategy #1 (b) OBTE strategy #2

Figure 10. Map of OBTE (%) for strategies #1 (a) and #2 (b).

The gap between BTE and OBTE for the strategy #1 was about 7% in the OBTE peak zone, and gradually decreased at higher speeds and low loads. A similar behavior characterizes the strategy #2, having in the peak zone a gap of about 3–4%. This is due to the power absorbed by the E-Comp, whose maps are reported in Figure 11. For strategy #1, the E-Comp power gets to the upper bound of 10 kW, which is about the 8% of the power rated by the engine, while for strategy #2, it only gets a maximum of about 7 kW.

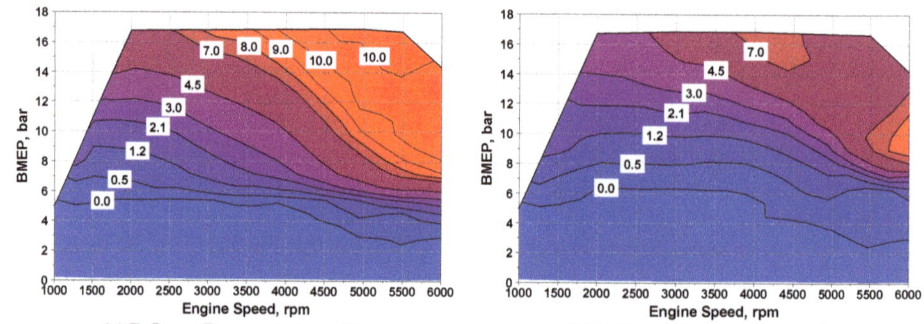

(**a**) E-Comp Power, strategy #1 (**b**) E-Comp Power, strategy #2

Figure 11. Map of E-Comp power consumption (kW) for strategies #1 (**a**) and #2 (**b**).

For the considered engine, featuring an ultra-lean air/fuel mixture, both strategies involved a very high plenum pressure, which is mandatory to achieve the load target. As an example, Figure 12a shows the boost pressure map for the strategy #1, which highlights a peak of 4.3 bar at the highest speeds and loads. The intake plenum pressure was conserved above 1.0 bar in most of the map (reduced pumping losses), and above 0.5 bar even in the area at lowest BMEP. The lambda target of 2 in the main chamber was reached in most of the operating domain, as shown in Figure 12b, except in the maximum power area. This is due to the attainment of the maximum E-Comp power, which obliges to reduce the mixture leaning ($\lambda_{MC} < 2$) to get the load. Similar boost pressure and lambda maps are also found for the strategy #2.

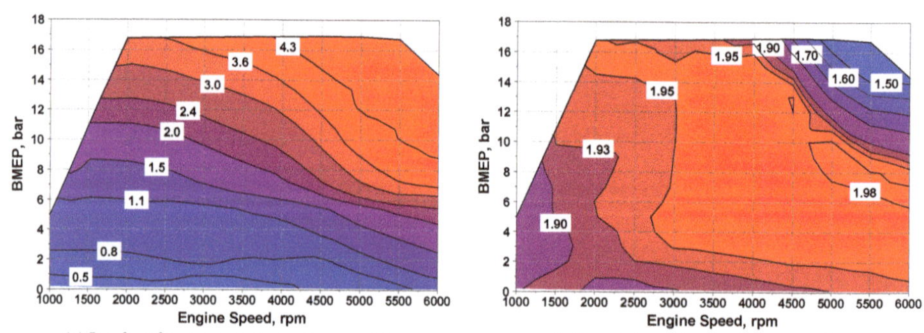

(**a**) Intake plenum pressure, strategy #1 (**b**) Lambda ratio, λ_{MC}, strategy #1

Figure 12. Map of intake plenum pressure (bar) (**a**) and (**b**) map of λ_{MC} (-) for strategy #1.

The predicted brake-specific emissions of NO_x and HC are shown in Figure 13a,b, respectively, still referring to strategy #1. With the aim to limit the temperature peaks in the burned zone of the combustion chamber preventing the formation of the above pollutants, the ultra-lean combustion concept was used. The NO_x production was very low in most parts of the operating map, except for the high speed and load zone. This is due to the lower mixture leaning in the above-mentioned zone, compared to the $\lambda_{MC} = 2$ target (see λ_{MC} contours in Figure 12b). Due to the higher burned temperature, if compared to the one in the MC, for this engine the NO_x production was prevailing in the pre-chamber. Despite the large excess air, the HC production is not negligible over the whole engine domain (Figure 13b). This is due to the lean burning concept, which, compared to a conventional stoichiometric engine, reduces the in-cylinder temperatures and consequently makes the post-oxidation phenomenon less effective. Looking to Figure 13b, the shape of the HC iso-contours is justified by a faster post-oxidation at increasing speed (less time for wall heat losses) and load (lower percent wall heat losses).

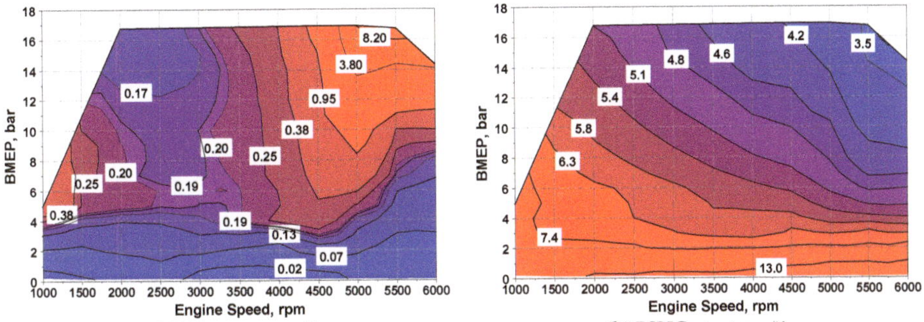

(**a**) BSNOx, strategy #1 (**b**) BSHC, strategy #1

Figure 13. Map of (**a**) brake specific NO_x emission (g/kWh) and (**b**) map of brake-specific HC emission (g/kWh) for strategy #1.

7. Vehicle Simulations

7.1. Hybrid Vehicle Architecture

The main characteristics of the vehicle here investigated are reported in Table 2 [38]. The vehicle is a HEV, belonging to the C-class. It is characterized by a combined parallel/series powertrain, schematized in Figure 14. The powertrain is composed of an ICE, two electric motor/generator units, (EM and EG), a battery (Ba), three clutches (Cl_{1-3}) and two gear boxes (GB_{1-2}).

Table 2. Main characteristics of the tested HEV.

Vehicle	
Mass, kg	1730
Car aero drag, m^2	0.775
Tire rolling resistance coeff., -	0.008
Wheel diameter, m	0.723
Axle ratio, -	4.4
Axle inertia, kgm^2	1.5
Internal Combustion Engine	
Displacement, cm^3	1633.1
Max Power, kW	125
Inertia, kgm^2	0.35
Electric Generator	
Max Power, kW	55
Max Torque, Nm	165
Inertia, kgm^2	0.10
Electric Motor	
Max Power, kW	50
Max Torque, Nm	240
Inertia, kgm^2	0.10
Battery	
Internal Resistance, Ohm	0.375
Voltage, Volt	400.0
Energy density, Wh/kg	170.0
Usable battery sizing, kWh	0.50
SoC limits, -	0.2–0.9

Table 2. *Cont.*

Gear-Box$_1$	
Gear 1 Ratio, -	2.72
Gear 2 Ratio, -	1.64
Gear 3 Ratio, -	0.99
Gear 4 Ratio, -	0.60
Gear-Box$_2$	
Gear 1 Ratio, -	2.67
Gear 2 Ratio, -	1.03

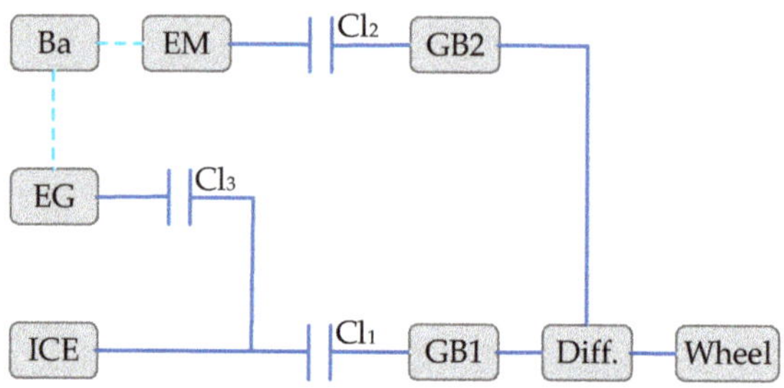

Figure 14. Powertrain schematic of the tested HEV.

7.2. Simulation Platform

The vehicle simulations were carried out by a "in-house developed" software, implemented in Fortran language (UniNa Vehicle Simulation-UNVS) [21]. It is based on a "backward-facing" (quasi-static) method [39]. All of the powertrain schematic components in Figure 14 were defined by a lumped-parameter method. The approach to describe the thermal unit was quasi-steady map based. In particular, the BSFC maps, derived from the calibration strategies #1 and #2, were implemented in the vehicle simulation, together with the curve of maximum shaft torque. In addition to the BSFC maps, the vehicle model handled the corresponding E-Comp power consumption maps. The electrical absorption of the E-Comp is considered as an additional load on the battery. The ICE thermal transient was neglected in the simulations, resulting in a null fuel consumption penalization at the cold start. The exhaust gas after-treatment system was not modeled, so the predicted emissions have to be considered as "engine-out" values. The electric unit efficiencies were described once again by a map-based approach, dependent on speed and shaft torque. The battery was treated by a conventional State of Charge (SoC) model [40]. In order to estimate the Joule-effect losses, an internal resistance was set, neglecting the possible variations in changing SoC and temperature. The mechanical losses in the gearboxes were evaluated assuming constant efficiencies (equal to 0.97).

7.3. Efficient Thermal Electric Skipping Strategy

The vehicle control strategy here adopted, if compared to the well assessed strategies (namely Pontryagin minimum principle [41], dynamic programming [42] or ECMS [43]), is designed to be simplified. The conventional power-split logics, typical of the abovementioned strategies, is substituted with an alternative utilization of the thermal and electric units for the vehicle driving (Efficient Thermal Electric Skipping Strategy–ETESS [21]). The basic concept behind ETESS is an alternate use of the electric units and thermal engine so to fulfill the power demand at the vehicle wheels. The choice between the two modes depends, at each time, on the comparison between the effective fuel rate of the thermal engine needed to fully satisfy the power demand at the vehicle wheels, and an equivalent fuel rate related to a pure electric driving of the vehicle. The ETESS is featured by a much-reduced computational effort if compared to most common optimization strategies, leading to similar performance in terms of kilometric consumption and emission [21]. The results shown in the next section were obtained by the off-line variant of the ETESS, where the control is tuned to realize the battery energy balance between the beginning and the end of the driving mission [44].

7.4. Simualtion Raster and Results Discussion

The vehicle simulations were carried out for two driving cycles, namely the WLTC and a real driving cycle (RDC) [45]. The results of CO_2 and pollutant emissions (grams per kilometer) are represented under the form of bar charts in Figures 15 and 16. Each couple of bars compares the results of the calibration strategy #1 (red) and strategy #2 (blue), reporting on the top percent difference.

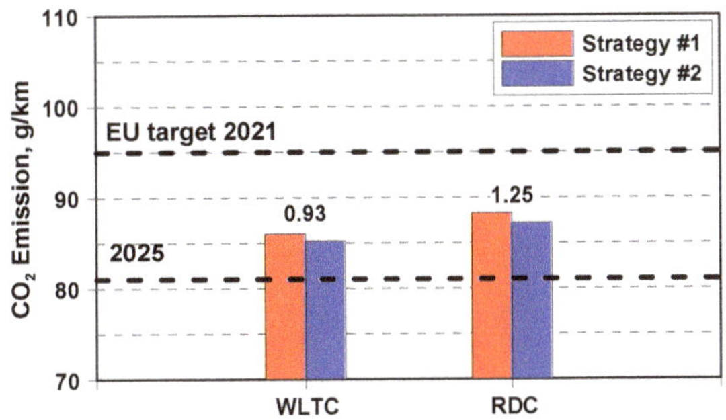

Figure 15. Strategies assessment of CO_2 emission and their percent difference along WLTC and RDC.

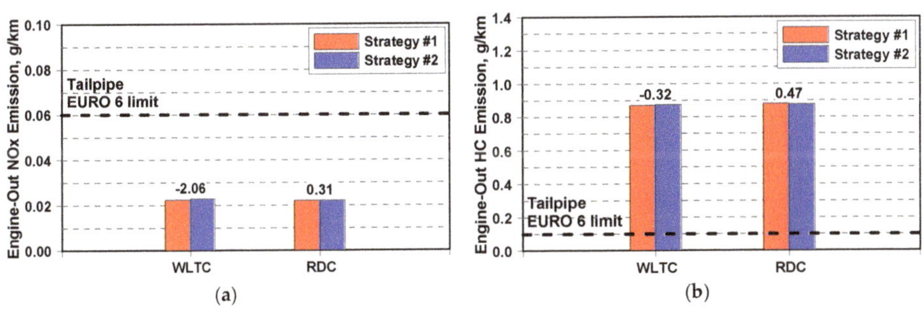

Figure 16. Strategies assessment of NO_x emission (**a**) and of HC emission (**b**) and their percent difference along WLTC and RDC.

Starting from the CO_2 results shown in Figure 15, a first consideration is that the emissions are below the target imposed by the European Union for the 2021 of 95 g/km, but above the one of the 2025, equal to about 81 g/km or whatever is the ICE control strategy. These results indicate that the HEV vehicle here analyzed, even equipped with a very efficient thermal unit, was not enough to match the EU 2025 target and a plug-in HEV variant is mandatory to get this. The pollutant emission outcomes, as stated above, refer to engine-out values compared to the tailpipe reference standards imposed by the legislation. The potential abatement due to an aftertreatment system is not explicitly considered. However, the comparison can also help to evaluate and design the most adequate exhaust after-treatment system (EATS) for the considered application. Looking at the values for the nitrogen oxides (Figure 16a), it can be seen that the emissions are within the bounds. The HC values (Figure 16b), for both the considered cycles and calibration strategies, are much higher than the limit imposed from the Euro 6d, highlighting the need for an oxidizing EATS for regulation compliance. Anyway, the technology of an oxidizing EATS at the current state-of-art is very robust, low-cost and reliable, not representing a substantial challenge for a real on-vehicle application. The engine calibration, either strategy #1 or #2, does not seem to exert a substantial influence on the pollutant emissions. It is worth noting that the ETESS control strategy was not conceived to minimize the pollutant emissions, but the fuel consumption and hence the CO_2 emissions.

The comparison presented underlines that the strategy #2 allows slightly lower CO_2 emissions for both the considered driving cycles. To clarify this result, the instantaneous trends of some representative control and performance parameters are plotted in Figure 17 with reference to the WLTC. The trends underline that the powertrain control is practically the same for both the ICE calibrations, resulting in almost superimposed traces of ICE power (b), EM power (c) and gear number (g). Some differences emerge for the trends of E-Comp power (d) and SoC (e), which reflect slight misalignments of fuel consumption (f) and battery power (h) profiles. Note that a positive value of battery power corresponds to an energy flux drained from the battery.

To better highlight the differences among the two engine calibrations, the plots of SoC, fuel rate and battery power are reproposed in Figure 18 with a magnified time scale in the most load-demanding portion of the driving cycle, namely between 1450–1750 s. During this period, the fuel rate is lower for the calibration strategy #1 if compared to the strategy #2, while the opposite occurs for the battery power consumption due to a higher E-Comp power absorption (Figure 17d). This reflects in a faster decrease of the SoC, which leads to reduced electric driving in the first portion of the cycle where the load demand is lower. For this reason, the ICE works for a longer time in the operating conditions, namely at low load/speed, where its efficiency is well below its maximum level. On the contrary, the calibration strategy #2 allows longer pure electric driving in the first portion of the cycle, avoiding for the thermal engine operations with a very reduced efficiency.

A similar behavior emerged for the RDC, as shown in Figure 19. This reports the vehicle speed profile (a) and the SoC trend (b). Note that the initial SoC level was chosen to get simultaneously the optimal control and battery energy balance along the cycle. Such a choice is a consequence of the simplified approach for the battery treatment, where the losses are not dependent on the instantaneous SoC level and on the possible stress when the extreme SoC bounds are attained.

Once again, during the most load-demanding portion of the cycle (after about 5400 s), the battery discharge for the calibration strategy #1 was faster, and this was compensated by a reduced use of electric driving in the first portion of the cycle. This occurrence globally penalizes the strategy #1 in comparison with the strategy #2.

Under a more general point of view, for a given battery capacity and vehicle mission, the engine calibration, which reduces the energy drain from the battery, allows for a more efficient choice of the electric driving phase along the cycle. In the presented cases, however, a limited benefit has been estimated. An explanation of such reduced fuel consumption differences can be drawn by analyzing the time evolution of the engine operating point. Referring to the RDC, the instantaneous engine operating points, sampled at each second, are shown in Figure 20. They are plotted over the map of OBTE difference between strategies #2 and #1. It can be observed that most of the samples are located in a speed range between 1500 and 2500 rpm and with loads below 8/9 bar BMEP, where the OBTE difference is lower than 0.6% and the E-Comp power is small (Figure 11). Changes between the two calibrations only emerge outside the above operating area where the percentage difference in terms of OBTE increases, contributing to reduce the fuel consumption of strategy #2. Similar considerations can be applied to the WLTC results. Different driving missions and more load demanding are expected to bring out more relevant differences between the considered strategies.

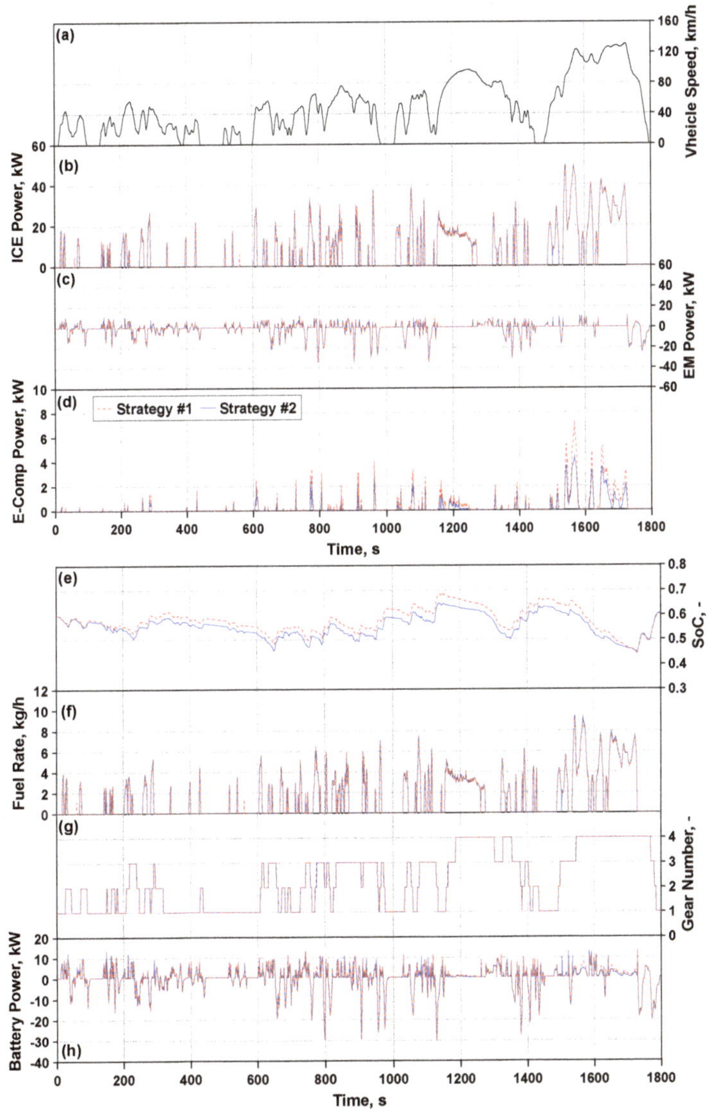

Figure 17. Instantaneous trends of vehicle speed (**a**), ICE power (**b**), EM power (**c**), E-Comp power (**d**), SoC (**e**), fuel rate (**f**), gear number (**g**), and battery power (**h**) along WLTC.

Even if the advantage of the strategy #2 cannot be considered yet as a general result due to the need of further verifications with different powertrain/vehicle types, it seems that a reduced battery consumption, even if the thermal engine will not give its maximum efficiency performance, is the path to improve the hybrid vehicle fuel economy. As stated above, further analyses will be performed in the next development of this activity to confirm the generality of this result.

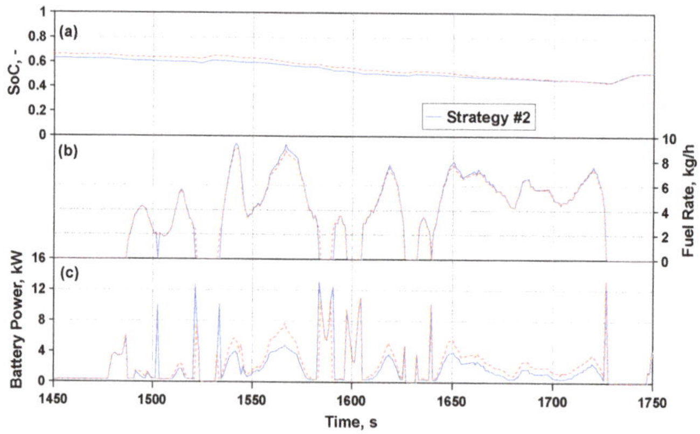

Figure 18. Instantaneous trends of SoC (**a**), fuel rate (**b**), and battery power (**c**) along the last portion of WLTC.

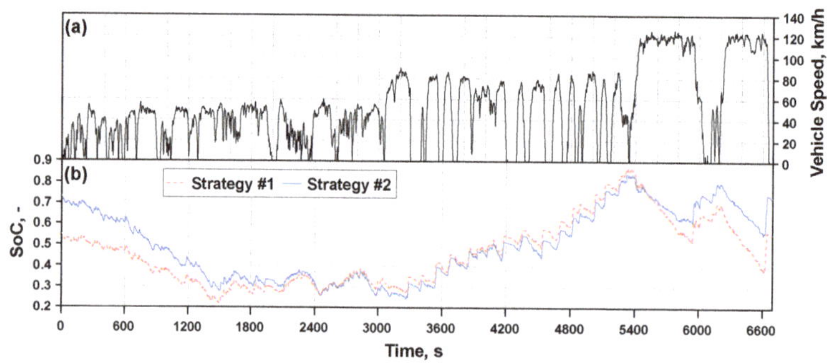

Figure 19. Instantaneous trends vehicle speed (**a**) and SoC (**b**) along the RDC.

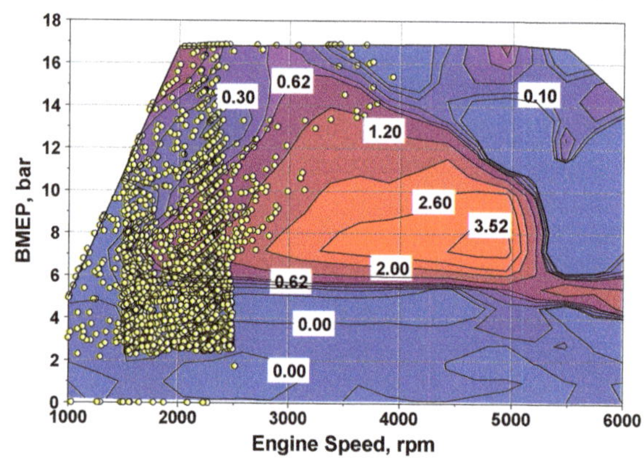

Figure 20. Instantaneous engine operating point along the RDC over the OBTE difference between strategy #2 and #1.

8. Conclusions

In this paper, a refined study of an innovative engine prototype was provided, aiming to estimate the impact of the engine calibration on CO_2 and pollutant emission of a hybrid vehicle equipped with the considered power unit, consisting in a two-stage boosted ultra-lean pre-chamber SI engine.

As a first step, a physical model for the prediction of the burn rate for this engine architecture was recalled, able to sense its main geometrical parameters and operating conditions. The combustion model, validated in previous authors' activities, was integrated with advanced sub-models of turbulence, heat transfer, knock, and engine-out pollutant emissions formation.

The model was employed to estimate complete engine performance maps once a rule-based numerical pre-calibration had been set up. In particular, two variants of the calibration procedure were proposed, aiming to maximize either the BTE or the OBTE; this last option also considered the E-Comp power consumption. Despite the great number of control parameters, the RB approach was successfully verified against the results of a more complex and time-consuming procedure where the 1D engine model was coupled to an external optimizer built on a genetic algorithm (MOGA-II).

The estimated performance maps revealed the potential of this advanced engine architecture to reach very high efficiency levels. Depending on the calibration strategy, a 49% BTE peak was estimated at 2000 rpm (full load), and a 43% OBTE peak was also gained at a lower load. The analyses also underline low cylinder-out specific NO_x emissions, thanks to the ultra-lean combustion concept and relevant cylinder-out specific HC emissions, due to an ineffective post-oxidation process.

The computed engine performance maps were employed in vehicle simulations to assess these strategies along two representative driving cycles, namely WLTC and an RDC. The comparisons underline that the strategy #2 allows slightly lower CO_2 emissions for both the considered driving cycles. This is due to a more extended pure electric driving during the low-speed portion of the driving cycles, allowed by a reduced electric consumption in the high-speed driving mission. In this way, the time in which the engine operates with a reduced efficiency was shortened. Either control strategies do not substantially show a clear impact on the pollutant emissions.

The discussed methodology represents a potential powerful tool to forecast both engine and vehicle performance. Once tested and validated against experimental data, this procedure will contribute to supporting and driving the development phase of a high-efficient HEV.

Author Contributions: Conceptualization, F.B. and V.D.B.; Methodology, F.B. and V.D.B.; Software, E.M.; Validation, E.M., L.T. and D.T.; Data Curation, E.M., L.T. and D.T.; Writing—Original Draft Preparation, F.B., V.D.B., E.M., L.T. and D.T.; Writing–Review & Editing, V.D.B. and E.M.; Visualization, E.M., L.T. and D.T.; Supervision, F.B. and V.D.B.; Project Administration, V.D.B.; Funding Acquisition, F.B. and V.D.B. All authors have read and agreed to the published version of the manuscript.

Funding: This project has received funding from the European Union's Horizon 2020 research and innovation programme under grant agreement No 724084.

Conflicts of Interest: The authors declare no conflict of interest.

Definitions/Abbreviations

0D/1D	Zero/One Dimensional
AFTDC	After Top Dead Center
AI	Auto Ignition
Ba	Battery
BB	Base Boost
BEV	Battery Electric Vehicles
BMEP	Brake Mean Effective Pressure
BSFC	Brake Specific Fuel Consumption
BSHC	Brake Specific HC
$BSNO_X$	Brake Specific NO_X
BTDC	Before Top Dead Center
BTE	Brake Thermal Efficiency
CAD	Crank Angle Degree
Cl	Clutch
DI	Direct Injection
DP	Dynamic Programming
EATS	Exhaust After-Treatment System

EG	Electric Generator
EGR	Exhaust Gas Recirculation
EM	Electric Motor
EML	Extreme Miller Lift
ETESS	Efficient Thermal Electric Skipping Strategy
EVC	Exhaust Valve Closure
EU	European Union
FL	Full Load
GB	Gear Box
HCCI	Homogeneous Charge Compression Ignition
HEV	Hybrid Electric Vehicle
HP	High-Pressure
HPC	High-Pressure Compressor
ICE	Internal Combustion Engine
IVC	Intake Valve Closure
LHV	Low Heating Value
LP	Low-Pressure
LPC	Low-Pressure Compressor
LPT	Low-Pressure Turbine
MC	Main Chamber
MFB	Mass Fraction Burned
OBTE	Overall Brake Thermal Efficiency
PC	Pre Chamber
PFI	Port Fuel Injection
PHEV	Plug-in Hybrid Electric Vehicle
PID	Proportional Integral Derivative
PMP	Pontryagin Minimum Principle
RB	Rule Based
RDC	Real Driving Cycle
RDE	Real Driving Emission
SACI	Spark Assisted Compression Ignition
SI	Spark Ignition
SML	Standard Miller Lift
SoC	State of Charge
UNVS	UniNa Vehicle Simulation
VVT	Variable Valve Timing
WLTC	Worldwide harmonized Light vehicles Test Cycle

Symbols

A	Area
A_L	Laminar Flame Area
c_{jet}	Tuning constant
k	Turbulent kinetic energy
K	Mean flow kinetic energy
m_b	Burned mass
$m_{b,entr}$	Burned portion of entrained mass
m_{entr}	Current entrained mass
$\dot{m}_f$	Total Fuel Flow Rate
$\dot{m}_{jet}$	Mass flow rate from the Pre-chamber
n	Engine rotational speed
P	Power
S	Swirl angular momentum
S_L	Laminar Flame Speed

t	Time
T	Tumble angular momentum
V	Volume

Greeks

η	Efficiency
λ	Relative air-fuel ratio
Λ_T	Taylor length scale
ρ	Density
Σ	Turbulence-induced flame wrinkling
τ_{AI}	Auto-ignition Time

References

1. Nemeth, K.; Carvalho, J.M. Dieselgate and Consumer Law: Repercussions of the Volkswagen scandal in the European Union. *J. Eur. Consum. Mark. Law* **2017**, *6*, 35.
2. Sperling, D. *Future Drive: Electric Vehicles and Sustainable Transportation*; Island Press: Washington, DC, USA, 2013; ISBN 9781559633277.
3. Helms, H.; Pehnt, M.; Lambrecht, U.; Liebich, A. Electric vehicle and plug-in hybrid energy efficiency and life cycle emissions. In Proceedings of the 18th International Symposium Transport and Air Pollution, Empa, Dübendorf, Switzerland, 18–19 May 2010; p. 113.
4. Senecal, P.K.; Leach, F. Diversity in transportation: Why a mix of propulsion technologies is the wayforward for the future fleet. *Results Eng.* **2019**, *4*. [CrossRef]
5. Thiel, C.; Perujo, A.; Mercier, A. Cost and CO_2 aspects of future vehicle options in Europe under new energy policy scenarios. *Energy Policy* **2010**, *38*, 7142–7151. [CrossRef]
6. Millo, F.; Rolando, R.; Fuso, R.; Zhao, J. Development of a new hybrid bus for urban public transportation. *Appl. Energy* **2015**, *157*, 583–594. [CrossRef]
7. Roberts, M. Benefits and Challenges of Variable Compression Ratio (VCR). *SAE Tech. Pap.* **2003**. [CrossRef]
8. Francqueville, L.; Michel, J. On the Effects of EGR on Spark-Ignited Gasoline Combustion at High Load. *SAE Int. J. Engines* **2014**, *7*, 1808–1823. [CrossRef]
9. Bozza, F.; De Bellis, V.; Teodosio, L. Potentials of cooled EGR and water injection for knock resistance and fuel consumption improvements of gasoline engines. *Appl. Energy* **2016**, *169*, 112–125. [CrossRef]
10. Bozza, F.; De Bellis, V.; Teodosio, L.; Tufano, D.; Malfi, E. Techniques for CO_2 emission reduction over a WLTC. A numerical comparison of increased compression ratio, cooled EGR and water injection. *SAE Tech. Pap.* **2018**. [CrossRef]
11. Germane, G.; Wood, C.; Hess, C. Lean Combustion in Spark-Ignited Internal Combustion Engines—A Review. *SAE Tech. Pap.* **1983**, 831694. [CrossRef]
12. Rapp, V.; Killingsworth, N.; Therkelsen, P.; Derek, D.-R. Lean Combustion, 2nd Edition. In *Lean-Burn Internal Combustion Engines*; Elsevier: Amsterdam, The Netherlands, 2016; pp. 111–146. [CrossRef]
13. Drake, M.; Fansler, T.; Lippert, A. Stratified-charge combustion: Modeling and imaging of a spray-guided direct-injection spark-ignition engine. *Proc. Combust. Inst.* **2005**, *30*, 2683–2691. [CrossRef]
14. Fajardo, C.; Sick, V. Flow field assessment in a fired spray-guided spark-ignition direct-injection engine based on UV particle image velocimetry with sub crank angle resolution. *Proc. Combust. Inst.* **2007**, *31*, 3023–3031. [CrossRef]
15. Stanglmaier, R.; Roberts, C. Homogeneous charge compression ignition (HCCI): Benefits, compromises, and future engine applications. *SAE Trans.* **1999**, 2138–2145. [CrossRef]
16. Okita, R. Mazda SKYACTIV-G Engine with New Boosting Technology. In Proceedings of the Advanced Clean Cars Symposium: The Road Ahead, Air Resources Board, Sacramento, CA, USA, 27–28 September 2016.
17. Sens, M.; Binder, E.; Benz, A.; Kramer, L.; Schultalbers, M.; Blumenröde, K. Pre-chamber ignition as a Key Technology for Highly Efficient SI Engines-New Approaches and Operating Strategies. In Proceedings of the 39th International Vienna Motor Symposium, Vienna, Germany, 26–27 April 2018.

18. Mueller, C.; Morcinkowski, B.; Habermann, K.; Uhlmann, T.; Schernus, C. Development of a pre-chamber for spark ignition engines in vehicle applications. In Proceedings of the 4th International Conference on Ignition Systems for Gasoline Engines, Berlin, Germany, 6–7 December 2018.
19. Bozza, F.; De Bellis, V.; Tufano, D.; Malfi, E.; Muller, C.; Habermann, K. A Quasi-Dimensional Model of Pre-Chamber Spark-Ignition Engines. *SAE Tech. Pap.* **2019**. [CrossRef]
20. Bozza, F.; De Bellis, V.; Tufano, D.; Malfi, E.; Muller, C.; Habermann, K. 1D Numerical and Experimental Investigations of an Ultra lean Pre-Chamber Engine. *SAE Int. J. Engines* **2020**. [CrossRef]
21. De Bellis, V.; Malfi, E.; Tufano, D.; Bozza, F. Efficient Thermal Electric Skipping Strategy applied to the Control of Series/Parallel Hybrid Powertrain. *SAE Tech. Pap.* **2020**. [CrossRef]
22. Serrano, D.; Zaccardi, J.M.; Müller, C.; Libert, C.; Habermann, K. Ultra-Lean Pre-Chamber Gasoline Engine for Future Hybrid Powertrains. *SAE Tech. Pap.* **2020**. [CrossRef]
23. Millo, F.; Bernardi, M.G.; Delneri, D. Computational Analysis of Internal and External EGR Strategies Combined with Miller Cycle Concept for a Two Stage Turbocharged Medium Speed Marine Diesel Engine. *SAE Int. J. Engines* **2011**, *4*, 1319–1330. [CrossRef]
24. Gottschalk, W.; Lezius, U.; Mathusall, L. Investigations on the Potential of a Variable Miller Cycle for SI Knock Control. *SAE Tech. Pap.* **2013**. [CrossRef]
25. De Bellis, V.; Severi, E.; Fontanesi, S.; Bozza, F. Hierarchical 1D/3D approach for the development of a turbulent combustion model applied to a VVA turbocharged engine. Part II: Combustion model. *Energy Procedia* **2014**, *45*, 1027–1036. [CrossRef]
26. North, G.L.; Santavicca, D.A. The fractal nature of premixed turbulent flames. *Combust. Sci. Technol.* **1990**, *72*, 215–232. [CrossRef]
27. Blizard, N.; Keck, J. Experimental and Theoretical Investigation of Turbulent Burning Model for Internal Combustion Engines. *SAE Tech. Pap.* **1974**, 740191. [CrossRef]
28. Hiraoka, K.; Nomura, K.; Yuuki, A.; Oda, Y.; Kameyama, T. Phenomenological 0-Dimensional Combustion Model for Spark-Ignition Natural Gas Engine Equipped with Pre-Chamber. *SAE Tech. Pap.* **2016**. [CrossRef]
29. Bozza, F.; Teodosio, L.; De Bellis, V.; Fontanesi, S.; Lorio, A. A Refined 0D Turbulence Model to Predict Tumble and Turbulence in SI Engines. *SAE Int. J. Engines* **2019**, *12*, 15–30. [CrossRef]
30. Bozza, F.; De Bellis, V.; Teodosio, L. A Tabulated-Chemistry Approach Applied to a Quasi-Dimensional Combustion Model for a Fast and Accurate Knock Prediction in Spark-Ignition Engines. *SAE Tech. Pap.* **2019**. [CrossRef]
31. Bounaceur, R.; Herbinet, O.; Fournet, R.; Glaude, P.-A.; Battin-Leclerc, F.; da Cruz, A.P.; Yahyaoui, M.; Truffin, K.; Moreac, G. Modeling the Laminar Flame Speed of Natural Gas and Gasoline Surrogates. *SAE Tech. Pap.* **2010**. [CrossRef]
32. Lavoie, G.; Heywood, J.; Keck, J. Experimental and theoretical study of nitric oxide formation in internal combustion engines. *Combust. Sci. Technol.* **1970**, *1*, 313–326. [CrossRef]
33. Heywood, J. *Internal Combustion Engine Fundamentals*; Mcgraw-Hill: New York, NY, USA, 1988; Volume 930, ISBN 007028637X.
34. Kaplan, J.; Heywood, J. Modeling the spark ignition engine warm-up process to predict component temperatures and hydrocarbon emissions. *SAE Tech. Pap.* **1991**, 361–376. [CrossRef]
35. Lavoie, G. Correlations of combustion data for SI Engine calculations–laminar flame speed, quench distance and global reaction rates. *SAE Tech. Pap.* **1978**, 1015–1033. [CrossRef]
36. Bozza, F.; Tufano, D.; Malfi, E.; Teodosio, L.; Libert, C.; De Bellis, V. Performance and Emissions of an advanced Multi-cylinder SI engine operating in ultra-lean conditions. *SAE Tech. Pap.* **2019**. [CrossRef]
37. Teodosio, L.; Pirrello, D.; Berni, F.; De Bellis, V.; Lanzafame, R.; D'Adamo, A. Impact of intake valve strategies on fuel consumption and knock tendency of a spark ignition engine. *Appl. Energy* **2018**, *216*, 91–104. [CrossRef]
38. Tufano, D.; De Bellis, V.; Malfi, E. Development of an on-line energy management strategy for hybrid electric vehicle. *Energy Procedia* **2018**, *148*, 106–113. [CrossRef]
39. Wipke, K.B.; Cuddy, M.R.; Burch, S.D. ADVISOR 2.1: A user-friendly advanced powertrain simulation using a combined backward/forward approach. *IEEE Trans. Veh. Technol.* **1999**, *48*, 1751–1761. [CrossRef]
40. Sciarretta, A.; Guzzella, L. Control of hybrid electric vehicles. *IEEE Control. Syst. Mag.* **2007**, *27*, 60–70. [CrossRef]

41. Kim, N.; Cha, S.; Peng, H. Optimal control of hybrid electric vehicles based on Pontryagin's minimum principle. *IEEE Trans. Control. Syst. Technol.* **2011**, *19*, 1279–1287. [CrossRef]
42. Patil, R.M.; Filipi, Z.; Fathy, H.K. Comparison of Supervisory Control Strategies for Series Plug-In Hybrid Electric Vehicle Powertrains Through Dynamic Programming. *IEEE Trans. Control. Syst. Technol* **2014**, *22*, 502–509. [CrossRef]
43. Paganelli, G.; Delprat, S.; Guerra, T.M.; Rimaux, J.; Santin, J.J. Equivalent consumption minimization strategy for parallel hybrid powertrains. In Proceedings of the Vehicular Technology Conference IEEE 55th Vehicular Technology Conference VTC Spring, Birmingham, AL, USA, 7 August 2002; pp. 2076–2081. [CrossRef]
44. Commission Regulation EU, 2018/1832 of 5 November 2018, Official Journal of the European Union L 301/1. Available online: https://eur-lex.europa.eu/legal-content/EN/TXT/PDF/?uri=CELEX:32018R1832&from=FR (accessed on 2 November 2019).
45. Valverde, V.; Michael, C.; Pierre, B.; Barouch, G.; Massimo, C.; Marcos, O.G.; Carsten, G.; Georgios, F.; Jrlica, P.; Giorgio, M.; et al. *Joint Research Centre 2018 Light-Duty Vehicles Emissions Testing—Contribution to the EU Market Surveillance: Testing Protocols and Vehicle Emissions Performance EUR29897 EN*; Publications Office of the European Union: Luxembourg, 2019; ISBN 978-92-76-12333-0. [CrossRef]

energies

MDPI

Article

Validation of a RANS 3D-CFD Gaseous Emission Model with Space-, Species-, and Cycle-Resolved Measurements from an SI DI Engine

Stefania Esposito [1,*,†], Max Mally [1,‡], Liming Cai [2], Heinz Pitsch [2] and Stefan Pischinger [1]

1 Institute for Combustion Engines (VKA), RWTH Aachen University, 52074 Aachen, Germany; mally@fev.com (M.M.); pischinger_s@vka.rwth-aachen.de (S.P.)

2 Institute for Combustion Technology (ITV), RWTH Aachen University, 52056 Aachen, Germany; l.cai@itv.rwth-aachen.de (L.C.); h.pitsch@itv.rwth-aachen.de (H.P.)

* Correspondence: s.esposito@itv.rwth-aachen.de

† Currently at Institute for Combustion Technology (ITV), RWTH Aachen University, 52056 Aachen, Germany.

‡ Currently at FEV Europe GmbH, 52078 Aachen, Germany.

Received: 19 July 2020; Accepted: 8 August 2020; Published: 19 August 2020

Abstract: Reynolds-averaged Navier–Stokes (RANS) three-dimensional (3D) computational fluid dynamics (CFD) simulations of gaseous emissions from combustion engines are very demanding due to the complex geometry, the emissions formation mechanisms, and the transient processes inside the cylinders. The validation of emission simulation is challenging because of modeling simplifications, fundamental differences from reality (e.g., fuel surrogates), and difficulty in the comparison with measured emission values, which depend on the measuring position. In this study, detailed gaseous emission data were acquired for a spark ignition (SI) direct-injection (DI) single-cylinder engine (SCE) fueled with a toluene reference fuel (TRF) surrogate to allow precise comparison with simulations. Multiple devices in different sampling locations were used for the measurement of average emission concentration, as well as hydrocarbon (HC) cycle- and species-resolved values. A RANS 3D-CFD methodology to predict gaseous pollutants was developed and validated with this experimental database. For precise validation, the emission comparison was performed in the exact same locations as the pollutants were measured. Additionally, the same surrogate fuel used in the measurements was defined in the simulation. To focus on the emission prediction, the pressure and heat release traces were reproduced by calibrating a *G*-equation flame propagation model. The differences of simulation results with measurements were within 4% for CO_2, while for O_2 and NO, the deviations were within 26%. CO emissions were generally overestimated probably because of inaccuracies in mixture formation. For HC emissions, deviations up to 50% were observed possibly due to inexact estimation of the influence of the piston-ring crevice geometry. The reasonable prediction accuracy in the RANS context makes the method a useful framework for the analysis of emissions from SI engines, as well as for mechanism validation under engine relevant conditions.

Keywords: internal combustion engine; combustion; emission; RANS simulation

1. Introduction

The modeling of pollutant emissions from internal combustion engines is of great importance for the development of future low-emission powertrains. The emission analysis within 3D-CFD simulation allows the optimization of the combustion system [1] or the operating strategy [2]. Additionally, 3D-CFD simulation can provide insights into the emission formation mechanisms as a function of in-cylinder phenomena such as fuel distribution, residual gas fraction, etc. [3–11]. The simulation of gaseous pollutants for spark ignition (SI) direct-injection (DI) engines in 3D-CFD require the correct prediction of many physical and chemical aspects, above all mixture formation, flame propagation, and chemical kinetics. Various studies have shown the usage of 3D-CFD to analyze gaseous emission formation with 3D-CFD simulation in SI DI engines [6,12–17].

The validation of emission simulation models with experiments is very challenging due to some fundamental differences. One aspect regards the surrogate fuel used in CFD that, due to its simpler composition, fails in the prediction of all the chemical and physical characteristics of market real fuels (e.g., distillation curve) [18]. For this reason, a few simulation studies used surrogate fuels with a large number of components to reproduce the evaporation curve of real fuels [8]. Additionally, validation requires a comparison in the location where the emission measurements are performed. It was shown in previous works [18,19] that the emission measuring position has a strong impact on the average emission level and cycle-resolved trends. However, validation studies reported in the literature rarely provide details on how the computed emissions were evaluated when compared to experiments. Often, the simulation domain does not cover the emission measuring position, and little information is given on what kind of computed values (e.g., average, instantaneous at a specific time) are evaluated where (e.g., in-cylinder, in the whole exhaust region, in a specific location) for the comparison with measurement data.

The implementation of chemical kinetics is fundamental for emission predictions. RANS cannot fully describe the turbulence-flame interaction, and the calculation of flame propagation only by solving species transport equations with detailed chemistry is often unsuccessful, as shown by Yang et al. [20]. Thus, a flame propagation model that takes into account the turbulence enhancement of the burning velocity is necessary. Among the different approaches [21], the level-set method (*G*-equation) formulated by Peters [22] for turbulent flames proved to be reliable for combustion simulation of SI engines [7,8,17,20,23,24]. To allow the interaction with chemical kinetics, Yang and Reitz [20] presented a criterion based on the ratio between turbulent and chemical time-scales to select whether fuel oxidation should follow the *G*-equation or the kinetics.

This study aims to provide a rigorous validation of a RANS CFD methodology for gaseous emission predictions. The simulations reproduce a specific experimental setup at a single-cylinder engine test bench. A toluene reference fuel (TRF) surrogate with ethanol (TRF+E) was considered in both experiments and simulation to minimize the impact of fuel on the validation results. The CFD geometry covers the emission sampling positions, allowing the comparison of computed emission values with experimental data at the same locations. The extensive comparison involves not only average gaseous emissions, but also species-, space-, and cycle-resolved hydrocarbon (HC) measurements. In this study, the chemical kinetics is used in combination with the *G*-equation combustion model, and a novel approach for their interaction is presented. Additional information on the described methodology can be found in [25].

2. Methodology

2.1. Engine

In this study, the measurement data came from a specific experimental setup, which was already presented by the authors in [18,19]. The main specifications of the single-cylinder engine are reported in Table 1.

Both the DI injector and the spark plug are in a central position in the combustion chamber. The spark plug is located between the exhaust valves, while the fuel injector is between the intake valves. The intake ports have a symmetrical high-tumble design.

Table 1. Single-cylinder engine technical data and operating parameter.

Parameter	Value	Unit	Parameter	Value	Unit
Displacement	400	cm^3	Intake and exhaust event length (@1 mm)	186	° CA
Bore/stroke	75/90.5	mm	Intake valve opening (IVO) (@1 mm)	380	° CA $aTDC_F$
Connecting rod length	152	mm	Exhaust valve closing (EVC) (@1 mm)	356	° CA $aTDC_F$
Compression ratio	11.8	-	Injection pressure	200	bar
Valves per cylinder	4	-	Injection angle	−280	° CA $aTDC_F$

CA = crank angle; $aTDC_F$ after firing top dead center.

2.2. Test-Bench Instrumentation

2.2.1. Thermodynamic Measurements

The cylinder pressure was measured with two pressure transducers (Kistler 6045B), flush-mounted in the combustion chamber roof, between the intake and the exhaust valve seat rings. The cylinder pressure signal was sampled via Kistler 5064 charge amplifiers and acquired with an FEV indication system (FEVIS) with a 0.1° crank angle (CA) resolution. The cycle-resolved intake and exhaust pressures were measured with Kistler 4045 A5 pressure transducers, sampled via Kistler 4665 and Kistler 4603 charge amplifiers with 0.1° CA resolution as well. For each operating point, one-thousand consecutive cycles were acquired. The static pressures and temperatures were measured with standard pressure transducers and thermocouples. The values were averaged over an interval of 30 s. The oil and coolant conditioning systems allowed steady-state operation. The intake air was conditioned to 25 °C upstream of the throttle flap. The pressure upstream of the throttle flap and in the exhaust manifold was controlled to 1.013 bar during throttled operation. During boosted operation, to simulate turbocharging, the pressure in the exhaust manifold was imposed equal to the pressure upstream of the throttle flap. The engine coupling with an eddy-current brake and an electric dynamometer allowed maintaining the desired engine speed with an accuracy of ±1 1/min, independently of the operating point. An ultrasonic air mass meter and a Coriolis-type mass flow sensor were used to measure the intake air mass flow and the fuel mass flow, respectively.

2.2.2. Emission Measurements

In these experiments, detailed emission measurements were conducted in two sampling positions in the engine exhaust manifold and with multiple devices. One position was in the exhaust port (Pos. 1) and the second position (Pos. 2) further downstream. As measurement devices, a fast flame ionization detector (FFID) was used for cycle-resolved measurements of total-HC (THC) emissions (THC are referenced to C_3H_8), while an ion molecule reaction mass spectrometer (IMR-MS) was applied to measure average concentration of selected HC species. The FFID device used was a Cambustion HFR500, which was equipped with two probes of 210 mm, with an internal diameter of 1.07 mm (0.042 inches). The two sampling lines were heated to 200 °C and had a length of 45 cm to the FID measuring head and had an estimated response time ($t_{10-90\%}$) of 1.8–2 ms. The FFID signal output (0–10 V) was connected to the FEV indication system (FEVIS) system and acquired with the resolution of 0.1° CA. The raw FFID signals need to be reconstructed for delay in the probe and the filtering of the dynamics inside the probe [26]. More details about the FFID signal reconstruction can be found in [19]. The FFID accuracy was expected to be lower and the drift higher than conventional FID devices due to the selected larger scales (due to high instantaneous values especially in Pos. 1) and pressure sensitivity.

The IMR-MS used in these investigations was a V&F TwinSense, which is a soft-ionization mass spectrometer. This ionizes the sample with a lower ionization energy (IE) in comparison to traditional electron-ionization mass spectrometry [27–30]. The lower IE results in lower fragmentation of HC molecules and makes this kind of mass spectrometry more suitable for application to engine exhaust gas. The IMR-MS has two channels available and can measure in two different positions at once, as the FFID. The sampling lines were insulated metallic capillaries of 0.5 mm of diameter, heated to 100–120 °C, approximatively 1 m long. The device measured single species concentration, and the results were averaged over the total measuring time (≥1000 cycles).

The standard gaseous emission measurements were performed by means of an FEV emission rate (FEVER) measurement system. This device contains analyzers to measure cycle-averaged concentrations of main emission species in the exhaust gases. The emission measurements were performed on a partial mass flow of exhaust gases, which was sampled downstream of the cylinder head flange. The sampling line was heated to 193 °C. The technical details of the FEVER system are reported in Appendix A in Table A1.

2.3. Fuel

Kinetic mechanisms and fuel surrogates are used in 3D-CFD simulation to emulate the combustion behavior of real fuels. The surrogate formulations are usually calculated in order to match relevant fuel chemical indicators (e.g., knocking tendency). However, emission formation mechanisms can be affected from the surrogate formulation, especially regarding HC emissions. Thus, the comparison of 3D-CFD simulated emissions with emission measurement data with real fuel (e.g., gasoline) can lead to deviations. The authors verified [18] that the average global emissions (CO, CO_2, NO, total-HC) of a TRF+E surrogate as defined in Cai et al. [31] are almost identical to that of its corresponding RON95E5 gasoline. However, high discrepancies were observed in terms of specific HC species, especially for the species that were also fuel components (toluene and n-heptane). In this study, in order to allow the comparison of CFD simulation with the species-resolved HC emissions, measurements at the single-cylinder engine with a TRF+E surrogate were taken into account. The surrogate composition in mass fraction was the following [18]: 51.8% i-C_8H_{18}, 13.3% n-C_7H_{16}, 28.8% C_7H_8, and 6.1% C_2H_5OH. The same fuel was used in 3D-CFD.

2.4. RANS 3D-CFD Simulation

2.4.1. Geometry and Meshing

The simulations were performed with the commercial software CONVERGE CFD (v2.4). The computational mesh domain was divided into three regions: the intake region, which included the intake runners and ports, the combustion chamber, and the exhaust region, which included exhaust ports and runners. The combustion chamber geometry included also the geometry of the piston top-land crevice up to the first compression-ring. The cold-geometry diameter (D) gap on the piston crown was 1 mm, with D_{liner} of 75 mm and D_{piston} of 74 mm. The gap in warm operation can be reduced due to the thermal expansion. However, the amount of thermal expansion is difficult to estimate and depends on the operating points. For this reason, in Section 3.5.2, a sensitivity analysis of the computed emissions on the piston-liner gap is reported. The CFD geometry included also the tips of all the sampling probes applied in the two emission measurement positions, in order to take into account possible flow disturbances. Figure 1 shows the CFD geometry including the emission sampling locations for the different devices.

The software CONVERGE has an automatic meshing at simulation time. The base size of the structured Cartesian mesh uses 2 mm cells in the intake and exhaust region and 1 mm in the cylinder region. Mesh refinements are implemented in the valve gaps during gas exchange and around the spark plug electrode gap at spark timing. Additionally, an automatic mesh refinement (AMR) reduces the mesh size up to 0.5 mm in the intake and exhaust regions if temperature and/or velocity gradients overcome a selected maximum value.

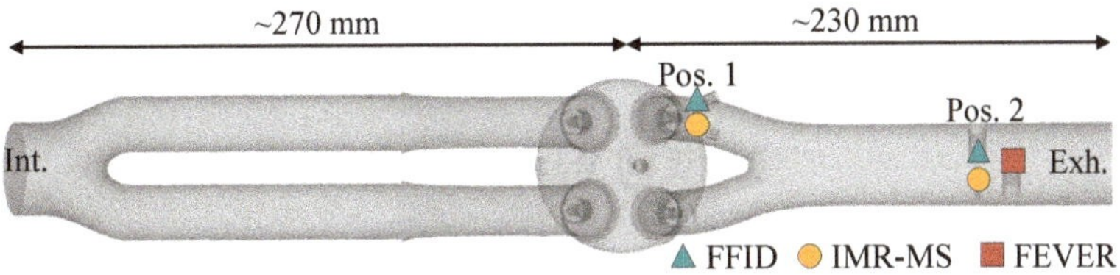

Figure 1. Top view of the CFD geometry with an overview of the emission sampling positions (Pos.) and length of the intake (Int.) and exhaust (Exh.) regions. FFID, fast flame ionization detector; IMR-MS, ion molecule reaction mass spectrometer; FEVER, FEV emission rate.

More details on the meshing refinement settings, boundary and initial conditions, and standard models adopted in the simulation are provided in Appendix A.2.

2.4.2. Operating Points

Four operating points (OPs) were selected for simulation. A load sweep (LS) at $n_E = 2500$ 1/min was simulated, as well as a high power (HP) operating point at $n_E = 4000$ 1/min. Table 2 reports the operating parameters of the selected operating points.

Table 2. Simulated operating points (OPs). LS, load sweep; HP, high power.

Operating Point (OP)	LS1	LS2	LS3	HP
Engine speed/1/min	2500	2500	2500	4000
Indicated mean effective pressure p_{mi}/bar	3	8	16	14
Relative air-to-fuel ratio λ/-	1.0	1.0	1.0	1.0

2.4.3. Injection and Wall-Film Model

For the spray break-up, the Kelvin–Helmholtz/Rayleigh–Taylor (KH-RT) hybrid model [32] was adopted. The model was calibrated in order to achieve the same spray penetration measured with static injector optical measurements. The calibration was performed by changing the time constant parameter [33] of the KH-RT model [32]. Typical values of this constant are in the range 5–100 [33]. In this work, it was calibrated to a value of 30. For the film splash, the O'Rourke model [34] was selected.

A strong accumulation of the fuel wall film in the piston top-land area was observed, especially in the case of the simulation of multiple consecutive cycles. Results of the film accumulation in the piston top-land area are shown in Figure 2a, while a schematization of the problem is reported in Figure 2b.

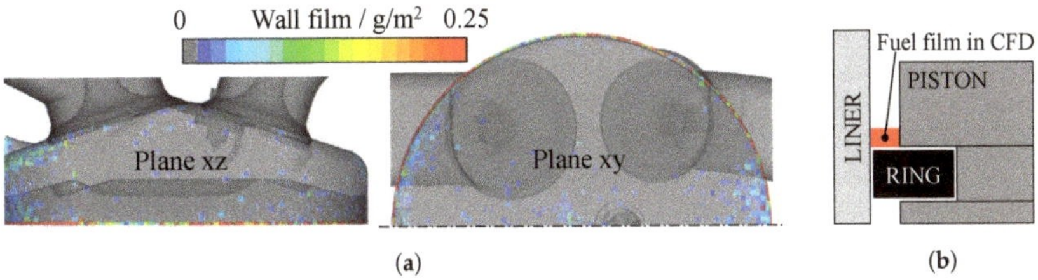

Figure 2. Fuel-film distribution in the combustion for the OP HP at 20° CA before TDC$_F$. (**a**) CFD rusults of the fuel-film formation on different planes, (**b**) schematization of the film positioning in the piston top-land crevice [25].

This effect had a strong impact on the HC emissions, increasing them by almost a factor of two. This accumulation effect was likely overestimated because the fuel-film in the piston-ring crevice would be mainly incorporated in the oil-film, transported in the ring pack, or reduced by the effect of the blow-by in the real engine operation. This was considered here by removing the residual wall film mass at $-20°$ CA before top dead center (TDC).

2.4.4. Chemical Kinetics and Combustion Model

The kinetic solver available in CONVERGE CFD (SAGE) [35] is active when a cell temperature exceeds 500 K and when the molar concentration of HC plus CO is higher than 0.1 ppm. The gasoline surrogate mechanism from [36] was taken into account in this study. Since the focus is on gasoline combustion in SI engines and the prediction of its knocking behavior is out of scope, the mechanism was reduced for improved computational cost by removing the low temperature chemistry of long-chain components, i-C_8H_{18} and n-C_7H_{16}. The reduced mechanism consisted of 239 species and 1068 reactions, including the detailed NO_x formation mechanism from Lamoureux et al. [37]. It was validated against the experimental laminar burning velocities available in the literature for gasoline fuels and their surrogates. An example of the mechanism validation is reported in Appendix A (Figure A1), and the kinetic mechanism itself is provided in the Supplementary Materials.

In order to focus on the prediction of the emissions, an accurate reproduction of the heat release and pressure trace is necessary, and it was achieved with the calibration of a G-equation flame propagation model. The model from Gülder [38] was selected for the calculation of the laminar flame speed. In this work, a novel methodology for the interaction of G-equation and kinetics with the scope of correct emission simulation was developed. As presented from Yang and Reitz [20], it is important to detect if the combustion is turbulence- or chemistry-dominated, especially in the late combustion phases. Indeed, the usage of the G-equation when the combustion is no longer turbulence-controlled does not allow for a correct evaluation of HC and CO emissions, since the flame propagation continues in the expansion stroke as long as G burnt ($G > 0$) and unburnt ($G < 0$) cells exist. An alternative to the approach of Yang and Reitz [20] was developed, in which the G-equation model was deactivated in the last phase of the combustion, when the turbulence level was low and further oxidation was kinetically controlled. The deactivation was achieved by means of a re-initialization of the G-scalar to negative values in the cylinder, and the later burn-out phase was then calculated solely by the SAGE solver. The timing at which the G-equation was deactivated was selected on the basis of the considerations of the heat release profile. The point at which the heat release rate slowed down was when the flame front touched the liner walls. This was when the flame propagation stopped being essentially turbulence-dominated and the combustion rate was expected to be mainly chemistry-dominated. This point was analytically determined on the basis of the measured heat release profiles. A sensitivity analysis on the effect of the G-equation deactivation point on the emission results will be presented in Section 3.5.2. The crank angle taken as deactivation angle α_{G_0} is where the highest maximum of the second derivative ($\dot{x}''_b$) of the heat release rate ($\dot{x}_b$) is reached in the second half of the combustion duration. Figure 3 reports graphical examples of the determination of α_{G_0} for LS2 and LS3.

In Figure 3, it can be observed how this analytical procedure identifies the point at which the heat release slows down for the two operating points.

Due to the selected methodology, the SAGE kinetic solver is used not only in the burnt zone, but also in the unburnt zone for the entire simulation time. Even if the low-temperature chemistry was not included, this was done for two reasons. First, after the G-equation deactivation and re-initialization of G to negative values, the whole combustion chamber is considered as the unburnt zone, and the SAGE solver is necessary to continue the calculation of the heat release. Additionally, the simulation of multiple consecutive cycles results in pollutants being present in the unburnt zone. The use of the SAGE solver allows the kinetic determination of species change in the unburnt zone during the time when the G-equation flame propagation model is active.

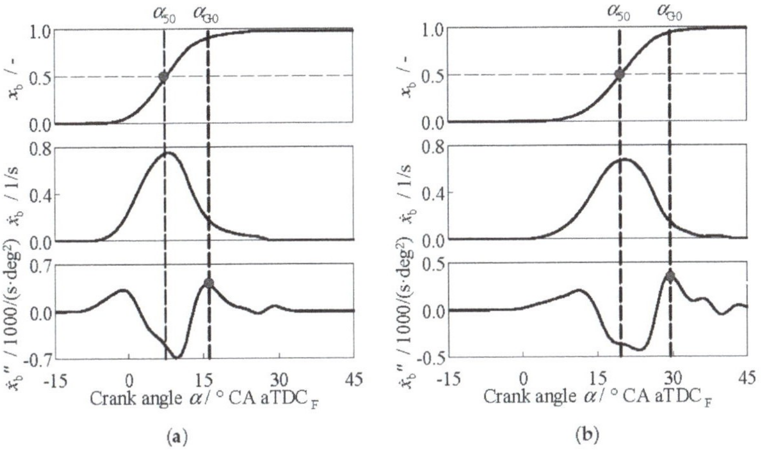

Figure 3. Example of G-equation deactivation angle calculation on the basis of the measured (TPA) burning functions (x_b) for two operating points [25]. (**a**) LS2; (**b**) LS3.

At the flame front (cells with $G = 0$), the species concentration is changed to burned conditions represented by chemical equilibrium to allow feasible computational times. This chemical equilibrium chemistry is only active as long as the G-equation model is active, which is for a relatively short time of about 30° CA over the entire cycle. The assumption of equilibrium chemistry during the main part of combustion, in which the highest temperature is reached, is a good approximation for almost all the species relevant for this study. The only exception is NO because it has longer chemical time scales, also at high temperatures. It has been verified that this approximation is acceptable for stoichiometric operation because the kinetically determined NO concentration come close to equilibrium levels during combustion, especially at higher loads, as also mentioned by Heywood [39]. This aspect was investigated for the selected operating points with 0D chemistry calculation in the burnt zone with a two-zone combustion chamber model in GT-POWER (v2019). These results are reported in Figure A2 in the Appendix A. For OPs LS2, LS3, and HP, the overestimation of NO concentration with equilibrium chemistry in comparison to chemical kinetics was approximately 15% at the time of G-equation deactivation. For OP LS1, a stronger overestimation of NO, over 50%, was observed due to lower combustion temperature. However, the magnitude of these deviations observed with 0D-chemistry cannot be directly carried to 3D-CFD. In conclusion, the adopted combustion methodology can show an overestimation tendency regarding NO that is expected to be quite limited in stoichiometric operation at mid-high loads, higher at stoichiometric operation at low loads, and very strong at lean operation.

To start the combustion with the G-equation model, the variable G was set to positive values at spark timing in a sphere of 0.6 mm in radius centered in the spark gap. This simplified ignition model resulted in a shorter early kernel development phase. Thus, the spark timing was slightly retarded to compensate for this aspect. The calibration of the combustion speed was achieved by modification of the b_1 parameter of the G-equation model ($b_1 = 2.0$ according to [22]). In Table 3, the parameters resulting from the combustion calibration are reported.

It is possible to note how the calibrated b_1 is on a similar level for all the operating points. The burnt fuel fraction for G-equation $x_b(\alpha_{G_0})$ was around 0.9 for all the operating points, with slightly higher values for the higher loads. In Table 3, the difference between the simulation and measurement ST is reported in degrees CA ($\Delta\alpha_{ST}$) and in ms ($t_{\Delta\alpha_{ST}}$). The values are consistent with the hypothesis that the ST retardation needs to compensate for the burn delay, which is artificially shortened from the ignition methodology. Indeed, $t_{\Delta\alpha_{ST}}$ follows a physical trend, as it increases at lower loads and at higher speeds.

Table 3. Calibrated b_1 values, G-equation deactivation angle α_{G_0}, and fuel burnt mass fraction at G-equation deactivation $x_b(\alpha_{G_0})$; spark timing (ST) in simulation versus the one in the experiments for all operating points, together with the difference in CA ($\Delta\alpha_{ST}$) degree as well as in ms ($t_{\Delta\alpha_{ST}}$).

Op. point (OP)	LS1	LS2	LS3	HP
b_1	2.75	2.75	2.75	3.00
$\alpha_{G_0}/°$ CA aTDC$_F$	16.0	15.5	29.5	29.0
$x_b(\alpha_{G_0})/1$	0.88	0.89	0.94	0.93
ST simulation/° CA aTDC$_F$	−10.0	−8.5	0.8	−1.0
ST measurement/° CA aTDC$_F$	−16.0	−11.8	−2.0	−8.5
$\Delta\alpha_{ST}$/CA	6.0	3.3	2.8	7.5
$t_{\Delta\alpha_{ST}}$/ms	0.40	0.22	0.19	0.31

2.4.5. Emission Post-Processing

Regarding the emission post-processing, monitoring points were introduced at the tips of all the modeled measurement probes. The average emission measurements (FEVER, IMR-MS, and FFID average values) were compared with the time averaged results from simulations in the same points as measured. In order to compare the emission values from CFD to the measurements, it is necessary to properly post-process them. For emission species like O_2, CO_2, and CO that are measured as dry (without water vapor), it is necessary to correct the CFD emissions (wet) according to:

$$Y_{i,\text{dry}} = \frac{Y_{i,\text{wet}}}{1 - Y_{H_2O}} \tag{1}$$

in which Y_i is the molar fraction of a generic emission species i and Y_{H_2O} is the molar fraction of water vapor.

Regarding HC emissions, an online evaluation of the total-HC concentration was done during the simulation. Indeed, the recording of all the species at each time step and a successive post-processing would have been unfeasible. In particular, C_3-equivalent HC values were calculated as user-defined passive quantities, in order to allow the direct comparison with the test bench results. The theoretical FID factor $f_{\text{thFID},i}$ for the generic HC species i can be defined as:

$$f_{\text{thFID},i} = \frac{c_i}{3} \tag{2}$$

where c_i represents the carbon atoms of the species i. The total-HC C_3-equivalent concentration Y_{THC} including all the HC species i can be calculated according to:

$$Y_{\text{THC}} = \sum_i Y_i \cdot f_{\text{thFID},i} \tag{3}$$

The real FID factor of each species is different from the theoretical one $f_{\text{thFID},i}$, considered in the post-processing, especially for oxygenated species. A comparison between real and theoretical factors was experimentally verified by the author for selected species and is reported in Appendix A in Table A4. However, even if a few real factors were known, the majority of them could not be estimated. For this reason, it was preferred to use the theoretical factors for all species. This was considered a possible source of deviations between CFD and measurements in terms of total-HC emissions.

3. Results and Discussion

3.1. Mixture Formation

Since the mixture formation affects the emissions strongly, the simulation results of the mixture formation are analyzed first. In Figure 4, the relative air-to-fuel ratio λ distribution at $-20°$ CA before TDC for all the simulated operating point is depicted. Rich tails can be observed for all the operating points. In the piston-ring crevice zone, a richer mixture is usually observed since the injected fuel droplets enter that region.

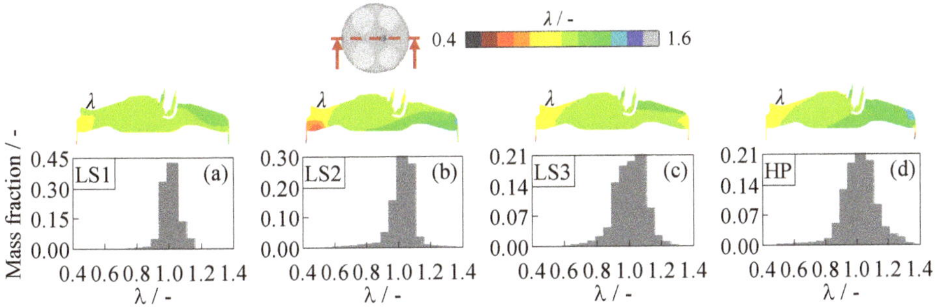

Figure 4. Lambda distribution for all the operating points at 20° CA before TDC_F: mass fraction distribution over λ-bins and central section of the combustion chamber. Operating points: (**a**) LS1, (**b**) LS2, (**c**) LS3, (**d**) HP.

As the starting value for the fuel mass, the injected mass is calculated on the basis of the measured fuel mass flow. However, since the cylinder filling and the trapped air mass in simulation can slightly differ from the trapped air of the real engine, the injected fuel mass was adjusted accordingly to achieve stoichiometric operation. Additionally, the simulation of multiple consecutive cycles is affected by CFD model inaccuracies (wall-film, injection, etc.), and small deviations in the simulation average λ up to 1.2% are observable. However, the accuracy of the λ determination in the measurements is typically around 1%. Figure 5 shows the deviation in terms of average λ for the different operating points.

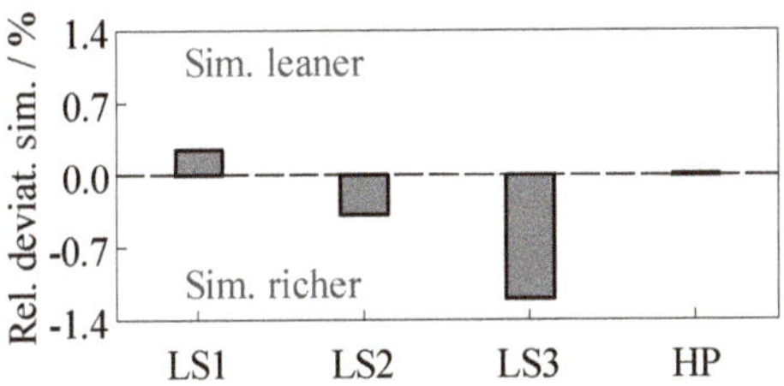

Figure 5. Relative deviation between simulation average λ and measured λ (according to Spindt [40]).

3.2. Combustion and Heat Release

Figure 6 shows the comparison of the measured and simulated pressure curves (p_{cyl}) and the heat release rate ($\dot{Q}_{cyl}$) for all simulated operating points. In all the cases, a good agreement in the pressure traces and heat release is achieved. The above-described *G*-equation deactivation analytical procedure results in a deactivation time at a point where the burn rate slows down. The heat release rate is accurately predicted even after the deactivation of the *G*-equation model, which is an indication of the validity of the proposed methodology.

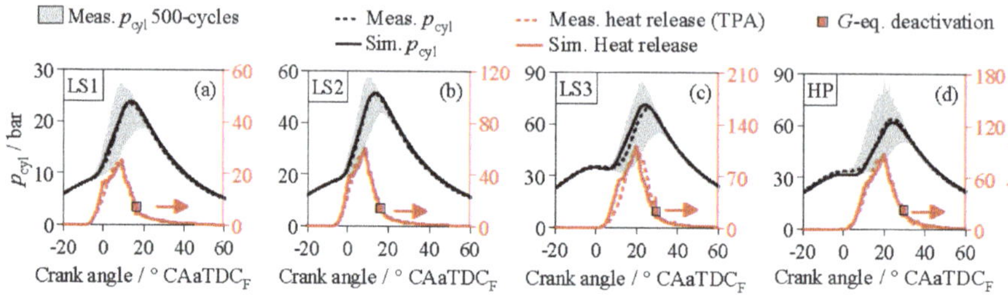

Figure 6. Comparison of measured and simulated pressure curves (p_{cyl}), as well as the heat release rate ($\dot{Q}_{cyl}$) obtained from three-pressure analysis (TPA) [41] and simulation. Operating points: (**a**) LS1, (**b**) LS2, (**c**) LS3, (**d**) HP.

3.3. Average Emissions

Figure 7 illustrates the comparison of the simulated average emission concentration with the FEVER measurements in Pos. 2. Together with the absolute values, the relative deviation of the simulation in comparison to the measurements is plotted.

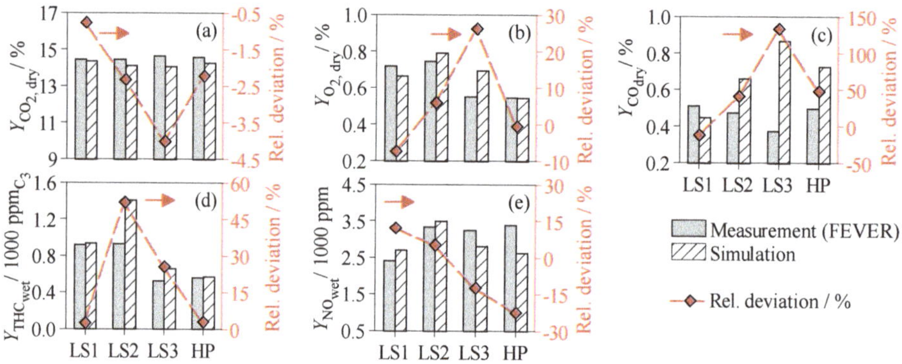

Figure 7. Comparison of average emissions in Pos. 2 between simulation and FEVER measurements (bars, left axis) with relative deviation (line plot, right axis): (**a**) CO_2, (**b**) O_2, (**c**) CO, (**d**) THC, and (**e**) NO.

Regarding CO_2 results (Figure 7a), a very high accuracy in simulation is achieved, with a maximum deviation of 4%. The O_2 (Figure 7b) content in the exhaust is in relatively good agreement as well, with a maximum deviation below 10%, with the exception of OP LS3, where the deviation reaches about 25%. The relative deviation trends of CO_2, O_2, CO are consistent with each other and suggest a mismatch of the calculated mixture formation with the real one. Since the injector was calibrated only with static measurements, it is possible that the spray-tumble interaction at different operating points is not correctly predicted and leads to inaccurate predictions of the λ-field. Above all, CO (Figure 7c) is the most sensitive species to mixture formation, especially around stoichiometric operation and hence shows overestimation up to 130% and incorrect trends over the operating points. However, the deviation in CO emissions can be partly explained with the trend of the average λ deviation of Figure 5.

The best overall agreement for CO_2, O_2, CO is observed for OP LS1, which shows a more homogeneous mixture in Figure 4a than the other OPs. In Section 3.5.1, a sensitivity analysis on global λ for OP LS3 will be presented in order to evaluate the impact on the CO emissions.

Regarding THC emissions (Figure 7d), for three out of four points, a relatively good agreement with deviations within 30% is observed. The LS2 value is overestimated by about 50%. In the CFD simulations, the main source of HC emissions is from the crevices, since the mesh is not fine enough to resolve the flame-wall quenching. Furthermore, fuel wall film is removed, and other HC mechanisms are not modeled. Thus, a possible reason for the overestimation of HC can result from a too large piston-ring crevice volume. In Section 3.5.2, a sensitivity analysis of the piston-ring crevice volume will be presented for OP LS2. A further reason for the possible overestimation could be a too rich mixture in the crevice, e.g., due to inaccuracies in the calculation of the mixture formation. Indeed, a richer mixture in the piston-ring crevice results in higher HC emissions, since the mass in the crevice expands back in the main combustion chamber during the expansion phase. Figure 4 shows that the OP LS2 (b) has the highest inhomogeneity among the simulated points and a very rich tail that corresponds partly to the mixture in the piston-ring crevice.

As far as NO is concerned, good predictions are achieved in all operating points with maximum deviations of 23%. The overestimation observed for LS1 can result from the combustion methodology and the expected overestimation tendency of the chemical equilibrium calculation at low loads. For the other operating points, specific deviations are difficult to be traced back to single sources. Since the λ-sensor used in the measurements has a certain inaccuracy (normally $\approx 1\%$) as well, a deviation in global λ could be a possible source. The sensitivity analysis on λ presented in Section 3.5.1 will give an evaluation of the variability of NO with λ.

Figure 8 shows the comparison of the non-standard emission measurement devices (FFID and IMR-MS) in Pos. 1 and Pos. 2 for two operating points (LS1 and LS2). The selected quantities are the average FFID THC values (a, e) and three selected HC species measured with the IMR-MS, C_2H_2 (b, f), C_6H_6 (c, g), and C_7H_{16} (d, h).

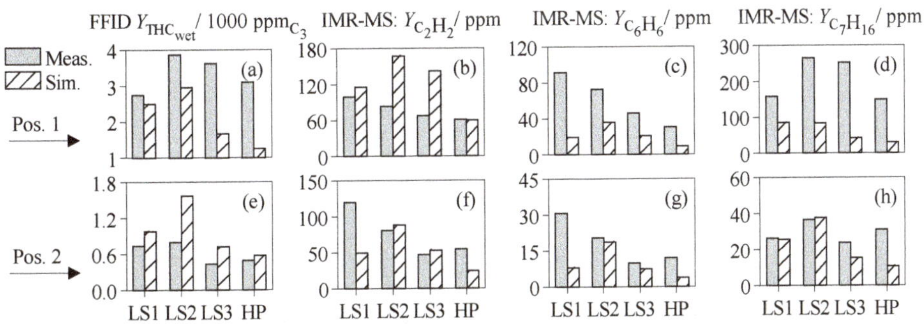

Figure 8. Comparison of average HC-emissions in Pos. 1 and in Pos. 2 between simulation (hatched bars) and measurements (FFID and IMR-MS) for all the operating points. FFID: **(a)** Pos. 1, **(e)** Pos. 2. IMR-MS: C_2H_2 **(b)** Pos. 1 and **(f)** Pos. 2, C_6H_6 **(c)** Pos. 1 and **(g)** Pos. 2, C_7H_{16} **(d)** Pos. 1 and **(h)** Pos. 2.

In comparison to the average FFID values, the simulated THC are underestimated in Pos. 1 and overestimated in Pos. 2 (similarly to what is observed with the FEVER THC comparison in Figure 7c). The THC concentration in Pos. 1 is found to be strongly dependent on the fuel-film in the piston-ring crevice, and elimination of this film resulted in a reduction of the THC concentration in Pos. 1.

Regarding the HC species measured with the IMR-MS, overall similar levels in simulations are observed. This aspect by itself is a very positive indication that from one side, the experimental methodology to measure HC species with the IMR-MS is reliable. Indeed, the correct determination of HC species concentration in the engine exhaust is challenging [18,19]. From the other side, similar levels of some specific HC species (fuel components like C_7H_{16} and partly oxidized species) are a further confirmation that the CFD methodology (combustion model, kinetics) is reproducing realistic results. Higher deviations are observed in Pos. 1 due to the high sensitivity to the piston-ring crevice mechanism. In Pos. 2, the deviation is reduced, and in several cases, a very high accuracy is achieved.

3.4. Crank Angle-Resolved HC Curves

Another detailed validation step is provided by the comparison of simulation results with the crank angle-resolved FFID measurements. Indeed, this can verify if the dynamics of the THC emissions in the CFD simulation are similar to what was measured on the test-bench. In Figure 9, the simulated instantaneous THC concentration is compared with the measured FFID signals in Pos. 1 and Pos. 2 for the OP LS2. The raw FFID signals (all 500 measured cycles and the average) are shown together with the signal reconstructed as explained in [19]. The exhaust valve opening and closing angles (EVO and EVC) are indicated as well.

Regarding the comparison in Pos. 1, a quite good agreement in trend and absolute values is observed. The simulated value during the time when the exhaust valve are closed has a high influence on the time-averaged value. The value appears to be lower than the average FFID signal (even if still in the scatter-band), while with the exhaust valve open, similar or slightly higher values are observed. In Pos. 2, a similar trend in simulation can be qualitatively observed even if the absolute values are higher overall. The timing of simulated and measured THC peak value are close to each other. This peak results from the passage of the HC stored in the exhaust port from the previous cycle, shortly after EVO, as visible in the 3D picture of Figure 9 at 170° CA aTDC. However, the peak shape and intensity differ between simulation and measurement, possibly because of the peak smoothing effect of the FFID probe.

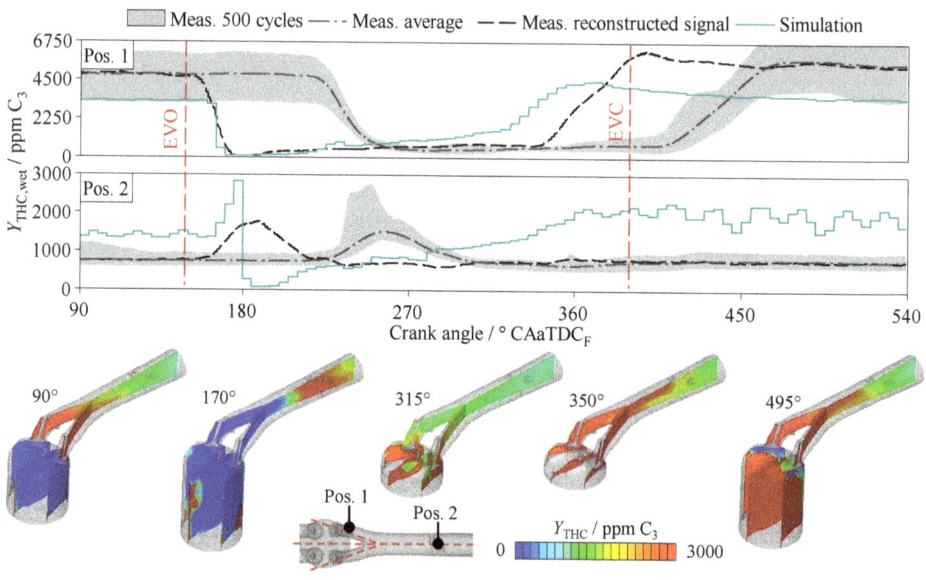

Figure 9. Top (diagram): comparison between FFID cycle-resolved measurements and simulated THC concentration in Pos. 1 and Pos. 2; FFID raw signals (all 500 measured cycles and the average) and the reconstructed curve, compared to the CFD prediction. Bottom: CFD pictures of the total-HC (THC) distribution in the combustion chamber and in the exhaust system (see the pictogram) at different crank angles. Operating point LS2.

3.5. *Sensitivity Analyses*

In this section, sensitivity analyses of specific parameters identified as relevant for the gaseous emission predictions are reported.

3.5.1. Average Relative Air-to-Fuel Ratio λ

O_2, CO, and NO emissions are supposed to be very sensitive to the average λ value. This is verified in OP LS3, where the highest deviation in CO is observed. Three simulations were calculated starting from 20° CA before TDC_F from the same 3D solution, in which the average λ was adjusted by means of a scaling of the fuel components in order to achieve $\lambda = 1.00$ (base value), $\lambda = 0.98$, and $\lambda = 1.02$. In Figure 10, the effect of this variation on the emission concentration in Pos. 2 is evaluated during the exhaust stroke.

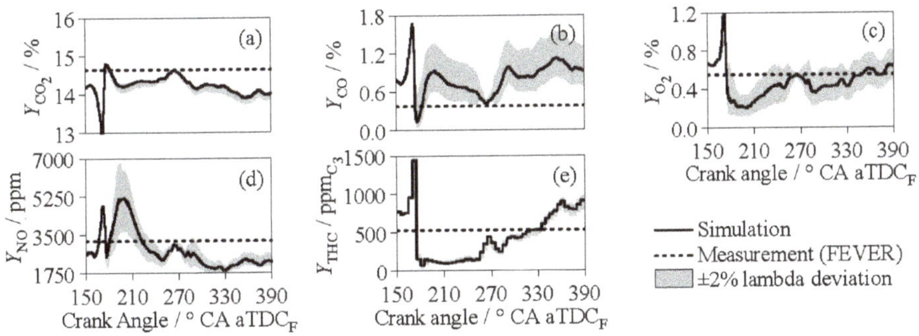

Figure 10. Effect of ±2% λ on simulated emissions concentrations in Pos. 2 compared with measurements (FEVER): (**a**) CO_2, (**b**) CO, (**c**) O_2, (**d**) NO, and (**e**) THC. Operating point LS3.

It can be observed that while CO_2 and THC are weakly influenced by the average λ, the effect on the other emission species is strongly non-linear, and a scatter-band up to ±30% is shown. This is a confirmation that matching the average λ in a very narrow band is of great importance for emission simulations and a great challenge due to the interaction of the measurement inaccuracies with all the CFD models (spray, wall film, evaporation).

3.5.2. Crevice Volume and *G*-Equation Deactivation

The sensitivity to the *G*-equation deactivation angle and to the crevice volume were investigated for OP LS2. To verify that the developed combustion methodology does not affect the emissions, a simulation in which the *G*-equation was deactivated 4.5° CA later (20 instead of 15.5° CA aTDC) was performed. Additionally, the effect of a 20% reduction of the crevice volume (by reducing the piston-liner gap from 1 mm to 0.8 mm) was investigated. The effect of these two changes was evaluated by calculating the relative change in the mass of a certain pollutant species released through the exhaust valves during the exhaust stroke. Figure 11 shows the results of these investigations.

The effect of later *G*-equation deactivation is negligible. This is a confirmation that the adopted combustion methodology is robust.

The reduction of 20% of the piston-liner gap results in a decrease of ≈15% for CO, O_2 and HC. This result correlates with the less than linear dependency of HC emissions with piston-liner gap verified by Min and Cheng [42].

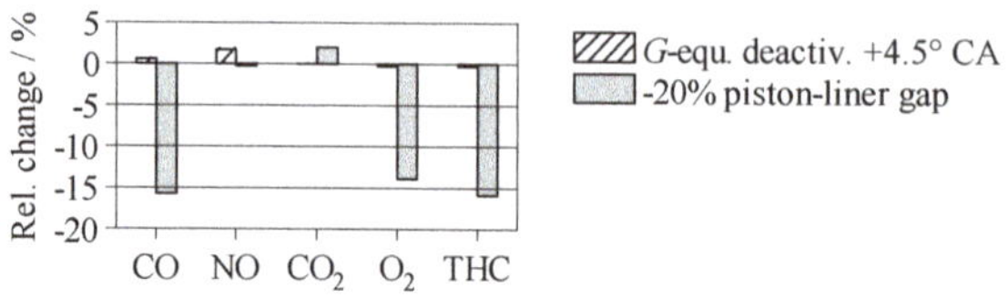

Figure 11. Relative change in species mass flows through the exhaust valves in the exhaust stroke in case of: later *G*-equation deactivation (*G*-equ. deactiv.) (hatched bars) and -20% piston-liner gap (grey bars). Operating point LS2.

4. Conclusions

In this study, a RANS 3D-CFD simulation model was used to reproduce a specific experimental setup and to enable precise validation with space-, species-, and cycle-resolved emission measurements. A newly developed approach for the interaction of the calibrated *G*-equation flame propagation model with kinetic mechanism was shown to be robust and to successfully reproduce measured in-cylinder pressure and heat release. The validation of emission predictions was based on four operating points with varying load and engine speed. The deviation of CO_2 was below 4%, O_2 below 26%, and NO below 23% for the investigated points. For CO, a high deviation at $n_E = 2500$ 1/min and $p_{mi} = 16$ bar of 130% was calculated, together with a high deviation of 25% in O_2. This effect suggests a mismatch in the simulated mixture field in comparison to the test-bench. Regarding THC, two operating points showed deviations below 3%, while for mid-high loads and lower speed deviation, up to 50% was observed. This behavior can derive from additional HC formation mechanisms not modeled in the simulation (e.g., fuel-film, quenching), which affect by a different magnitude the operating points. Additionally, an overestimation of the piston-ring crevice contribution to the THC formation is not excluded.

The comparison with single HC-species measurements showed that the simulated values were in the same order of magnitude, with better agreement in the measurement position further downstream the exhaust valve. Although deviations were present, the combined experimental-simulative methodology shows potential for further validation of kinetic mechanisms under engine conditions. Lastly, the transient trace of THC emissions was validated by the measurement results, which showed a similar shape and timings in two sampling points. Sensitivity analyses showed that the effect of the timing at which the *G*-equation model was deactivated was negligible, while the global λ led to changes in CO, O_2, and NO of up to $\pm 30\%$ when varied by $\pm 2\%$. The reduction of the crevice volume resulted in a less than linear reduction of THC, CO, and O_2 emissions.

Supplementary Materials: The following are available online at http://www.mdpi.com/1996-1073/13/17/4287/s1. The kinetic mechanism in CHEMKIN format is made available as Supplementary Material.

Author Contributions: Conceptualization, S.E., M.M., S.P.; Methodology, S.E., M.M., L.C.; Software, S.E. and M.M.; Validation, S.E. and L.C.; Formal Analysis, S.E.; Investigation, S.E.; Writing—Original Draft Preparation, S.E. and M.M.; Writing—Review & Editing, S.E., M.M., L.C., H.P. and S.P.; Visualization, S.E. and M.M.; Supervision, L.C., H.P. and S.P. All authors have read and agreed to the published version of the manuscript.

Funding: This work was funded by "Deutsche Forschungsgemeinschaft" (DFG)—GRK 1856.

Acknowledgments: Simulations were performed with computing resources granted by RWTH Aachen University under the project "rwth0239". The authors are thankful to Convergent Science Inc. for providing licenses for CONVERGE.

Conflicts of Interest: The authors declare no conflicts of interest.

Appendix A

Appendix A.1. Standard Emission Analyzer (FEVER)

Table A1. Specifications of the standard gaseous exhaust analyzer (FEVER) [18].

Property	FEVER Device					
	CLD [1]	NDIR [2]		PMD [3]	FID [4]	
Species	NO_x/ppm	CO/%	CO_2/%	O_2/%	THC / ppm_{C_3} [5]	CH_4/ppm_{C_1}
Range	0–10,000	0–10	0–20	0–25	10–10,000	10–3000
Accuracy	1% of measured value					
Linearity	2% of measured value					
Response [6]	<2 s				<4 s	

[1] Chemiluminescence detector; [2] non-dispersive infrared detector; [3] para-magnetic detector; [4] flame-ionization detector; [5] total-HC, referred to propane (C_3); [6] time 10 to 90% ($t_{10-90\%}$).

Appendix A.2. Additional Information on the Simulation Methodology

Mesh refinements The mesh refinements in the CFD model, both user-defined and AMR (automatic mesh refinement), consist of a halving of base cell size δ of n_{ref} times, with n_{ref} the refinement degree ($\delta_{\text{ref}} = \delta/2^{n}_{\text{ref}}$) [33]. Table A2 summarizes the meshing parameters of the simulation.

Table A2. Geometry meshing parameters. Base mesh size $\delta = 2$ mm. AMR, automatic mesh refinement; EVO, exhaust valve opening; EVC, exhaust valve closing.

Mesh Embedding			AMR				
Area	Phase	n_{ref}	Region	AMR-Variable	Phase	Δ [b]	n_{ref}
Cylinder region	permanent	1	Cylinder	Temperature	700–800° CA	2.5 K	2
Spark plug $r = 1$ mm [a]	7° CA @ST	4	Intake	Velocity	IVO-IVC	1 m/s	1
Spark plug $r = 3$ mm [a]	7° CA @ST	3	Exhaust	Velocity	EVO-EVC	1 m/s	1
Intake valves gap	IVO-IVC	1					
Exhaust valves gap	EVO-EVC	1					

[a] Embedding sphere radius centered in the spark gap; [b] maximum difference between adjacent cells allowed from AMR.

Boundary and initial conditions, heat transfer, and wall temperatures The pressure and temperature boundary conditions were generated on the basis of the post-processing of the measured indicated pressures through a three-pressure-analysis (TPA) [41] with the software GT-POWER (v2019). Regarding the gas composition at the boundaries, on the intake side, fresh air 76.8% in mass of N_2 and 23.2% of O_2 were set, while on the exhaust side, the gas composition was calculated based on emission measurements. Since the real HC composition was unknown, the measured THC concentration was set at the exhaust boundary as C_3H_8.

The simulations were started at the second half of the exhaust stroke. The initial composition in the intake region was fresh air. In the combustion chamber and in the exhaust region, the gas composition of measured engine exhaust gas was used. However, for each simulation point, multiple consecutive cycles ($\geq$3) were simulated in order to minimize the possible influence of initial conditions on simulated emission species.

For the heat transfer, the O'Rourke and Amsden model [43] was adopted. For the description of the temperature and velocity boundary layer, the law of the wall [44] was applied. The imposed wall temperatures on the simulation surfaces are reported in Table A3.

Table A3. Assumed wall temperatures in K for the different simulated operating points.

Op. point (OP)	LS1	LS2	LS3	HP
Intake runners	310	310	310	320
Intake ports	363	363	363	363
Intake valves	400	420	450	500
Cylinder liner	400	410	420	420
Cylinder head	435	440	450	450
Piston	445	450	460	470
Spark plug	780	790	800	800
Exhaust valves	550	600	650	700
Exhaust ports and runner	420	430	450	470

Flow and evaporation modeling The RANS equations were solved by using the standard $k-\epsilon$ [45] within CONVERGE [33]. The thermodynamic quantities of the gas were calculated as a function of temperature. The liquid fuel was considered as incompressible. The droplet evaporation was assumed as ideal and described by the Frossling correlation [46].

Appendix A.3. Kinetic Mechanism Validation

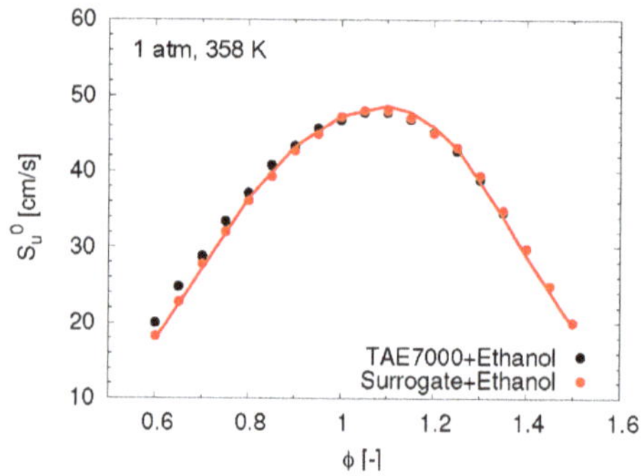

Figure A1. Burning velocities of a commercial gasoline TAE7000and its surrogate (13.7% n-heptane, 42.9% iso-octane, 43.4% toluene) with air mixtures with 15% ethanol addition. Symbols denote the experimental measurements by Dirrenberger et al. [47] at 1 atm and 358 K. Solid lines show the numerical results for the present kinetic model. Comparison analogous to that presented by the authors in [31].

Appendix A.4. Comparison of NO Production between Kinetics and Equilibrium Chemistry

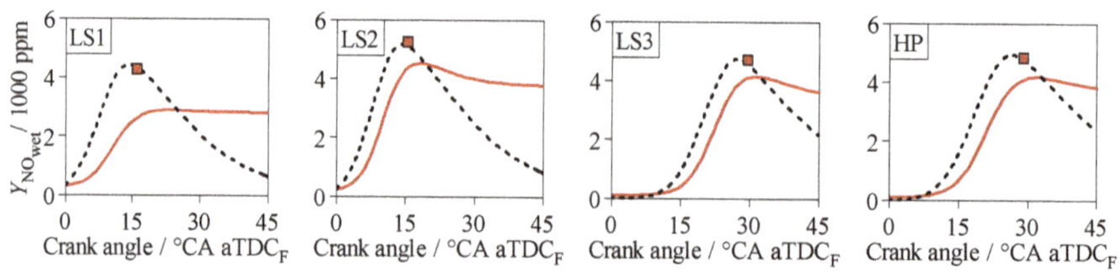

Figure A2. Comparison of in-cylinder NO-production prediction with a 0D two-zone GT-POWER model (v2019) using chemical kinetics (same mechanism used in 3D-CFD and provided in the Supplementary Materials, red continuous line) and chemical equilibrium calculations (from [48]).

Appendix A.5. Experimentally Determined FID Factors

Table A4. FID response factors measured for selected species for the FEVER device and the FFID device (two lines). The theoretical response factor $f_{thFID,i}$ (s.Equation (2)) is compared with the measured one, and the relative deviation is reported [25].

Species			FEVER		FFID Line 1		FFID Line 2	
		$f_{thFID,i}$	$f_{FID,i}$	Deviation	$f_{FID,i}$	Deviation	$f_{FID,i}$	Deviation
Methane	CH_4	0.33	0.35	4.8%	0.36	8.5%	0.38	13.6%
Acetylene	C_2H_2	0.67	1.04	55.4%	0.90	34.3%	0.91	36.5%
Ethane	C_2H_6	0.67	0.66	−0.8%	0.69	3.7%	0.68	1.5%
Methanol	CH_3OH	0.33	0.22	−34.9%	0.26	−21.3%	0.26	−21.9%
Furan	C_4H_4O	1.33	1.04	−21.7%	1.04	−22.2%	1.02	−23.3%
Benzene	C_6H_6	2.00	2.14	6.9%	1.96	−1.9%	1.98	−0.9%
Propene	C_3H_6	1.00	0.98	−2.2%	0.95	−5.5%	0.88	−11.8%
Ethanol	C_2H_5OH	0.67	0.51	−23.0%	0.53	−20.7%	0.53	−20.7%
1.3-Butadiene	C_4H_6	1.33	1.29	−3.0%	1.30	−2.6%	1.27	−4.4%
n-Butane	C_4H_{10}	1.33	1.34	0.2%	1.36	1.8%	1.35	1.1%
n-Heptene	C_7H_{14}	2.33	2.31	−1.2%	2.30	−1.4%	2.27	−2.7%
n-Pentane	C_5H_{12}	1.67	1.71	2.9%	1.74	4.7%	1.56	−6.7%
n-Hexane	C_6H_{14}	2.00	2.00	0.0%	2.03	1.7%	2.01	0.5%
Toluene	C_7H_8	2.33	2.42	3.8%	2.23	−4.5%	2.21	−5.3%
n-Heptane	C_7H_{16}	2.33	2.32	−0.6%	2.33	−0.3%	2.31	−1.2%
Xylene	C_8H_{10}	2.67	2.76	3.7%	2.57	−3.8%	2.53	−5.0%

References

1. Morcinkowski, B.; Adomeit, P.; Mally, M.; Esposito, S.; Walter, V.; Yadla, S. Emissionsvorhersage in der Entwicklung ottmotorischer EU7-Antriebe. In *Experten-Forum Powertrain: Ladungswechsel und Emissionierung 2019*; Springer Vieweg: New York, NY, USA, 2020.
2. Boldaji, M.R.; Gainey, B.; Lawler, B. Thermally stratified compression ignition enabled by wet ethanol with a split injection strategy: A CFD simulation study. *Appl. Energy* **2019**, *235*, 813–826. [CrossRef]
3. Addepalli, S.K.; Saw, O.P.; Mallikarjuna, J.M. *Effect of Mixture Distribution on Combustion and Emission Characteristics in a GDI Engine—A CFD Analysis*; SAE Technical Paper Series; SAE: Warrendale, PA, USA, 2017.
4. Cao, L.; Su, H.; Mosbach, S.; Kraft, M.; Bhave, A.; Kook, S.; Bae, C. *Studying the Influence of Direct Injection on PCCI Combustion and Emissions at Engine Idle Condition Using Two Dimensional CFD and Stochastic Reactor Model*; SAE Technical Paper 2008-01-0021; SAE: Warrendale, PA, USA, 2008.
5. Coskun, G.; Soyhan, H.S.; Demir, U.; Turkcan, A.; Ozsezen, A.N.; Canakci, M. Influences of second injection variations on combustion and emissions of an HCCI-DI engine: Experiments and CFD modelling. *Fuel* **2014**, *136*, 287–294. [CrossRef]
6. Jadhav, P.D.; Mallikarjuna, J.M. *Effect of EGR on Performance and Emission Characteristics of a GDI Engine—A CFD Study*; SAE Technical Paper 2017-24-0033; SAE: Warrendale, PA, USA, 2017.
7. Jiao, Q.; Reitz, R.D. *Modeling of Equivalence Ratio Effects on Particulate Formation in a Spark-Ignition Engine under Premixed Conditions*; SAE Technical Paper 2014-01-1607; SAE: Warrendale, PA, USA, 2014.
8. Jiao, Q.; Reitz, R.D. The Effect of Operating Parameters on Soot Emissions in GDI Engines. *Sae Int. J. Engines* **2015**, *8*, 1322–1333. [CrossRef]
9. Kavuri, C.; Tiry, M.; Paz, J.; Kokjohn, S.L. Experimental and computational investigation of soot production from a premixed compression ignition engine using a load extension injection. *Int. J. Engine Res.* **2016**, *18*, 573–590. [CrossRef]
10. Distaso, E.; Amirante, R.; Tamburrano, P.; Reitz, R.D. Understanding the role of soot oxidation in gasoline combustion: A numerical study on the effects of oxygen enrichment on particulate mass and number emissions in a spark-ignition engine. *Energy Convers. Manag.* **2019**, *184*, 24–39. [CrossRef]

11. Roberts, J.; Chuahy, F.D.F.; Kokjohn, S.L.; Roy, S. Laminar burning velocity of gasolines with addition of ethanol. *Fuel* **2019**, *237*, 522–535. [CrossRef]
12. Duclos, J.M.; Zolver, M. 3D Modeling of Intake, injection and combustion in a DI-SI engine under homogeneous and stratified operating conditions. In Proceedings of the 4th International Symposium on Diagnostics and Modeling of Combustion in Internal Combustion Engines (COMODIA 1998), Kyoto, Japan, 20–23 July 1998; Volume 98, pp. 287–292.
13. Kosmadakis, G.M.; Rakopoulos, C.D.; Demuynck, J.; de Paepe, M.; Verhelst, S. CFD modeling and experimental study of combustion and nitric oxide emissions in hydrogen-fueled spark-ignition engine operating in a very wide range of EGR rates. *Int. J. Hydrog. Energy* **2012**, *37*, 10917–10934. [CrossRef]
14. Kosmadakis, G.M.; Rakopoulos, D.C.; Rakopoulos, C.D. Investigation of nitric oxide emission mechanisms in a SI engine fueled with methane/hydrogen blends using a research CFD code. *Int. J. Hydrog. Energy* **2015**, *40*, 15088–15104. [CrossRef]
15. Kosmadakis, G.M.; Rakopoulos, D.C.; Rakopoulos, C.D. Methane/hydrogen fueling a spark-ignition engine for studying NO, CO and HC emissions with a research CFD code. *Fuel* **2016**, *185*, 903–915. [CrossRef]
16. Reddy, A.A.; Mallikarjuna, J.M. *Parametric Study on a Gasoline Direct Injection Engine—A CFD Analysis*; SAE Technical Paper 2017-26-0039; SAE: Warrendale, PA, USA, 2017.
17. Liu, J.; Szybist, J.; Dumitrescu, C. *Choice of Tuning Parameters on 3D IC Engine Simulations Using G-Equation*; SAE Technical Paper 2018-01-0183; SAE: Warrendale, PA, USA, 2018.
18. Esposito, S.; Cai, L.; Günther, M.; Pitsch, H.; Pischinger, S. Experimental comparison of combustion and emission characteristics between a market gasoline and its surrogate. *Combust. Flame* **2020**, *214*, 306–322. [CrossRef]
19. Esposito, S.; Mauermann, P.; Lehrheuer, B.; Günther, M.; Pischinger, S. *Effect of Engine Operating Parameters on Space- and Species-Resolved Measurements of Engine-Out Emissions from a Single-Cylinder Spark Ignition Engine*; SAE Technical Paper 2019-01-0745; SAE: Warrendale, PA, USA, 2019.
20. Yang, S.; Reitz, R.D.; Iyer, C.O.; Yi, J. A Transport Equation Residual Model Incorporating Refined G-Equation and Detailed Chemical Kinetics Combustion Models. *Sae Int. J. Engines* **2009**, *1*, 1028–1044. [CrossRef]
21. Tan, J.Y.; Bonatesta, F.; Ng, H.K.; Gan, S. Developments in computational fluid dynamics modelling of gasoline direct injection engine combustion and soot emission with chemical kinetic modelling. *Appl. Therm. Eng.* **2016**, *107*, 936–959. [CrossRef]
22. Peters, N. *Turbulent Combustion*; Cambridge monographs on mechanics; Cambridge University Press: Cambridge, UK; New York, NY, USA, 2000.
23. Tan, Z.; Kong, S.C.; Reitz, R.D. *Modeling Premixed and Direct Injection SI Engine Combustion Using the G-Equation Model*; SAE Technical Paper 2003-01-1843; SAE: Warrendale, PA, USA, 2003.
24. Liang, L.; Reitz, R.D. *Spark Ignition Engine Combustion Modeling Using a Level Set Method with Detailed Chemistry*; SAE Technical Paper 2006-01-0243; SAE: Warrendale, PA, USA, 2006. [CrossRef]
25. Esposito, S. Analysis of Gaseous Pollutant Emissions from Spark-Ignition Engines. Ph.D. Thesis, RWTH Aachen University, Aachen, Germany, 2020.
26. Cheng, W.K.; Summers, T.; Collings, N. The fast-response flame ionization detector. *Prog. Energy Combust. Sci.* **1998**, *24*, 89–124. [CrossRef]
27. Hoffmann, E.d.; Stroobant, V. *Mass Spectrometry: Principles and Applications*, 3rd ed.; J. Wiley: Chichester, UK; Hoboken, NJ, USA, 2007.
28. Gross, J.H. *Mass Spectrometry*; Springer: Berlin/Heidelberg, Germany, 2011. [CrossRef]
29. Munson, M.S.B.; Field, F.H. Chemical Ionization Mass Spectrometry. I. General Introduction. *J. Am. Chem. Soc.* **1966**, *88*, 2621–2630. [CrossRef]
30. Frischmuth, M. Massenspektrometrie mit Chemischer Ionisierung (CI-MS) zur dynamischen Erfassung limitierter und nicht-limitierter Komponenten im Automobilabgas. Ph.D. Thesis, Universität Lüneburg, Lüneburg, Germany, 2001.
31. Cai, L.; Pitsch, H. Optimized chemical mechanism for combustion of gasoline surrogate fuels. *Combust. Flame* **2015**, *162*, 1623–1637. [CrossRef]
32. Reitz, R.D.; Beale, J.C. Modeling spray atomization with the Kelvin-Helmholtz/Rayleigh-Taylor hybrid model. *At. Sprays* **1999**, *9*, 623–650. [CrossRef]
33. Richards, K.J.; Senecal, P.K.; Pomraning, E. Available online: https://www.scribd.com/document/367228444/Converge-2-4-Manual (accessed on 19 July 2020).

34. O'Rourke, P.J. Collective Drop Effects on Vaporizing Liquid Sprays. Ph.D. Thesis, Princeton University, Princeton, NJ, USA, 1981.
35. Senecal, P.K.; Pomraning, E.; Richards, K.J.; Briggs, T.E.; Choi, C.Y.; Mcdavid, R.M.; Patterson, M.A. *Multi-Dimensional Modeling of Direct-Injection Diesel Spray Liquid Length and Flame Lift-off Length Using CFD and Parallel Detailed Chemistry*; SAE Technical Paper 2003-01-1043; SAE: Pennsylvania, PA, USA, 2003.
36. Cai, L.; Ramalingam, A.; Minwegen, H.; Heufer, A.K.; Pitsch, H. Impact of exhaust gas recirculation on ignition delay times of gasoline fuel: An experimental and modeling study. *Proc. Combust. Inst.* **2019**, *37*, 639–647. [CrossRef]
37. Lamoureux, N.; Desgroux, P.; El Bakali, A.; Pauwels, J.F. Experimental and numerical study of the role of NCN in prompt-NO formation in low-pressure CH4–O2–N2 and C2H2–O2–N2 flames. *Combust. Flame* **2010**, *157*, 1929–1941. [CrossRef]
38. Gülder, Ö.L. *Correlations of Laminar Combustion Data for Alternative S.I. Engine Fuels*; SAE Technical Paper 841000; SAE: Warrendale, PA, USA, 1984.
39. Heywood, J.B. *Internal Combustion Engine Fundamentals*; McGraw-Hill series in mechanical engineering; McGraw-Hill: New York, NY, USA, 1988.
40. Spindt, R.S. *Air-Fuel Ratios from Exhaust Gas Analysis*; SAE Technical Paper 650507; SAE: Warrendale, PA, USA, 1965.
41. *GT-SUITE 2018: Engine Performance Application Manual*; Gamma Technologies: Westmont, IL, USA, 2018.
42. Min, K.; Cheng, W.K. Oxidation of the Piston Crevice Hydrocarbon during the Expansion Process in a Spark Ignition Engine. *Combust. Sci. Technol.* **1995**, *106*, 307–326. [CrossRef]
43. Amsden, A.A. KIVA-3V: A Block Structured KIVA Program for Engines with Vertical or Canted Valves. Available online: https://www.osti.gov/servlets/purl/505339-tlyFln/webviewable/ (accessed on 19 July 2020).
44. Warsi, Z.U.A. *Fluid Dynamics: Theoretical and Computational Approaches/Z.U.A. Warsi*, 3rd ed.; Taylor & Francis: Boca Raton, FL, USA; London, UK, 2006.
45. Ferziger, J.H.; Perić, M. *Numerische Strömungsmechanik*; Springer: Berlin/Heidelberg, Germany, 2008.
46. Amsden, A.A. KIVA-II: A Computer Program ofr Chemically Reactive Flows with Sprays. Available online: https://www.osti.gov/biblio/6228444-kiva-ii-computer-program-chemically-reactive-flows-sprays (accessed on 19 July 2020)
47. Dirrenberger, P.; Glaude, P.A.; Bounaceur, R.; Le Gall, H.; da Cruz, A.P.; Konnov, A.A.; Battin-Leclerc, F. Laminar burning velocity of gasolines with addition of ethanol. *Fuel* **2014**, *115*, 162–169. [CrossRef]
48. Olikara, C.; Borman, G.L. *A Computer Program for Calculating Properties of Equilibrium Combustion Products with Some Applications to I.C. Engines*; SAE Technical Paper 750468; SAE: Pennsylvania, PA, USA, 1975.

energies

MDPI

Article

Methodological Approach for 1D Simulation of Port Water Injection for Knock Mitigation in a Turbocharged DISI Engine

Federico Millo *, Fabrizio Gullino and Luciano Rolando

Energy Department, Politecnico di Torino, Corso Duca degli Abruzzi 24, 10129 Torino, Italy; fabrizio.gullino@polito.it (F.G.); luciano.rolando@polito.it (L.R.)

* Correspondence: federico.millo@polito.it; Tel.: +39-011-090-4517

Received: 19 July 2020; Accepted: 9 August 2020; Published: 19 August 2020

Abstract: In the upcoming years, more challenging CO_2 emission targets along with the introduction of more severe Real Driving Emissions limits are expected to foster the development and the exploitation of innovative technologies to further improve the efficiency of automotive Spark Ignition (SI) engines. Among these technologies, Water Injection (WI), thanks to its knock mitigation capabilities, can represent a valuable solution, although it may significantly increase the complexity of engine design and calibration. Since, to tackle such a complexity, reliable virtual development tools seem to be mandatory, this paper aims to describe a quasi-dimensional approach to model a Port Water Injection (PWI) system integrated in a Turbocharged Direct Injection Spark Ignition (T-DISI) engine. Through a port-puddling model calibrated with 3D-CFD data, the proposed methodology was proven to be able to properly replicate transient phenomena of water wall film formation, catching cycle by cycle the amount of water that enters into the cylinder and is therefore available for knock mitigation. Moreover, when compared with experimental measurements under steady state operating conditions, this method showed good capabilities to predict the impact of the water content on the combustion process and on the knock occurrence likelihood.

Keywords: turbocharged gasoline engine; port water injection; predictive combustion model; knock mitigation; engine performance; virtual test rig

1. Introduction

Environmental sustainability represents one of the main drivers for the future development of the automotive industry. As a matter of fact, in the European Union (EU), to guarantee a significant reduction in the contribution of passenger cars to greenhouse gas emissions, more challenging targets will be introduced. For instance, the target of 95 grams of CO_2 per kilometer issued for 2020 [1] is expected to be reduced by 15% and 37.5% by 2025 and 2030, respectively [2]. Moreover, the introduction of Real Driving Emissions (RDE) for vehicle tests will force automotive manufacturers to keep criteria pollutant emissions low throughout the whole engine operating map [3].

In this scenario, downsized and turbocharged gasoline engines are currently increasing their market share thanks to their noteworthy potential for fuel consumption reduction [4]. As a matter of fact, engine displacement reduction allows the shifting of load points towards higher efficiency regions of the operating map, while performance is preserved or even enhanced thanks to turbocharging [5–7]. Nevertheless, the search for continuously increasing boost levels [8] has led to a dramatic increment of the knock likelihood, requiring quite often spark timing retards, mixture enrichment and/or compression ratio reductions, which can jeopardize the expected CO_2 emissions benefits [9]. Therefore, the development of effective knock mitigation techniques is crucial to extend efficiency improvements enabled by downsizing at high loads. Several innovative techniques have already been investigated to meet this target, such as Miller cycles [10–12], Variable Compression Ratios (VCR) [13–15], cooled Exhaust Gas Recirculation (EGR) [16–18] or the employment of alternative fuels with higher octane numbers [19–21].

Besides the abovementioned technologies, Water Injection (WI) is also nowadays rising as one of the most promising solutions for knock mitigation. WI, taking advantage of the high latent heat of vaporization of the water introduced into the cylinders, allows effective cooling of the charge, thus reducing knock tendency. Moreover, thanks to the increase in the heat capacity of the cylinder charge, an additional cooling effect can be exploited during compression and combustion.

Water Injection technology has been thoroughly examined in the last decade, as has emerged from the literature, where various experimental activities and numerical studies are available. For instance, several papers have assessed the pros and cons of direct and indirect water injection strategies [22–24]. Indirect water injection, thanks to its simplicity and cost-effectiveness, seems to be the best candidate for series production. Several studies have focused on the optimization of the Port Water Injection system (PWI): the careful optimization of both operating and design parameters such as water spray Sauter Mean Diameter (SMD), spray targeting, injection timing and pressure has been demonstrated to play a key-role in reducing water consumption [22,25–29]. In a previous work of the authors [30], issues related to both liquid film formation and oil dilution were discussed. Paltrinieri et al. [31] analyzed through both a numerical simulation and experimental activities the effects on engine performance of several calibrations of a Port Water Injection system integrated in a single cylinder high specific power engine. The potential of a PWI system was also demonstrated in [32], where a two cylinder turbocharged spark ignition engine could be operated at a stoichiometric Air-to-Fuel (A/F) ratio in a wide range of operating conditions at high loads with a significant reduction in the Indicated Fuel Consumption (ISFC). Moreover, Khatri et al. [33] showed the capability of water injection to counterbalance the engine performance reduction related to the use of low octane gasolines, while Wilson [34] estimated a considerable Brake Specific Fuel Consumption (BSFC) reduction coupling PWI and increased Compression Ratio (CR). Moreover, both Cazzoli et al. [35] and Bozza et al. [36] studied the impact of water vapors on Laminar Flame Speed and auto-ignition delay time, while Gern et al. [37] investigated the effect of water injection on Particulate Matter (PM) emissions.

Based on the above literature overview, 3D-CFD models proved to be very useful in better understanding and in predicting with high accuracy the influence of water on the in-cylinder phenomena. Nevertheless, in order to foresee possible synergies between WI and other promising technologies (e.g., high CR configuration and Miller cycle) by means of a virtual test rig, a reliable 1D model of the whole engine appears to be the most suitable numerical tool. In the scientific literature, different studies on WI technology have been carried out by adopting a 1D approach, which affords the achievement of a reliable prediction of the whole system behavior with reduced computational effort. For example, Cavina et al. [38] investigated through 1D-CFD simulations the effects of Port Water Injection and Direct Water Injection (DWI) on exhaust gas temperature and knock intensity, subsequently validating their numerical results with an experimental campaign. Furthermore, Bozza et al. developed a 1D model to generate BSFC maps of a downsized turbocharged 2-cylinder engine under wet and dry operating conditions; the computed maps were then embedded in a vehicle model with the aim of estimating the potential CO_2 reduction achievable through WI over a worldwide harmonized light-duty test procedure (WLTP) cycle [36,39] and to define the proper sizing of the water tank [40].

The 1D approach for WI modelling requires reliable sub-models that allow for a physically based reproduction in turbulence, combustion, wall film formation, heat transfer, Cycle to Cycle Variability (CCV) and knock phenomena. Only in this way can a preliminary engine calibration with an enhanced knock resistance be numerically identified, together with realistically expected performance improvements.

Based on the above considerations, this paper aims to describe a quasi-dimensional approach to model a PWI system integrated in a state-of-the-art Turbocharged Direct Injection Spark Ignition engine [41]. After a brief introduction to the experimental set-up, the authors describe the main features of the numerical model, which was preliminary calibrated and validated in dry operating conditions. Afterwards, the methodology for PWI modelling is presented and discussed in detail, in three sub-sections. The first describes the integration of the port-puddling model, which allows the proper replication of transient phenomena of water wall film formation in the intake ports. The second focuses on the water evaporation model, calibrated with the aim to correctly account for the fraction of in-cylinder water that evaporates by absorbing heat from walls. The third examines the combustion model, which requires a further refinement in order to accurately predict the effects of water vapors on the burn rate. A final section, in addition to reporting the PWI model validation, assesses the capability of the proposed approach to create a virtual test bench for the optimization of the engine operating parameters at different values of water flow rate.

2. Case Study

The research activities described in this paper were carried out on a state-of-the-art 3-cylinder 1.0L Turbocharged Direct Injection Spark Ignition engine, the main characteristics of which are reported in Table 1 (see also [41] for more details). The engine features a very high Compression Ratio for efficiency maximization and is equipped with an electro-hydraulic Variable Valve Actuation (VVA) MultiAir system [42] on the intake side for gas exchange optimization. A Miller cycle strategy with Late Intake Valve Closing (LIVC) is also exploited both at low and high engine speed to decrease the knock likelihood and to reduce the exhaust gas temperature, respectively.

Table 1. Engine specifications.

Specification	Value
N. of cylinders [-]	3
Displacement [cm^3]	999
Bore [mm]	70
Stroke [mm]	86.5
Compression Ratio [-]	13
VVA System	MultiAir
Injection System	GDI
Full rated power [kW]	88 at 5500 rpm
Full rated torque [N-m]	190 at 2000 rpm

For the experimental campaign, the engine was installed on the test bench in its standard configuration. It was equipped with three piezoelectric pressure transducers integrated in the spark-plugs to detect in-cylinder pressure signal within each cylinder. A linear lambda sensor was mounted downstream of the turbine to measure the relative Air-to-Fuel ratio. Furthermore, several piezo-resistive pressure transducers and K-type thermocouples were used to monitor pressure and temperature along both intake and exhaust lines. Knocking cycles were experimentally detected on the basis of the value of their knock intensity, measured by means of the Maximum Amplitude of Pressure Oscillations (MAPO) knock index [43]: cycles for which MAPO was exceeding a proper threshold value were classified as knocking cycles. Knock Limited Spark Advance (KLSA) was experimentally identified as the spark timing for which approximately 2–3% of engine cycles were knocking.

A Port Water Injection system was then integrated into this base layout (see Figure 1).

The water used for the experiments was de-ionized distilled water stored in a tank at 25 °C. A recirculating pump delivered water at a pressure of 6 bar to the rail, which in turn was connected to three injectors, which were located in the intake runners, one for each cylinder.

During all the tests, the engine was operated at the stoichiometric Air-to-Fuel ratio and was fueled with gasoline with a Research Octane Number (RON) of 95 and a Lower Heating Value (LHV) of 42.5 MJ/kg. Furthermore, water injection timing was kept constant with an End of Injection (EOI) equal to 360 a Crank Angle degree (i.e., equal to gas exchange Top Dead Center). For all measurements, experimental data were recorded and evaluated over 300 consecutive engine cycles, in order to gather a significant statistical sample for knock-limited spark advance operating conditions, where few percent of engine cycles are typically knocking.

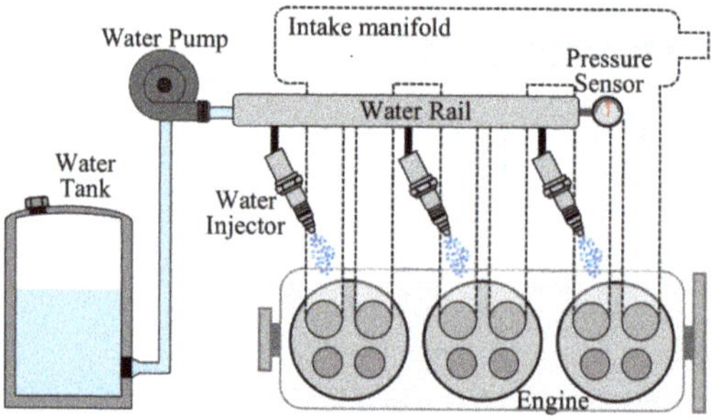

Figure 1. Schematic layout of Port Water Injection system.

Experimental Measurements

A summary of the experimental campaign is depicted in Figure 2. At first, a set of tests were carried out in dry conditions (i.e., with a Water-to-Fuel ratio (W/F) equal to 0) at Wide Open Throttle (WOT) to define the engine reference performance (red circles). Afterwards, two-part load conditions, namely 2500 rpm–18 bar Brake Mean Effective Pressure (BMEP) and 4000 rpm–17 bar BMEP (green rhombuses), were chosen to analyze the effects of water injection on the combustion process. In particular, different values of Water-to-Fuel ratio, from 0% to 90% at steps of 10%, were tested: for each level of W/F, the spark timing was advanced until knock onset, and the boost level was adjusted in order to keep the engine load constant. Finally, the increase in engine performance enabled by PWI knock suppression capabilities were assessed at WOT exploiting a Water-to-Fuel ratio between 50% and 60% (blue squares).

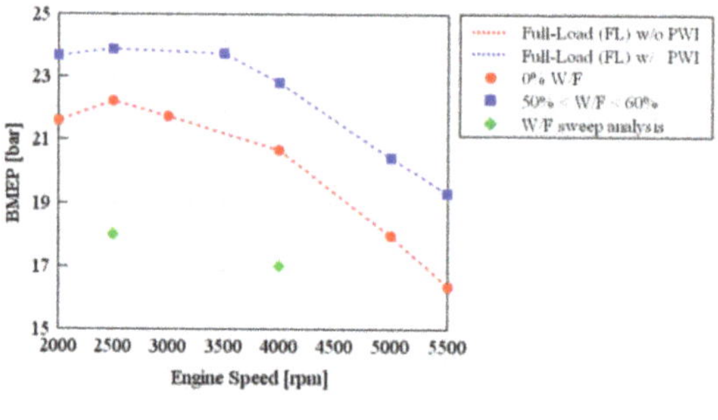

Figure 2. Experimental test matrix.

3. Numerical Set-Up

At first, a detailed 1D-CFD model of the reference engine was built by means of GT-POWER^{TM}, a one-dimensional fluid-dynamic code developed by Gamma Technologies Inc. for engine performance simulation. In particular, a simple injector model was adopted to control the injected fuel mass in order to guarantee a stoichiometric combustion. Concerning water injection, it was performed via multiple profile injectors connected to the intake runners, while the injected water mass was imposed equal to the experimental one. A model-based controller was then introduced to actuate the waste-gate opening diameter in order to match the experimental boost pressure.

After a thorough calibration of the gas-exchange processes, a predictive combustion model, the so-called SITurb [44–46], was implemented in the code for the prediction of the burn rate at various operating conditions. Knock detection was instead accomplished through a Kinetics-Fit model, which had been tuned on the basis of a CCV-replicating model.

The following paragraphs contain a detailed description of SITurb, CCV and Knock model calibration processes, which were carried out considering only the experimental tests without the Port Water Injection system.

3.1. Calibration of the Combustion Model

Following a well-assessed hierarchical 1D/3D approach, as a first step a turbulence sub-model was calibrated on the basis of 3D-CFD results obtained at various valve strategies and engine speeds [44]. The second step of the tuning procedure concerned the optimization of the SITurb combustion model constants. The main characteristics of the SITurb model are briefly summarized in the following sub-section. More details can be found in [45,46].

3.1.1. SITurb Model

The entrained mass (M_e) rate of the unburned air-fuel mixture in the flame front is dependent on the flame area (A_f), the unburned gas density (ρ_u) and the sum of the Turbulent and Laminar Flame Speeds, as expressed by Equation (1):

$$\frac{dM_e}{dt} = \rho_u \cdot A_f \cdot (S_T + S_L) \tag{1}$$

S_T, namely the Turbulent Flame Speed, is calculated through Equation (2), where R_f, u' and L_t are, respectively, flame radius, turbulent intensity and turbulent length scale.

$$S_T = C_S \cdot u' \cdot \left(1 - \frac{1}{1 + \frac{C_k \cdot R_f^2}{L_t^2}}\right) \tag{2}$$

In Equation (2), two calibration parameters are present: the Flame Kernel Growth Multiplier (C_k), which scales the flame front evolution, and the Turbulent Flame Speed Multiplier (C_S), which globally scales the Turbulent Flame Speed.

A further calibration parameter, the Dilution Effect Multiplier (C_D), is instead adopted to tune the effect of diluents such as exhaust residuals, EGR or water on the Laminar Flame Speed (S_L), which is calculated by Equation (3):

$$S_L = \left(B_m + B_\phi \cdot (\phi - \phi_m)^2\right) \cdot \left(\frac{T_u}{T_0}\right)^\alpha \cdot \left(\frac{p}{p_0}\right)^\beta \cdot \left(1 - 0.75 \cdot C_D \cdot \left(1 - (1 - 0.75 \cdot C_D \cdot [D_i])^7\right)\right) \tag{3}$$

where ϕ is the in-cylinder equivalence ratio, T_u is the temperature of the unburned gas, p is the in-cylinder pressure and D_i is the mass fraction of the diluents in the unburned zone. B_m, B_ϕ, ϕ_m, T_0, p_0, α and β are model constants.

Finally, Equation (4) is used to determine the burn rate of the entrained unburned mass (M_u):

$$\frac{dM_b}{dt} = \frac{M_u}{\tau} = \frac{M_e - M_b}{\tau} \tag{4}$$

where τ is assumed to be the time needed by the Laminar Flame Speed to cover the Taylor Microscale (λ) of turbulence, expressed by Equation (5). C_λ, the Taylor Length Scale Multiplier, is a further calibration parameter inversely proportional to the burnup rate.

$$\tau = \frac{\lambda \cdot C_\lambda}{S_L} \tag{5}$$

3.1.2. SITurb Model Calibration Results

Six engine operating points at Full-Load, the ones shown with red circles in Figure 2, were chosen for the tuning of the four SITurb calibration parameters (see Section 3.1.1). By imposing the experimental spark timing, a Design of Experiment (DoE) methodology, integrated with a Genetic Algorithm optimum search, was used to identify the best set of SITurb constants aimed at minimizing the error between experimental and predicted burn rates for the selected operating conditions [44]. Once the optimal SITurb calibration parameters have been defined (see Table 2), the results of the complete engine model were compared with experimental data in order to assess its accuracy and reliability. The values of BMEP, BSFC and Volumetric Efficiency (referred to the air conditions in the intake manifold), depicted in Figure 3a–c, respectively, highlight a very good agreement with an error below 2%. Moreover, the simulated pressure cycle and burn rate profile reported in Figure 3d also show a satisfactory correlation with experimental measurements.

Table 2. Optimized values of the calibration parameters of the SITurb combustion model.

Calibration Parameter	Value
C_k	4.112
C_S	0.625
C_λ	0.789
C_D	0.872

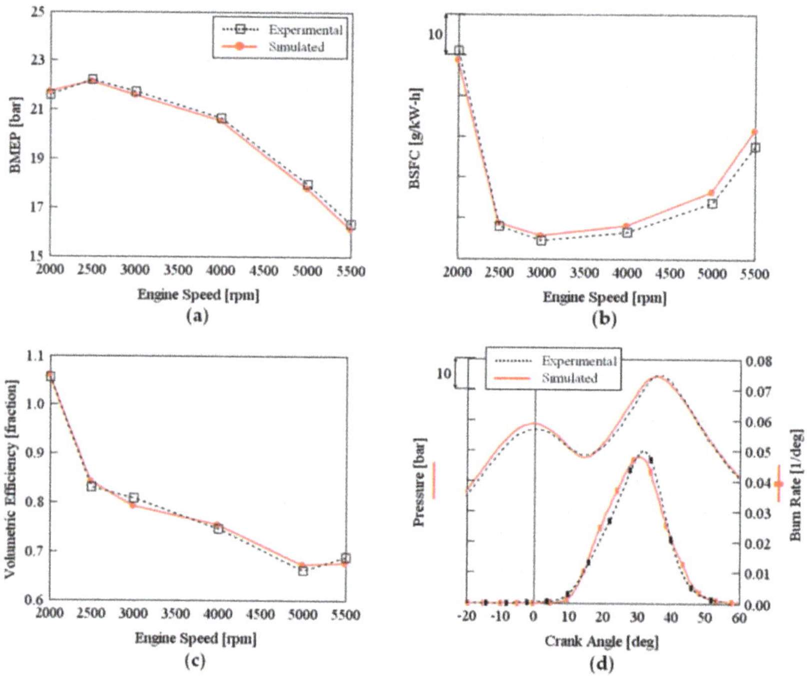

Figure 3. (**a**) Brake Mean Effective Pressure (**b**) Brake Specific Fuel Consumption (**c**) Volumetric Efficiency (Manifold) (**d**) Pressure Cycle (left) and Burn rate (right) at 3000 rpm obtained at Full-Load operating conditions without Port Water Injection; simulated (red solid line) and experimental (black dashed line).

3.2. Calibration of CCV Model

Since knock is heavily dependent on Cycle-to-Cycle variability (CCV), its occurrence cannot be predicted by analyzing the average cycle only. However, to physically reproduce the CCV, it would be necessary to accurately model the stochastic behavior of in-cylinder charge motion, the local inhomogeneities of the Air-to-Fuel ratio inside the cylinder, etc. as discussed in detail in [47]. As a result, the adoption of a 0D phenomenological combustion model does not allow a rigorous modelling of CCV, which instead must be "injected" into the model using a statistically-based approach [48]. In this research work, the GT CCV model [49] developed by Gamma Technologies was adopted. Such an approach, after the simulation of some preliminary cycles to achieve the convergence of the flow solution, perturbs two variables of the SITurb combustion model, namely the Turbulent Flame Speed Multiplier C_S and the Taylor Length Scale Multiplier C_λ, for 300 cycles in a row. C_S and C_λ assume random values from Gaussian distributions whose standard deviations represent the calibration parameters of the CCV model. Similarly to the approach adopted for SITurb calibration, a DoE methodology integrated with a Genetic Algorithm optimum search was used to calculate the standard deviations of C_S and C_λ that minimized the error between the experimental and simulated Coefficient of Variation (COV) of the Indicated Mean Effective Pressure (IMEP) (see Table 3). Afterwards, the average IMEP and its COV resulting from the CCV-run were compared against experimental outcomes. As it can be seen in Figure 4, the CCV model provides a good estimation of these quantities for almost the entire engine speed range, with the only exception of the 2000 rpm operating condition, for which the simulation underestimates the COV. The root cause of this deviation could be attributed to an overestimation of the scavenging air, due to the particular intake valve lift profile adopted at 2000 rpm; as a consequence, the residuals are underestimated and so is the COV.

Table 3. Optimized values of the calibration parameters for the Cycle to Cycle Variability CCV model.

Calibration Parameter	Value
Standard deviation for C_S	0.048
Standard deviation for C_λ	0.167

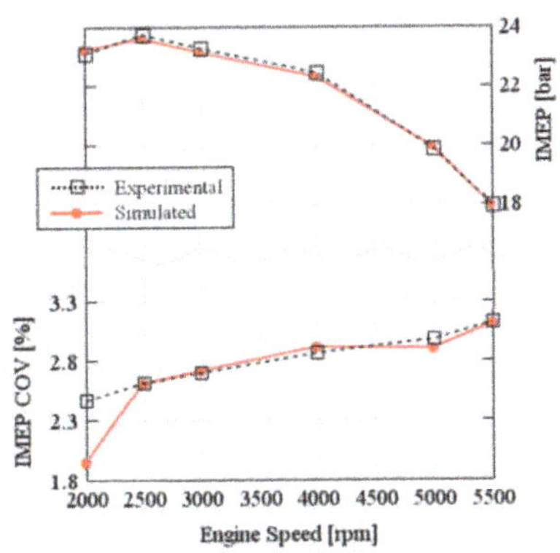

Figure 4. Coefficient of Variation of the Indicated Mean Effective Pressure (top) and average Indicated Mean Effective Pressure over 300 CCV-cycles (bottom) at Full-Load without Port Water Injection PWI; simulated (red solid line) and experimental (black dashed line).

3.3. Calibration of Knock Model

Most of the models developed for knock prediction are based on the approach proposed by Livengood and Wu [50] which states that the auto-ignition of the end gas occurs when:

$$\int_{t=0}^{t_{knock}} \frac{1}{\tau} dt = 1 \tag{6}$$

where τ is the induction time of the Air-Fuel mixture and t_{knock} is the time corresponding to the auto-ignition instant (computed from the start of compression).

Several empirical relationships for induction time calculation have been developed in the scientific literature [51–53]. Most of them, however, do not take into account induction time dependence from mixture composition in terms of diluents, such as water, and are not capable of reproducing the Negative Temperature Coefficient (NTC) behavior [54,55]. For these reasons, the so-called Kinectics-Fit model, proposed by Gamma Technologies [49], was preferred for this work; it calculates the induction time on the basis of Equation (7):

$$\tau_i = M_1 \cdot a_i \cdot \left(\frac{ON}{100}\right)^{b_i} \cdot [Fuel]^{c_i} \cdot [O_2]^{d_i} \cdot [D_i]^{e_i} \exp\left(\frac{f_i}{M_2 T_u}\right) \tag{7}$$

where ON is the fuel Octane Number, [Fuel] and $[O_2]$ are the concentrations in [mol/m^3] of fuel and oxygen, respectively; T_u is the temperature of the unburned gas; $[D_i]$ is the sum of N_2, CO_2, and H_2O concentrations in [mol/m^3]. The knock Induction Time Multiplier (M_1) and the Activation Energy Multiplier (M_2) are the knock calibration parameters while quantities from a_i to f_i are model constants.

After the set-up of the artificial CCV model described in the previous section, which led to the simulation of Cycle to Cycle Variability as shown in Figure 5a, the knock model was calibrated through a Genetic Algorithm optimum search. M_1 and M_2 were tuned (see Table 4) so that the percentage of knocking cycles detected by the model during the 300 CCV-cycles was coincident to the experimental one. Optimization results are presented in Figure 5b. The Knock model demonstrated the capability to reproduce the experimental knocking behavior, with the exception of the operating point at 2000 rpm. This is the direct consequence of CCV-model, which underestimates IMEP COV at 2000 rpm (see Figure 4), thus underestimating the knock occurrence likelihood.

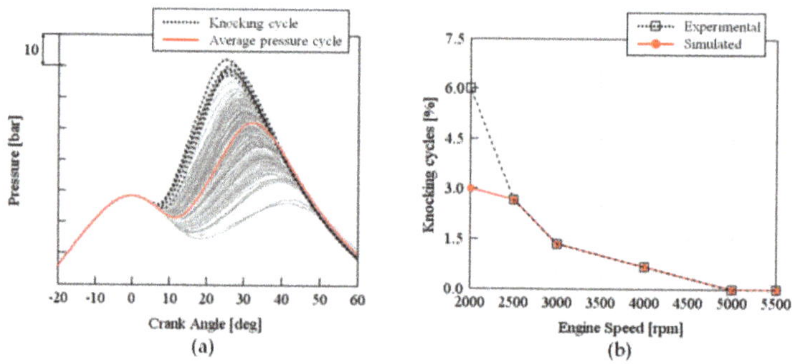

Figure 5. (**a**) Example of CCV simulation; (**b**) comparison between simulated (red solid line) and experimentally measured (black dashed line) knocking cycles percentage over 300 CCV-cycles at Full-Load without PWI.

Table 4. Optimized value of the calibration parameters for the Knock model.

Calibration Parameter	Value
M_1	0.860
M_2	0.916

4. Modelling of Port Water Injection

In this section, the numerical approach adopted for the modelling of Port Water Injection system is presented and discussed in detail, in three sub-sections. The first describes the integration of the port-puddling model, which allows the proper replication of transient phenomena of water wall film formation in the intake ports. The second focuses on the water evaporation model, calibrated to discern the water amount that is directly participating in the charge cooling from that which is evaporating on the walls. Finally, the last sub-section examines the SITurb combustion model, which requires a further refinement in order to accurately describe the effect of water vapor on the burn rate.

The calibration of the PWI model has been carried out considering the experimental outcomes obtained at the operating points of 2500 rpm–18 bar BMEP and 4000 rpm–17 bar BMEP, shown in Figure 2 with green rhombuses.

4.1. Calibration of Port-Puddling Model

A particular model capable of reproducing the formation of a water wall film was adopted for the intake ports downstream of water injectors. Afterwards, with the aim of describing the roles that water performs during each cycle, Equation (8) was introduced:

$$m_{inj}^{(n)} + m_{back}^{(n-1)} = m_{film}^{(n)} + m_{trap}^{(n)} + m_{back}^{(n)} \quad (8)$$

As represented in Figure 6, the entire water mass injected at cycle 'n' ($m_{inj}^{(n)}$) temporarily becomes part of the wall film. Subsequently, the code models the building-up of the liquid film (see Figure 6a), along with its transportation under the action of shear forces generated by intake air flow and its evaporation. The latter is governed by the heat transfer between the film and the gas flow and between the film and the port walls. However, because of the intimate contact of the thin fuel film with the pipe wall, the film temperature is close to the port wall temperature.

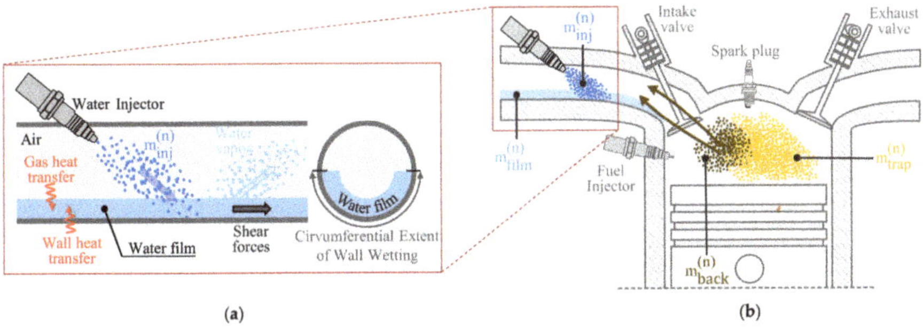

Figure 6. Schematic representation of (**a**) port-puddling model (**b**) injected water behavior.

The liquid film modelling is achieved by the two following parameters: the Circumferential Extent of Wall Wetting (CEWW) and the Puddle Velocity Multiplier (PVM). The first one, as shown in Figure 6a, defines the fraction of the intake port circumference that is covered by the water puddle: higher values imply a greater amount of water that can be retained by the wall film. On the other hand, PVM represents a multiplier to increase or decrease the effect of shear stresses on the transportation of the water puddle: higher values of PVM cause lower amount of water to be trapped in the wall film.

It follows that just a fraction of the injected mass ($m_{inj}^{(n)}$) remains trapped in the wall film ($m_{film}^{(n)}$), while the remaining part escapes from the film and enters into the cylinder. Once reached the combustion chamber, the liquid water starts evaporating by extracting heat either from the surrounding air or from the cylinder walls. Afterwards, some vapor can flow back into the intake ports ($m_{back}^{(n)}$)

due to the LIVC. $m_{back}^{(n)}$, as vapor, does not become part of the liquid film and it gets entrained in the cylinder during the following cycle. Finally, $m_{trap}^{(n)}$ represents the amount of water which is present in the cylinder during the combustion process.

With the aim of characterizing the evaporation process of the injected water, an index was defined through 3D-CFD simulations [30]. This index, as described by Equation (9), represents instant by instant the ratio between the mass of water vapor into the cylinder ($m_{H_2O,evap}$) and the maximum amount of water theoretically available at each engine cycle, that is the sum of the injected water during cycle 'n' ($m_{inj}^{(n)}$) and of the backflow water from the previous cycle 'n−1' ($m_{back}^{(n-1)}$).

$$\text{Evaporation Index (EI)} = \frac{m_{H_2O,evap}}{m_{inj}^{(n)} + m_{back}^{(n-1)}} \tag{9}$$

Figure 7 reports an example of Evaporation Index computation: after Intake Valve Opening (IVO), entrained water mass produces a subsequent growth of EI; then, from Bottom Dead Centre (BDC) on, a reduction in EI is observed as a consequence of the high water backflow generated by the Miller cycle (Late Intake Valve Closing). The Evaporation Index, due to the build-up of the wall film, varies from cycle to cycle and achieves its highest value, one, when the wall film is saturated and all the injected water enters into the combustion chamber.

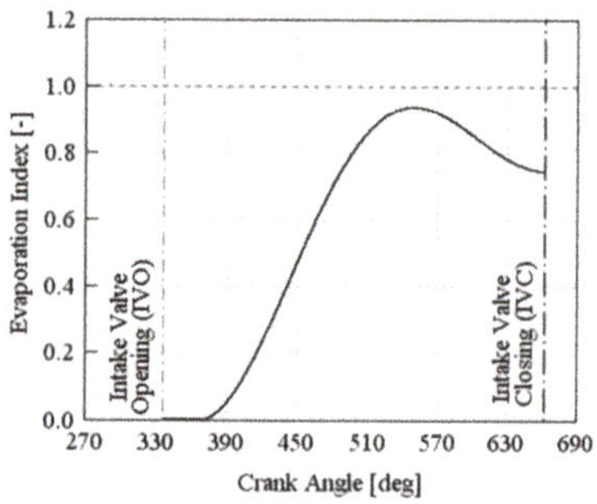

Figure 7. Example of Evaporation Index as a function of Crank Angle (CA).

To carry out the calibration of the water wall film model, the results of a previous work of the authors were considered [30]: in particular, the Evaporation Indexes computed for three different values of Water-to-Fuel ratios, namely 9%, 18% and 50%, at the operating point 4000 rpm–17 bar BMEP, were taken as reference. Such calculations showed that the higher the mass of water injected, the higher the fraction retained by the wall film. Moreover, as depicted in Figure 8a by dashed lines, none of the computed indexes reached the maximum value of one: indeed, they were recorded during the third cycle in a row of Port Water Injection and therefore during the building-up of the liquid film.

To replicate the 3D-CFD Evaporation Indexes, the water wall film model required a calibration process which was carried out through DoE methodology. In particular, for each Water-to-Fuel ratio, the optimal pairs of CEWW and PVM aimed at minimizing the error between 1D-CFD and 3D-CFD Evaporation Indexes were determined.

Solid lines of Figure 8a represent the results of the refined water wall film model, while Figure 8b reports the values of its calibration parameters as a function of the total mass of injected water per cylinder.

Considering the operating point 4000 rpm–17 bar BMEP and a Water-to-Fuel ratio of 50%, the different roles of the injected water were analyzed as a function of the consecutive engine cycles with PWI (see Equation (8)), as depicted in Figure 9a. Figure 9b shows instead the impact of the water wall film formation on PWI effectiveness in reducing knock tendency. It can be noticed that during the first cycle of port water injection, almost all injected water remains trapped in the wall film (Figure 9a), and therefore knock likelihood is not reduced (Figure 9b). On the other hand, after six cycles, film building-up is almost complete and practically all the injected water gets trapped in the cylinder, thus effectively diminishing knock tendency. Therefore, it can be stated that the PWI system, at the operating point 4000 rpm–17 bar BMEP and with a Water-to-Fuel ratio of 50%, becomes fully effective after six cycles of water injection. In steady state conditions, represented in Figure 9a,b by cycles 'k' and 'k+1', the fraction of injected water mass that gets trapped in the wall film is null (i.e., the amount of water that enters the film is equal to that which comes out). Moreover, the backflow water of cycle 'n−1' is equal to the one of cycle 'n'. It follows that, in steady state conditions, the amount of water that takes part in the combustion process ($m_{trap}^{(n)}$) is equal to the injected water ($m_{inj}^{(n)}$).

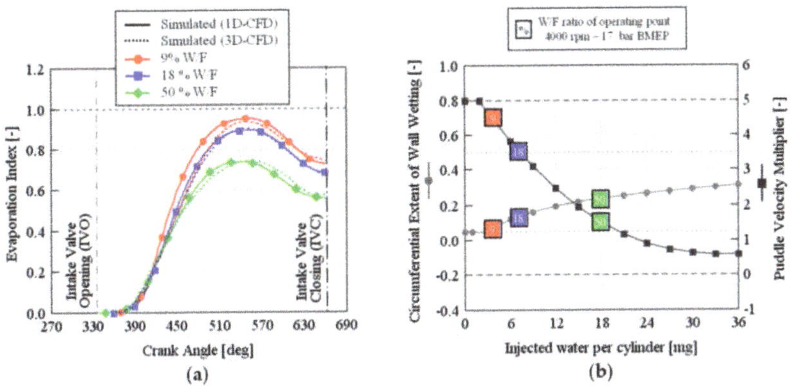

Figure 8. (**a**) Evaporation index measured during the 3rd cycle in a row of PWI at 4000 rpm—17 bar for Water-to-Fuel ratios of 9% (red line), 18% (blue line) 50% (green line); 1D simulation (solid line), 3D simulation (dashed line); (**b**) Optimized values of the calibration parameters for the port-puddling model as a function of the total injected water per cylinder.

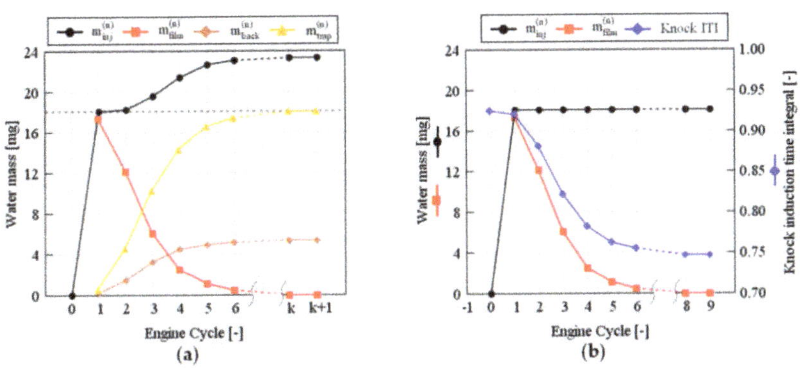

Figure 9. (**a**) The amount of water available before Intake Valve Opening (IVO), water trapped in the wall film, water backflow and water which is present into the cylinder during the combustion process as a function of the consecutive engine cycles with Port Water Injection (PWI); (**b**) The amount of injected water, trapped water in the wall film and the knock induction time integral as a function of the consecutive engine cycles with PWI; k ≥ 13 cycles.

4.2. Calibration of Evaporation Model

The engine model governs the water vaporization within the cylinder through a simple phenomenological evaporation model. However, a detailed investigation to quantify the charge cooling effect produced by the phase change of water needs to be carried out. As a matter of fact, when entering the cylinder, the water droplets can impact on either the piston or the liner, thus extracting the heat requested for evaporation from the walls instead of the surrounding gases. The share between these contributions can be tuned through a parameter known as Fraction of Vaporized Heat Taken from Walls (FHW). The latter accounts for the fraction of water whose vaporization occurs on the cylinder walls rather than within the cylinder gases. As shown in Figure 10a,b, where the in-cylinder temperature at Intake Valve Closing (IVC) and the Volumetric Efficiency are respectively reported, an excessively low value of FHW implies a larger amount of heat adsorbed from surrounding air, thus inducing a significant in-cylinder gases cooling with a consequent overestimation of the Volumetric Efficiency. The increase in FHW improves the calculation of engine airflow: however, a constant value for this parameter does not ensure a satisfactory correlation between numerical and experimental Volumetric Efficiency. Therefore, considering the operating point 4000 rpm–17 bar BMEP, the FHW was optimized as a function of the total amount of liquid water that enters the cylinder, with the objective being to match the experimental air flow rate (Equation (10)).

$$FHW_{opt} = 0.95 - 0.0028 \cdot \int \dot{m}_{H_2O,liq} \tag{10}$$

where $\dot{m}_{H_2O,liq}$ represents the instantaneous mass flow rate of liquid water that passes through the intake valves of a single cylinder.

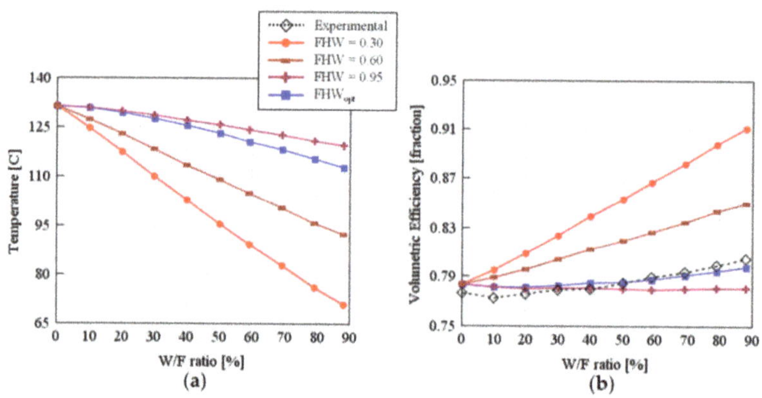

Figure 10. (**a**) In-cylinder temperature measured at Intake Valve Closing (IVC) and (**b**) Volumetric Efficiency as a function of the Water-to-Fuel ratio for different values of the Fraction of Vaporized Heat Taken from Walls (FHW) model calibration parameter; operating point 2500 rpm–18 bar.

From Equation (10) it was noticed that, considering a 10% Water-to-Fuel ratio, only around 6% of in-cylinder liquid water vaporizes within the cylinder gases. Furthermore, such a percentage rises along with the increase in the injected water and achieves a value around 15% with a Water-to-Fuel ratio equal to 90%. In accordance with previous findings [29,31,33], these results evidence a poor efficiency of PWI system in reducing the charge air temperature.

The robustness of Equation (10), which was defined for the operating point 4000 rpm—17 bar BMEP, was assessed at the operating point 2500 rpm–18 bar BMEP. As depicted by the blue solid line of Figure 10, the FHW optimization limits the reduction in in-cylinder temperature produced by the increase in the injected water, allowing the correct reproduction of the experimental Volumetric Efficiency for any Water-to-Fuel ratio.

4.3. Calibration of Dilution Effect Multiplier

Despite replicating the experimental values of boost pressure, spark timing, lambda and intake airflow, the engine model was not capable of satisfactorily predicting the evolution of the combustion process when water was injected. In particular, the higher the water content, the higher the deviation from experimental results: this trend is evident when observing Figure 11a, where both the center of gravity of the combustion and the in-cylinder pressure peak (p_{max}) are reported. This error was attributed to an incorrect estimation of the Laminar Flame Speed, the only parameter involved in burn rate calculation which depends on the in-cylinder mass fraction of water (see Equations (1) and (3)). As claimed by several authors [31,35,36], the water dilution of the air–fuel mixture drops the S_L, thus delaying the combustion duration. This effect, governed in the SITurb model by the C_D (see Equation (3)), seems to be underestimated, thus resulting in a more advanced MFB_{50} (Crank Angle corresponding to 50% of fuel burned) and a consequently higher in-cylinder pressure peak.

Based on the above considerations and similarly to the approach used for FHW, C_D was calibrated as a function of the total amount of in-cylinder water vapor present at spark timing, in order to minimize the error on the prediction of the combustion center of gravity. FHW optimization, carried out at operating point 4000 rpm–17 bar BMEP, led to definition of Equation (11):

$$C_{D,opt} = 0.872 + 0.0062 \cdot \left\{ m_{trap}[mg] \right\} \tag{11}$$

As expected, Equation (11) shows that the greater the water content, the higher the optimal C_D and therefore the impact on S_L slows, as described in the scientific literature. It is worth remembering that, at steady state conditions, the amount of water that takes part in the combustion process (m_{trap}) is equal to the amount of injected water, as already described in Section 4.1.

The robustness of Equation (11) was then assessed at the operating point 2500 rpm–18 bar BMEP: observing the green line of Figure 11a, the C_D optimization led to MFB_{50} and maximum in-cylinder pressure values well correlated with the experimental results for the entire Water-to-Fuel ratio sweep, with errors lower than 2%. Moreover, also the simulated pressure cycle and the burn rate profile reported in Figure 11b for a 90% W/F show a satisfactory correlation with experimental measurements.

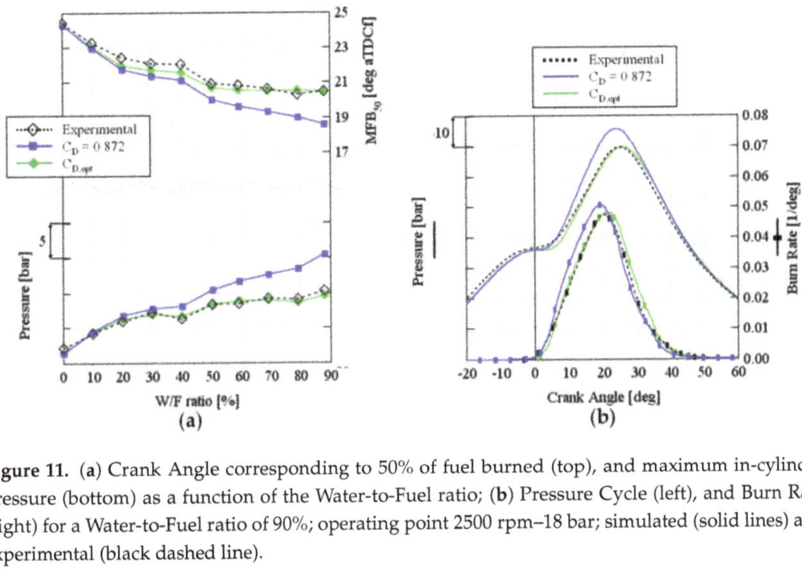

Figure 11. (**a**) Crank Angle corresponding to 50% of fuel burned (top), and maximum in-cylinder pressure (bottom) as a function of the Water-to-Fuel ratio; (**b**) Pressure Cycle (left), and Burn Rate (right) for a Water-to-Fuel ratio of 90%; operating point 2500 rpm–18 bar; simulated (solid lines) and experimental (black dashed line).

5. Results

In this section, the engine model validation against the experimental tests carried out at WOT with PWI is presented and discussed. Moreover, the capability of the proposed methodology to create a virtual test bench for the optimization of the engine operating parameters at different water flow rate values is assessed.

5.1. PWI Model Validation

At first, FHW_{opt} and $C_{D,opt}$ equations together with the port-puddling model were implemented in the engine model calibrated in dry conditions (see Section 3). The knock model and CCV model were not modified. The latter, which is not predictive, was kept unchanged since all the experimental tests carried out with water injection evidenced only a quite moderate increase in IMEP COV (see Figure 12).

Afterwards, by imposing the experimental boost pressure and by controlling the injected fuel mass in order to guarantee stoichiometric combustion, the experimental amount of water mass was injected in the intake runners. Subsequently, by simulating a significant number of engine cycles with the CCV model enabled, the spark timing was adjusted in order to achieve the same percentage of knocking cycles observed at Full-Load without PWI (see the red solid line of Figure 5). Since engine performance at 5000 rpm and 5500 rpm were not knock limited (see Figure 5), such analysis focused on the engine speed range 2000–4000 rpm.

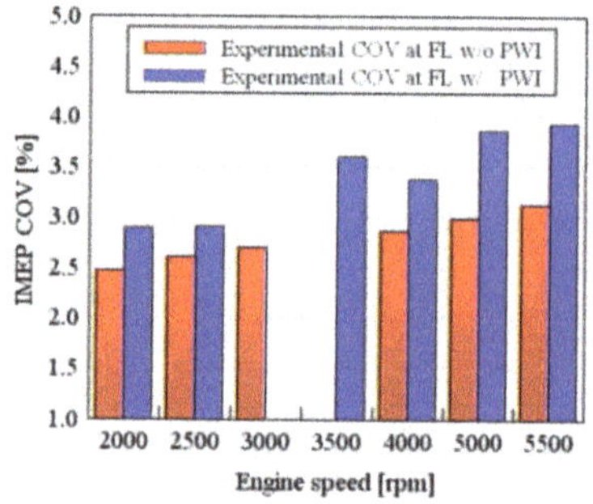

Figure 12. Experimental Coefficient of Variation of the Indicated Mean Effective Pressure at Full-Load in dry (red bars) and wet (blue bars) conditions.

The predicted Knock Limited Spark Advance, BMEP, as well as pressure cycles and the burn rates, which for the sake of conciseness have been presented only at 4000 rpm, showed very good agreement when compared to the experimental results (see Figure 13a–c, respectively). The error around 4% obtained in estimating the engine load at 2000 rpm–Full-Load with PWI was attributed to a knock tendency underestimation. Such an error, already evident in Figure 5, allowed the exploitation of a more advanced spark timing with a consequently higher BMEP.

5.2. Engine Virtual Calibration

The predictive capabilities of the proposed methodology were finally used to perform a virtual calibration of the engine. In a first stage, the relative BMEP improvement achievable with different levels of water mass flow rate in knock limited conditions was investigated. Subsequently, the same analysis was carried out by introducing an additional constraint related to the in-cylinder pressure peak, which was kept below 100 bar. Besides knock and maximum in-cylinder pressure, major limits to the engine performance enhancement could be represented by either excessive COV of IMEP or lubricant oil dilution, which can suddenly increase at high W/F ratios [30,31]. Unfortunately, these phenomena cannot be caught by the model since a blow-by model is not present and the CCV is "injected" into the model on the basis of the experimental data.

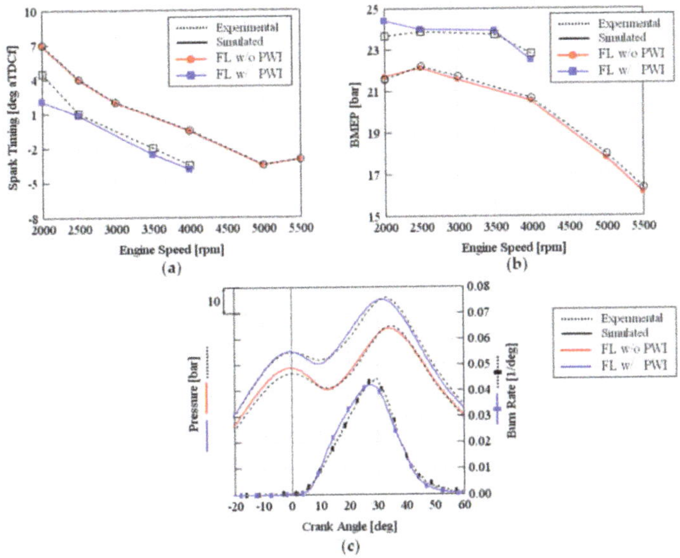

Figure 13. Experimental (dotted lines) and simulated (solid lines) without PWI (red lines) and with PWI (blue lines) at Full Load (FL) operating conditions: (**a**) Spark Timings vs. engine speed; (**b**) BMEP vs. engine speed; (**c**) in-cylinder pressure (left axis) and burn rate (right axis) vs. crank angle at 4000 rpm.

The results of both the abovementioned analyses, which were carried out by keeping constant the experimental boost pressure target and maintaining a stoichiometric combustion, are presented in Figure 14. The latter shows the BMEP improvements as a function of the water flow rate for each engine speed: each line belonging to a certain engine speed contains a specific symbol (e.g., round for 2000 rpm, square for 2500, etc.) which is repeated every time the injected water mass per cycle per cylinder (m_{inj}) increases by 10 mg. Moreover, the black hexagons of Figure 14 give the relative increase in BMEP achieved with the water flow rate exploited experimentally in WOT operating conditions.

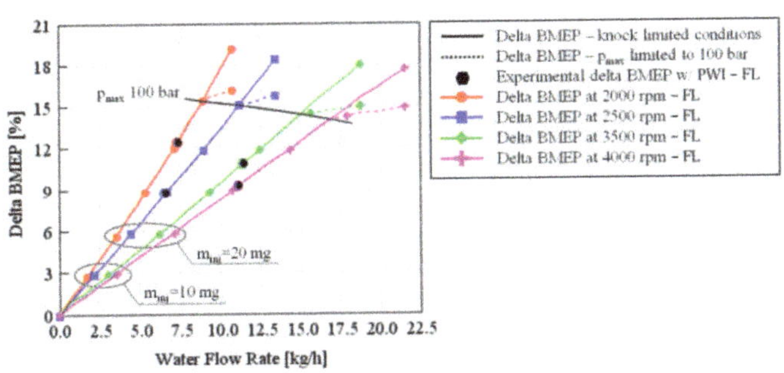

Figure 14. Relative improvement of Brake Mean Effective Pressure BMEP enabled by PWI at Full-Load as a function of the injected water flow rate.

At a constant water flow rate, as shown in Figure 14, greater improvement of engine BMEP can be achieved at lower engine speed. This happens because at equal water consumption, the lower the engine speed, the higher the injected water mass per cycle: the larger water content, thanks to its knock mitigation capability, enables a higher advance of spark timing with a consequently higher improvement of engine load. Considering instead the symbols within the two grey circles of Figure 14, it is noticeable that, at equal injected water mass per cycle, the relative increase in engine BMEP is almost constant independently from the engine speed. Moreover, if the engine is knock limited (solid lines), a linear correlation is found between the water mass flow and the BMEP rise, which exceeds 17% with an m_{inj} equal to 60 mg. On the other hand, by adding the constraint on the in-cylinder pressure peak (dashed lines), a significant reduction in delta BMEP increase is observed as soon as the maximum in-cylinder pressure achieves the value of 100 bar. However, as expected, even when engine performance is limited by the in-cylinder pressure peak, a further increase in the water mass flow rate results in a higher engine BMEP. Indeed, the additional water causes a further slowing of the Laminar Flame Speed and, in turn, of the burn rate (see Section 4.3): it follows a reduction in the maximum in-cylinder pressure, thus enabling an advance of the spark timing. The subsequent ignition advance, in addition to returning the pressure peak in the combustion chamber below the 100 bar threshold value, guarantees a small improvement of engine BMEP.

6. Conclusions

This paper describes a methodology, based on a quasi-dimensional approach, to model a PWI system and to assess its knock mitigation potential, thus significantly reducing the experimental effort for engine calibration.

In a first stage, experimental investigations were carried out at the engine test bench to characterize the engine behavior under "dry" (i.e., without WI) and "wet" (i.e., with WI) operation. Afterwards, the experimental outcomes obtained without PWI were used to calibrate the combustion (SITurb), CCV and knock models. In particular, knock model constants were tuned by simulating CCV over 300 consecutive engine cycles with the aim of matching the experimental percentage of knocking cycles. In a second stage, the results of 3D-CFD simulations of the PWI were used to calibrate a port-puddling model: the latter, once implemented in the reference engine model, was proven to properly replicate transient phenomena of water wall film formation, catching cycle by cycle the amount of water that enters into the cylinder and is therefore available for knock mitigation. Subsequently, the experimental results collected at medium loads with W/F sweeps were considered for the refinement of both water evaporation and SITurb model. The first one was calibrated to properly discern the fraction of the injected water that is producing the charge cooling through its evaporation from that which is evaporating due to the heating from the walls. Then, the SITurb model was tuned in order to carefully predict the impact of the injected water on the Laminar Flame Speed: in particular, a linear dependence between the in-cylinder vapor content and the Dilution Effect Multiplier was found. The PWI model was then validated against the experimental tests carried out at Wide Open Throttle with PWI, showing the capability to accurately predict the effects of water injection in terms of both burning rate slow down and knock tendency reduction. Finally, the proposed methodology was used to create a virtual test bench, which allowed the evaluation of engine performance improvement as a function of the injected water mass flow rate.

The same methodology will be adopted in a future work in order to determine the water consumption as well as the potential benefits in terms of CO_2 emission achievable over RDE cycles.

Author Contributions: Conceptualization, F.M.; methodology definition, F.M., F.G., L.R.; software, F.G., L.R.; validation, F.G., L.R.; formal analysis, F.G., L.R.; investigation, F.G., L.R.; writing—original draft preparation, F.G.; writing—review and editing, L.R.; supervision, F.M.; project administration, F.M.; funding acquisition, F.M. All authors have read and agreed to the published version of the manuscript.

Funding: This research activity was financially supported by FCA through the grant "Advanced combustion processes for high efficient Spark Ignition Engines 2016–2017".

Acknowledgments: The valuable support provided to the research activity by FCA is gratefully acknowledged. In particular, the authors would like to thank Caterina Venezia and Andrea Stroppiana for their precious and constant support as well as for their invaluable suggestions during the simulation activities.

Conflicts of Interest: The authors declare no conflict of interest.

Definitions/Abbreviations

1D	One Dimensional
3D	Three Dimensional
aTDCf	After Top Dead Center Firing
A/F	Air-to-Fuel ratio
BDC	Bottom Dead Centre
BMEP	Brake Mean Effective Pressure
BSFC	Brake Specific Fuel Consumption
CA	Crank Angle
CCV	Cycle-to-Cycle Variability
CEWW	Circumferential Extent of Wall Wetting
CFD	Computational Fluid Dynamic
COV	Coefficient Of Variation
CR	Compression Ratio
DoE	Design of Experiment
DWI	Direct Water Injection
ED_{50}	50% Evaporation Duration
EI	Evaporation Index
EOI	End Of Injection
EU	European Union
EGR	Exhaust Gas Recirculation
FHW	Fraction of vaporized Heat taken from Walls
FL	Full-Load
IMEP	Indicated Mean Effective Pressure
IVC	Intake Valve Closing
IVO	Intake Valve Opening
KLSA	Knock Limit Spark Advance
LHV	Lower Heating Value
LIVC	Late Intake Valve Closing
MFB_{50}	50% Mass Fraction Burned
N	Number
ON	Octane Number
PM	Particulate Matter
PVM	Puddle Velocity Multiplier
PWI	Port Water Injection
RDE	Real Driving Emissions
RON	Research Octane Number
rpm	revolutions per minute
SMD	Sauter Mean Diameter
t	time
t_{knock}	Auto-Ignition instant
T-DISI	Turbocharged Direct Injection Spark Ignition
VCR	Variable Compression Ratio
VE	Volumetric Efficiency
VVA	Variable Valve Actuation
w/	with
W/F	Water-to-Fuel ratio
w/o	without
WI	Water Injection
WOT	Wide Open Throttle
WLTP	Worldwide Harmonized Light-Duty Test Procedure

Symbols

ϕ	Equivalence Ratio
λ	Taylor Microscale
ρ_u	Unburned Gas Density
τ	Induction Time
ω	Engine Speed
A_f	Flame Area
C_λ	Taylor Length Scale Multiplier
C_D	Dilution Effect Multiplier
C_k	Flame Kernel Growth Multiplier
C_S	Turbulent Flame Speed Multiplier
D_i	Mass Fraction of Diluents
L_t	Turbulent Length Scale
M_1	Knock Induction Time Multiplier
M_2	Activation Energy Multiplier
M_b	Burned Mass
m_{back}	In-Cylinder Water Mass that flows back to the Intake Ports
M_e	Entrained Mass
m_{film}	Water Trapped in the Wall Film
$m_{H_2O,liq}$	In-cylinder Mass of Liquid Water
$m_{H_2O,evap}$	In-cylinder Mass of Water Vapor
m_{inj}	Injected Water Mass per Cylinder
m_{trap}	Water Mass Trapped in the Cylinder after IVC
M_u	Unburned Mass
$^{(n)}$	Cycle Number
p	In-cylinder Pressure
p_{max}	Maximum in-cylinder Pressure
R_f	Flame Radius
S_L	Laminar Flame Speed
S_T	Turbulent Flame Speed
T_u	Temperature of the Unburned Gas
u'	Turbulent Intensity

References

1. Regulation (EC) No 443/2009 of the European Parliament and of the Council of 23 April 2009. OJ L140, 5.6.2009. pp. 1–15. Available online: https://eur-lex.europa.eu/legal-content/EN/ALL/?uri=CELEX%3A32009R0443 (accessed on 11 August 2020).
2. Regulation (EU) 2019/631 of the European Parliament and of the Council of 17 April 2019. OJ L111, 25.4.2019. pp. 13–53. Available online: https://eur-lex.europa.eu/legal-content/EN/TXT/?uri=CELEX%3A32019R0631 (accessed on 11 August 2020).
3. Wanker, R. "Real Driving Emissions-New Requirements and Solutions for LD and HD Vehicles", presented at in PDiM—Product Development in Motion, Göteborg. 2016. Available online: https://www.avl.com/c/document_library/get_file?uuid=d6679ecd-c9f6-4bfe-8c1f-47e2be4c3afb&groupId=10138&download (accessed on 17 April 2019).
4. Isenstadt, A.; German, J.; Dorobantu, M.; Boggs, D.; Watson, T. *Downsized, Boosted Gasoline Engines*; Working Paper 2016-2; International Council on Clean Transportation: Washington, DC, USA, 1 October 2016.
5. Königstein, A.; Grebe, U.D.; Wu, K.-J.; Larsson, P.-I. Differentiated analysis of downsizing concepts. *MTZ Worldw.* **2008**, *69*, 4–11. [CrossRef]
6. Bozza, F.; Siano, D.; Torella, E. Cycle-by-cycle analysis, Knock modelling and spark- advance setting of a downsized spark-ignition turbocharged engine. *SAE Int. J. Engines* **2010**, *2*, 381–389. [CrossRef]
7. Marelli, S.; Capobianco, M. Steady and pulsating flow efficiency of a waste-gated turbocharger radial flow turbine for automotive application. *Energy* **2010**, *36*, 459–465. [CrossRef]

8. Neußer, H.-J. The car of the future will continue to fascinate people. In Proceedings of the 36th International Vienna Motor Symposium, Vienna, Austria, 7–8 May 2015.
9. Leduc, P.; Dubar, B.; Ranini, A.; Monnier, G. Downsizing of Gasoline Engine: An Efficient Way to Reduce CO_2 Emissions. *Oil Gas. Sci. Technol. Rev. IFP* **2003**, *58*, 115–127. [CrossRef]
10. Millo, F.; Luisi, S.; Borean, F.; Stroppiana, A. Numerical and experimental investigation on combustion characteristics of a spark ignition engine with an early intake valve closing load control. *Fuel* **2014**, *121*, 298–310. [CrossRef]
11. Luisi, S.; Doria, V.; Stroppiana, A.; Millo, F.; Mirzaeian, M. Experimental Investigation on Early and Late Intake Valve Closures for Knock Mitigation through Miller Cycle in a Downsized Turbocharged Engine. No. 2015-01-0760. *SAE Tech. Pap.* **2015**. [CrossRef]
12. Millo, F.; Accurso, F.; Zanelli, A.; Rolando, L. Numerical Investigation of 48 V Electrification Potential in Terms of Fuel Economy and Vehicle Performance for a Lambda-1 Gasoline Passenger Car. *Energies* **2019**, *12*, 998. [CrossRef]
13. Boretti, A.; Scalzo, J. Exploring the Advantages of Variable Compression Ratio in Internal Combustion Engines by Using Engine Performance Simulations. No. 2011-01-0364. *SAE Tech. Pap.* **2011**. [CrossRef]
14. Shelby, M.H.; Leone, T.G.; Byrd, K.D.; Wong, F.K. Fuel Economy Potential of Variable Compression Ratio for Light Duty Vehicles. No. 2017-01-0639S. *AE Tech. Pap.* **2017**. [CrossRef]
15. Tomazic, D.; Kleeberg, H.; Bowyer, S.; Dohmen, J.; Wittek, K.; Haake, B. Two-Stage Variable Compression Ratio (VCR) System to Increase Efficiency in Gasoline Powertrains. In Proceedings of the DEER Conference 2012, Dearborn, MI, USA, 16 October 2012.
16. Potteau, S.; Lutz, P.; Leroux, S.; Moroz, S.; Tomas, E. Cooled EGR for a Turbo SI Engine to Reduce Knocking and Fuel Consumption. *SAE Tech. Pap. SAE Int.* **2007**. [CrossRef]
17. Kumano, K.; Yamaoka, S. Analysis of Knocking Suppression Effect of Cooled EGR in Turbo-Charged Gasoline Engine. *SAE Tech. Pap. SAE Int.* **2014**. [CrossRef]
18. Francqueville, L.; Michel, J. On the Effects of EGR on Spark-Ignited Gasoline Combustion at High Load. *SAE Int. J. Engines* **2014**, *7*. [CrossRef]
19. Celik, M.B.; Ozdalyan, B.; Alkan, F. The use of pure methanol as fuel at high compression ratio in a single cylinder gasoline engine. *Fuel* **2011**, *90*, 1591–1598. [CrossRef]
20. Selim, M.Y.E. Sensitivity of dual fuel engine combustion and knocking limits to gaseous fuel composition. *Energy Convers Manag.* **2004**, *45*, 411–425. [CrossRef]
21. Saikaly, K.; Corre, O.L.; Rahmouni, C.; Truffet, L. Preventive knock protection technique for stationary SI engines fuelled by natural gas. *Fuel Process. Technol.* **2010**, *91*, 641–652. [CrossRef]
22. Pauer, T.; Frohnmaier, M.; Walther, J.; Schenk, P.; Hettinger, A.; Kampmann, S. Optimization of Gasoline Engines by Water Injection. In Proceedings of the 37th International Vienna Motor Symposium, Vienna, Austria, 28–29 April 2016.
23. Mingrui, W.; Sa, N.T.; Turkson, R.F.; Jinping, L.; Guanlun, G. Water Injection for Higher Engine Performance and Lower Emissions. *J. Energy Inst.* **2016**, 1–15. [CrossRef]
24. Netzer, C.; Franken, T.; Seidel, L.; Lehtiniemi, H.; Kulzer, A.C. Numerical Analysis of the Impact of Water Injection on Combustion and Thermodynamics in a Gasoline Engine using Detailed Chemistry. No. 2018-01-0200. *SAE Tech. Pap.* **2018**. [CrossRef]
25. Battistoni, M.; Grimaldi, C.; Cruccolini, V.; Discepoli, G.; De Cesare, M. Assessment of Port Water Injection Strategies to Control Knock in a GDI Engine through Multi-Cycle CFD Simulations. No. 2017-24-0034. *SAE Tech. Pap.* **2017**. [CrossRef]
26. Hoppe, F.; Thewes, M.; Baumgarten, H.; Dohmen, J. Water Injection for Gasoline Engines: Potentials, Challenges, and Solutions. *Int. J. Engine Res.* **2016**, *17*, 86–96. [CrossRef]
27. Hoppe, F.; Thewes, M.; Seibel, J.; Balazs, A.; Scharf, J. Evaluation of the Potential of Water Injection for Gasoline Engines. *SAE Int. J. Engines* **2017**, *10*. [CrossRef]
28. Berni, F.; Breda, S.; Lugli, M.; Cantore, G. A Numerical Investigation on the Potentials of Water Injection to Increase Knock Resistance and Reduce Fuel Consumption in Highly Downsized GDI Engines. *Energy Procedia* **2015**, *81*, 826–835. [CrossRef]
29. Zembi, J.; Battistoni, M.; Ranuzzi, F.; Cavina, N.; De Cesare, M. CFD Analysis of Port Water Injection in a GDI Engine under Incipient Knock Conditions. *Energies* **2019**, *12*, 3409. [CrossRef]

30. Millo, F.; Mirzaeian, M.; Rolando, L.; Bianco, A.; Postrioti, L. A Methodology for the Assessment of the Knock Mitigation Potential of a Port Water Injection System. *Fuel* **2020**, in press.
31. Paltrinieri, S.; Mortellaro, F.; Silvestri, N.; Rolando, L.; Medda, M.; Corrigan, D. Water Injection Contribution to Enabling Stoichiometric Air-to-Fuel Ratio Operation at Rated Power Conditions of a High-Performance DISI Single Cylinder Engine. No. 2019-24-0173. *SAE Tech. Pap.* **2019**. [CrossRef]
32. Tornatore, C.; Siano, D.; Marchitto, L.; Iacobacci, A.; Valentino, G.; Bozza, F. Water Injection: A Technology to Improve Performance and Emissions of Downsized Turbocharged Spark Ignited Engines. *SAE Int. J. Engines* **2017**, *10*. [CrossRef]
33. Khatri, J.; Denbratt, I.; Dahlander, P.; Koopmans, L. Water Injection Benefits in a 3-Cylinder Downsized SI-Engine. *SAE Int. J. Adv. Curr. Pract. Mobil.* **2019**, *1*, 236–248. [CrossRef]
34. Wilson, J.P. Effects of Water Injection and Increased Compression Ratio in a Gasoline Spark Ignition Engine. Master's Thesis, Mechanical Engineering, University of Idaho, Moscow, ID, USA, 2011.
35. Cazzoli, G.; Bianchi, G.M.; Falfari, S.; Ricci, M.; Forte, C. Evaluation of Water and EGR Effects on Combustion Characteristics of GDI Engines Using a Chemical Kinetics Approach. No. 2019-24-0019. *SAE Tech. Pap.* **2019**. [CrossRef]
36. Bozza, F.; De Bellis, V.; Giannattasio, P.; Teodosio, L.; Marchitto, L. Extension and Validation of a 1D Model Applied to the Analysis of a Water Injected Turbocharged Spark Ignited Engine at High Loads and over a WLTP Driving Cycle. *SAE Int. J. Engines* **2017**, *10*. [CrossRef]
37. Gern, M.S.; Vacca, A.; Bargende, M. Experimental Analysis of the Influence of Water Injection Strategies on DISI Engine Particle Emissions. No. 2019-24-0101. *SAE Tech. Pap.* **2019**. [CrossRef]
38. Cavina, N.; Rojo, N.; Businaro, A.; Brusa, A.; Corti, E.; De Cesare, M. Investigation of Water Injection Effects on Combustion Characteristics of a GDI TC Engine. *SAE Int. J. Engines* **2017**, *10*, 2209–2218. [CrossRef]
39. De Bellis, V.; Bozza, F.; Teodosio, L.; Valentino, G. Experimental and Numerical Study of the Water Injection to Improve the Fuel Economy of a Small Size Turbocharged SI Engine. *SAE Int. J. Engines* **2017**, *10*. [CrossRef]
40. Bozza, F.; De Bellis, V.; Teodosio, L.; Tufano, D.; Malfi, E. Techniques for CO_2 Emission Reduction over a WLTC. A Numerical Comparison of Increased Compression Ratio, Cooled EGR and Water Injection. No. 2018-37-0008. *SAE Tech. Pap.* **2018**. [CrossRef]
41. De Martino, C.; Maiorana, G.; Pallotti, P.; Quinto, S.; Sacco, D. The Global Small Engine 3 and 4 Cylinder Turbo: The New FCA's Family of Small High-Tech Gasoline Engines. In Proceedings of the 39th International Vienna Motor Symposium, Vienna, Austria, 26–27 April 2018.
42. Millo, F.; Mirzaeian, M.; Porcu, D. Knock Mitigation Techniques for highly boosted downsized SI engines. In Proceedings of the 29th SIA Powertrain Congress, Paris, France, 6–7 June 2017.
43. Millo, F.; Ferraro, C. Knock in S.I. Engines: A Comparison between Different Techniques for Detection and Control. *SAE Tech. Pap.* **1998**. [CrossRef]
44. Mirzaeian, M.; Millo, F.; Rolando, L. Assessment of the Predictive Capabilities of a Combustion Model for a Modern Downsized Turbocharged SI Engine. No. 2016-01-0557. *SAE Tech. Pap.* **2016**. [CrossRef]
45. Wahiduzzaman, S.; Morel, T.; Sheard, S. Comparison of Measured and Predicted Combustion Characteristics of a Four-Valve, S.I. Engine. *SAE Tech. Pap.* **1993**. [CrossRef]
46. *GT-SUITE Engine Performance Application Manual, Software User Manual*; Gamma Technologies: Westmont, IL, USA, 2019.
47. Zhao, L.; Moiz, A.A.; Som, S.; Fogla, N.; Bybee, M.; Wahiduzzaman, S.; Mirzaeian, M.; Millo, F.; Kodavasal, J. Examining the role of flame topologies and in-cylinder flow fields on cyclic variability in spark-ignited engines using large-eddy simulation. *Int. J. Engine Res.* **2017**. [CrossRef]
48. Millo, F.; Rolando, L.; Pautasso, E.; Servetto, E. A Methodology to Mimic Cycle to Cycle Variations and to Predict Knock Occurrence through Numerical Simulation. No. 2014-01-1070. *SAE Tech. Pap.* **2014**. [CrossRef]
49. *GT-SUITE Engine Templates Reference Manual, Software User Manual*; Gamma Technologies: Westmont IL, USA, 2019.
50. Livengood, J.C.; Wu, P.C. Correlation of Auto-ignition Phenomenon in Internal Combustion Engines and Rapid Compression Machines. In *Symposium (International) on Combustion*; Elsevier: Amsterdam, The Netherlands, 1955; Volume 5, Issue 1, pp. 347–356. [CrossRef]
51. By, A.; Kempinski, B.; Rife, J. Knock in Spark Ignition Engines. *SAE Tech. Pap.* **1981**. [CrossRef]
52. Belli, M.; Danieli, G.; Amelio, M.; Bova, S.; Fragiacomo, N. A Detonation Model in Spark-Ignition Engines: Preliminary Results on Engine Octane Requirement. *SAE Tech. Pap.* **1984**. [CrossRef]

53. Douaud, A.; Eyzat, P. Four-Octane-Number Method for Predicting the Anti-Knock Behavior of Fuels and Engines. *SAE Tech. Pap.* **1978**. [CrossRef]
54. Hahn, T. Ignition Study in Rapid Compression Machine. Bachelor's Thesis, MIT, Cambridge, MA, USA, 2006.
55. Schernus, C.; Nebbia, C.; Di Matteo, F.; Thewes, M. Application of the New Kinetic Knock Model to A Downsized TGDI Engine. In Proceedings of the Gamma Technology User Conference 2013, Frankfurt, Germany, 21–22 October 2013.

energies
MDPI

Article

Study of the Radar Cross-Section of Turbofan Engine with Biaxial Multirotor Based on Dynamic Scattering Method

Zeyang Zhou * and Jun Huang

School of Aeronautic Science and Engineering, Beihang University, Beijing 100191, China; junh@china.com
* Correspondence: zeyangzhou@buaa.edu.cn

Received: 29 September 2020; Accepted: 2 November 2020; Published: 5 November 2020

Abstract: With the continuous advancement of rotor dynamic electromagnetic scattering research, the radar cross-section (RCS) of turbofan engines has attracted more and more attention. In order to solve the electromagnetic scattering characteristics of a biaxial multirotor turbofan engine, a dynamic scattering method (DSM) based on dynamic simulation and grid transformation is presented, where the static RCS of the engine and its components is calculated by physical optics and physical theory of diffraction. The results show that the electromagnetic scattering of the engine is periodic when the engine is working stably, while the rotors such as fans and turbines are the main factors affecting the dynamic electromagnetic scattering and the ducts greatly increase the overall RCS level of the engine. The proposed DSM is effective and efficient for studying the dynamic electromagnetic scattering characteristic of the turbofan engine.

Keywords: engine fan; rotor blade; electromagnetic scattering characteristic; dynamic simulation; grid transformation

1. Introduction

Stealth characteristics of aircraft engine intake and exhaust systems have been research hotspots, including infrared radiation, electromagnetic scattering, and noise [1,2]. In the face of the need for a comprehensive stealth design, the study of radar scattering characteristics of the turbofan engine is also receiving attention [3]. Therefore, the calculation of engine electromagnetic scattering has important practical significance for the stealth performance evaluation of the aircraft in the front-rear direction.

Turbofan engines usually consist of a compressor, combustion chamber, high-pressure turbine, low-pressure turbine, and exhaust system [4,5], which makes its electromagnetic scattering characteristics extremely complicated [6]. The engine intake contributes to the forward radar cross-section (RCS) of the aircraft [7] and the entire airflow duct creates complex reflections of the electromagnetic waves [8,9]. In general, physical optics (PO) [3], iterative physical optics [6], and method of moment (MOM) [10] are used to solve the surface scattering of the inlet/outlet chambers. The method of equivalent current is used to solve the edge diffraction of the cavity [11,12]. The inlet complex terminal is divided into several equivalent terminals when calculating the radar scattering area inside the air intake of the aircraft by using the mode superposition method [13,14]. The finite difference method based on the concept of generalized scattering amplitude is applied to the scattering of electromagnetic waves by a two-dimensional airfoil where the wing here uses an improved NACA 4418 airfoil [15]. These studies are based on the electromagnetic scattering calculations when the target is static [16,17], while the engine has many high-speed rotating rotor parts [18,19], which obviously has a dynamic effect [20,21] on the electromagnetic scattering characteristics of the aircraft.

For the solution and calculation of rotor parts RCS, quasi-static methods are commonly used to simulate the rotation of the rotor [22–24]. This method is suitable for low-speed rotors or a small number of rotation angle sample points [18,22], and it is not enough for high-speed rotating targets including engine blades. The difficulty with this type of problem is that the rotor rotates at high speed and its support structure is still stationary or relatively stationary, which causes the grid data of the entire model to be constantly updated. The grid transformation method is widely used to solve the electromagnetic scattering problem of the target body [25,26], which could reduce a large amount of calculation time and experimental cost [27]. Component decomposition methods, improved mesh transformation methods, and joint transformation methods are used to solve the dynamic process of the particular problems [28–30]. For turbofan engines, the fan and air intake are the main sources of scattering for the forward RCS of the aircraft [3], the high-speed rotating fan is the main part of the aircraft's forward dynamic RCS [31–33], and the low-pressure turbine and tail nozzles contribute to the tail RCS of the aircraft. It could be seen that the method of component decomposition or mesh transformation could be used to realize the high-speed rotation of each rotor of the engine in the engine cavity according to the structure and motion characteristics of the turbofan engine itself [34,35], thereby solving the dynamic electromagnetic scattering characteristics of the entire target body.

These studies focus on the static electromagnetic scattering calculations of the inlet cavity and exhaust nozzles of turbofan engines while ignoring the dynamic effects of large-scale, high-speed rotating fans, and turbines on target stealth performance. Therefore, this paper attempts to establish the dynamic principle of simulating the high-speed rotation of engine blades and relies on the component decomposition combination and the grid transformation method to solve the dynamic radar cross-section of the engine. It can be seen that studying the dynamic electromagnetic scattering of the engine has practical significance and promotion value for the stealth characteristics of important directions such as the front and rear direction of the aircraft.

The structure of this manuscript is as follows: the research method is described in Section 2, the turbofan engine models are designed and provided in Section 3, the results are presented and discussed in Section 4, and finally, this article is summarized.

2. Dynamic Scattering Method

The dynamic electromagnetic scattering of a turbofan engine is described in Figure 1, where α is the azimuth of the radar observatory, β is the elevation angle, θ_{fan} is the rotation angle of the fan, $V_{tip,\,fan}$ is the linear velocity of the fan blade tip, t is the observation time, and the subscript 0 indicates the initial state.

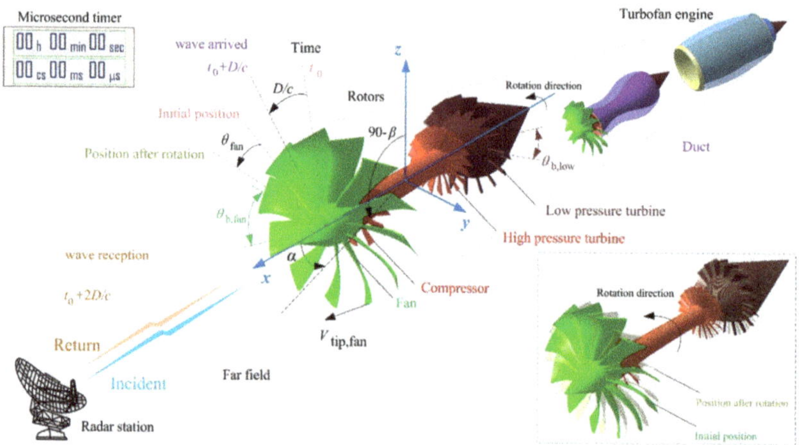

Figure 1. Schematic diagram of dynamic electromagnetic scattering of a biaxial turbofan engine.

For this biaxial turbofan engine, the fan and the low-pressure turbine share a low-speed rotating shaft, and the compressor and the high-pressure turbine share a high-speed rotating shaft, where the compressor, the high-pressure turbine and the low-pressure turbine are all in the inner duct of the engine, while the fan is located at the outer duct of the air inlet.

2.1. Electromagnetic Scattering Calculation

When the rotors of the engine are at rest, the electromagnetic scattering characteristics of the entire engine could be calculated by physical optics and physical theory of diffraction (PTD). In the physical optics method [8,18], an induced current is generated in the illuminated area when the incident wave is irradiated on the surface of the target body, while the magnetic vector position thereof could be described as follows:

$$A(r) = \frac{\mu}{4\pi}\iint\limits_{S} J_s(r')\frac{e^{-jkR}}{R}dS', \tag{1}$$

where $A(r)$ is the magnetic vector position generated by the surface-induced current, μ is the permeance coefficient, r' is the coordinate vector of the source point, and R is the distance between the field point and the source point:

$$R = \left|r - r'\right|, \tag{2}$$

where r represents the coordinate vector of the field point. According to the magnetic vector position, the electric field and the magnetic field can be, respectively, determined:

$$E(r) = \frac{1}{j\omega\varepsilon\cdot 4\pi}\iint\limits_{S'}\left[\frac{3-k^2R^2+j3kR}{R^5}e^{-jkR}R\times(R\times J_s(r')) + 2J_s(r')\frac{1+jkR}{R^3}e^{-jkR}\right]dS' \tag{3}$$

$$H(r) = \frac{1}{4\pi}\iint\limits_{S'}\frac{-1-jkR}{R^3}e^{-jkR}(R\times J_s(r'))dS', \tag{4}$$

where ω is the electromagnetic wave angular frequency and ε is the dielectric permittivity.

According to the assumptions of physical optics, there are:

$$J_s = \begin{cases} 2n\times H & ,Z_i \\ 0 & ,Z_d \end{cases}, \tag{5}$$

where n is expressed as the unit normal vector of the normal direction of r' at the surface of the scatterer, Z_i is the illuminated area, and Z_d is the dark area. According to the mirror principle:

$$E^s(r) = \frac{-k^2}{j\omega\varepsilon\cdot 2\pi}\iint\limits_{S'}\hat{R}\times\left[\hat{R}\times\left(\hat{n}(r')\times H^i(r')\right)\right]\frac{e^{-jkR}}{R}dS'. \tag{6}$$

When the incident wave is a plane wave, there are:

$$E^i(r') = |E_0|e^{-jk\cdot r'}, \tag{7}$$

where k refers to the wave vector. The resulting electric field formula is as follows:

$$E^s(r) = \frac{j}{\lambda r}|E_0|e^{-jk\cdot r}\iint\limits_{S'}\hat{r}\times\left\{\hat{r}\times\left[(\hat{n}(r')\cdot E_0)\,\hat{k} - \left(\hat{n}(r')\cdot\hat{k}\right)E_0\right]\right\}e^{-jk(-\hat{r}+\hat{k})\cdot r'}dS'. \tag{8}$$

The integral term could be written as:

$$I = \iint\limits_{S'}\hat{r}\times\left\{\hat{r}\times\left[(\hat{n}(r')\cdot E_0)\,\hat{k} - \left(\hat{n}(r')\cdot\hat{k}\right)E_0\right]\right\}e^{-jk(-\hat{r}+\hat{k})\cdot r'}dS'. \tag{9}$$

Then, we get the following RCS expression:

$$\sigma = \frac{4\pi}{\lambda^2}|I|^2. \tag{10}$$

When there are many edges or sharp points in the target, it is necessary to assist in solving the related theory of edge diffraction. The actual scattering field is the sum of the contribution of the physical optics method and the result of PTD, namely:

$$J_S = J_{PO} + J_{PTD}. \tag{11}$$

More statements about PTD could be found in the literature [8,18].

2.2. Dynamic Simulation Method

When the rotors of the engine begin to work, the dynamic simulation method is used to simulate their high-speed rotation process. Consider the linear circumferential velocity of the tip of each rotor blade of the engine:

$$\begin{cases} V_{\text{tip,fan}} = \omega_l \cdot R_{\text{fan}} \\ V_{\text{tip,comp}} = \omega_h \cdot R_{\text{comp}} \\ V_{\text{tip,high}} = \omega_h \cdot R_{\text{high}} \\ V_{\text{tip,low}} = \omega_l \cdot R_{\text{low}} \end{cases} \tag{12}$$

$$V_{\text{tip,max}} = \max\left\{V_{\text{tip,fan}}, V_{\text{tip,comp}}, V_{\text{tip,high}}, V_{\text{tip,low}}\right\}, \tag{13}$$

where R_{fan} is the radius of the fan, R_{comp} is the radius of the compressor, R_{high} is the radius of the high-pressure turbine, and R_{low} is the radius of the low-pressure turbine. Each time the fan is rotated, its model will be updated as follows:

$$m_{\text{fan}} = m_{\text{r,fan}} \left| \theta_{\grave{\text{fan}}}(m_{\text{fan}}) = \theta_{t,\text{fan}}(m_{\text{r,fan}}, t) \right., \tag{14}$$

where m_{fan} is the model of the fan, $m_{\text{r,fan}}$ is the fan model after rotation, and $\theta_{t,\text{fan}}$ is the rotating real-time angle of the fan. The rotation angle is a function of time:

$$\begin{cases} \theta_{\text{l}}(t) = \frac{180}{\pi}\omega_{\text{l}} \cdot t \\ \theta_{\text{h}}(t) = \frac{180}{\pi}\omega_{\text{h}} \cdot t \end{cases}. \tag{15}$$

The rotational motion of the engine blades has periodic characteristics, so consider a base pass time here, taking the fan as an example:

$$t_{\text{base,fan}} = \frac{\theta_{\text{fan}}}{\omega_{\text{l}}} \cdot \frac{\pi}{180}, \quad \theta_{\text{fan}} = \frac{360}{N_{\text{fan}}}, \tag{16}$$

where $t_{\text{base,fan}}$ is the base pass time during which the fan blade passes the angle between adjacent blades when the fan rotates. Consider the propagation time of electromagnetic waves:

$$\begin{cases} t_{\text{prop}} = D/c \\ \varepsilon = \min\left\{\left|t_{\text{base,fan}} - t_{\text{prop}}\right|, \left|t_{\text{base,comp}} - t_{\text{prop}}\right|, \left|t_{\text{base,high}} - t_{\text{prop}}\right|, \left|t_{\text{base,low}} - t_{\text{prop}}\right|\right\} \end{cases}, \tag{17}$$

where t_{prop} is the time from emitting electromagnetic waves to electromagnetic waves reaching the blade surface. ε is used to measure the difference between the basic passage time of each rotor and the electromagnetic wave propagation time. The smaller the ε, the more the dynamic RCS needs to be considered. After the time t, the model of the engine could be updated as:

$$\begin{cases} m_{\text{fan}} = m_{\text{r,fan}}|t = t_{\text{r}} \in [0, T_{\text{obs}}] \\ m_{\text{engine}} = \left[m_{\text{fan}}, m_{\text{comp}}, m_{\text{high}}, m_{\text{low}}, m_{\text{duct}}\right] \end{cases}, \tag{18}$$

where m_{engine} is the model of the engine and t_{r} is the real observation time when the rotors rotate.

$$t_{\text{base,max}} = \max\left\{t_{\text{base,fan}}, t_{\text{base,comp}}, t_{\text{base,high}}, t_{\text{base,low}}\right\} \tag{19}$$

$$T_{\text{obs}} = N_{\text{obs}} \cdot t_{\text{base,max}}, \tag{20}$$

where N_{obs} is a custom observation time multiple, which defaults to 1. The larger the N_{obs}, the more dynamic RCS changes can be monitored in order to find periodicity.

2.3. Grid Transformation Method

For the engine model at each observation time, the grid matrix of the whole model could be generated as follows:

$$\boldsymbol{M}_{\text{engine}} = \left[\boldsymbol{M}(m_{\text{fan}}); \boldsymbol{M}(m_{\text{comp}}); \boldsymbol{M}(m_{high}); \boldsymbol{M}(m_{low}); \boldsymbol{M}(m_{duct})\right], \tag{21}$$

where M_{engine} is the grid coordinate matrix of the engine model. When the rotors rotate around the x-axis, the real-time model of the fan could be updated to:

$$\boldsymbol{M}(m_{\text{fan}})\Big|_{t=t_{\text{r}}} = \begin{bmatrix} 1 & 0 & 0 \\ 0 & \cos\theta_{\text{r,fan}} & -\sin\theta_{\text{r,fan}} \\ 0 & \sin\theta_{\text{r,fan}} & \cos\theta_{\text{r,fan}} \end{bmatrix} \cdot \boldsymbol{M}(m_{\text{fan}})\Big|_{t=0}, \quad \theta_{\text{r,fan}} = \theta_{\text{l}}(t = t_{\text{r}}). \tag{22}$$

The other rotors model could be generated as follows:

$$\boldsymbol{M}(m_{\text{comp}})\Big|_{t=t_{\text{r}}} = \begin{bmatrix} 1 & 0 & 0 \\ 0 & \cos\theta_{\text{r,comp}} & -\sin\theta_{\text{r,comp}} \\ 0 & \sin\theta_{\text{r,comp}} & \cos\theta_{\text{r,comp}} \end{bmatrix} \cdot \boldsymbol{M}(m_{\text{comp}})\Big|_{t=0}, \quad \theta_{\text{r,comp}} = \theta_{\text{h}}(t = t_{\text{r}}) \tag{23}$$

$$\boldsymbol{M}(m_{\text{high}})\Big|_{t=t_{\text{r}}} = \begin{bmatrix} 1 & 0 & 0 \\ 0 & \cos\theta_{\text{r,high}} & -\sin\theta_{\text{r,high}} \\ 0 & \sin\theta_{\text{r,high}} & \cos\theta_{\text{r,high}} \end{bmatrix} \cdot \boldsymbol{M}(m_{\text{high}})\Big|_{t=0}, \quad \theta_{\text{r,high}} = \theta_{\text{h}}(t = t_{\text{r}}) \tag{24}$$

$$\boldsymbol{M}(m_{\text{low}})\Big|_{t=t_{\text{r}}} = \begin{bmatrix} 1 & 0 & 0 \\ 0 & \cos\theta_{\text{r,low}} & -\sin\theta_{\text{r,low}} \\ 0 & \sin\theta_{\text{r,low}} & \cos\theta_{\text{r,low}} \end{bmatrix} \cdot \boldsymbol{M}(m_{\text{low}})\Big|_{t=0}, \quad \theta_{\text{r,low}} = \theta_{\text{l}}(t = t_{\text{r}}), \tag{25}$$

where the duct model of the engine is considered to be stationary, so its mesh matrix is not transformed.

The dynamic scattering method (DSM) is verified by PO+MOM/MLFMM (multilevel fast multipole method) in FEKO as shown in Figure 2, where f_{R} is the radar wave frequency. The RCS-α curve determined by DSM is generally consistent with the results of FEKO, where the mean value of the DSM RCS-α curve is 1.59 dBm^2 smaller than that of FEKO and there are errors in some local peaks, because different mesh types and RCS algorithms are used. The quasi-static principle (QSP) is used to discretize the high-speed rotation state of the fan at a few time sample points. The calculation results show that the DSM values in the given time range are consistent with the discrete calculation values, showing that DSM is accurate and efficient.

The above calculation results also show that the RCS of the individual fan has changed more than 21.92 dBm^2 in a very short time, so it is necessary to study the dynamic electromagnetic scattering of the engine blade. More information about the RCS algorithm test is shown in Figure A1. In the following calculations, if not specified, the radar wave frequency is always 6 GHz, horizontal polarization.

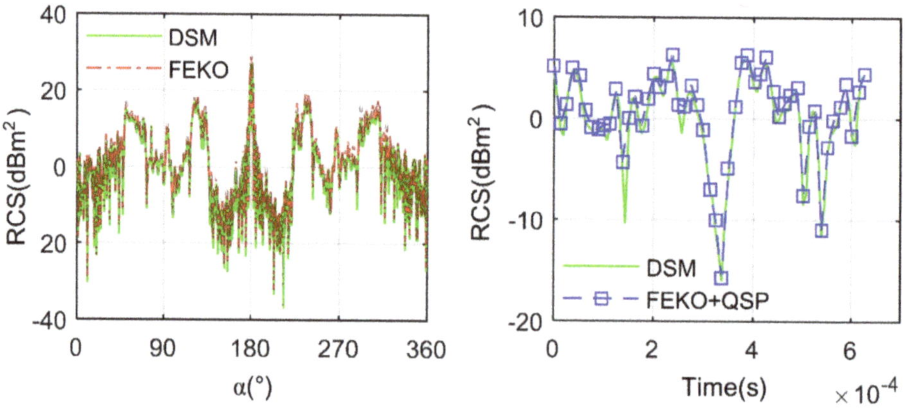

(a) $t = 1.25 \times 10^{-4}$ s (b) $\alpha = 30°$

Figure 2. Verification of the dynamic scattering method (DSM) with m_{fan}, $\beta = 0°$, and $f_R = 6$ GHz.

3. Model

The geometric model of a turbofan engine is conceived by the full factorial design (FFD) as shown in Figure 3, where L_{duct} is the length of the duct, A_{cut} is the cutting angle of the end face of the duct inlet, R_{duct} is the outer radius of the duct, noting that the duct includes the outer duct, the inner ducts, and the associated support components.

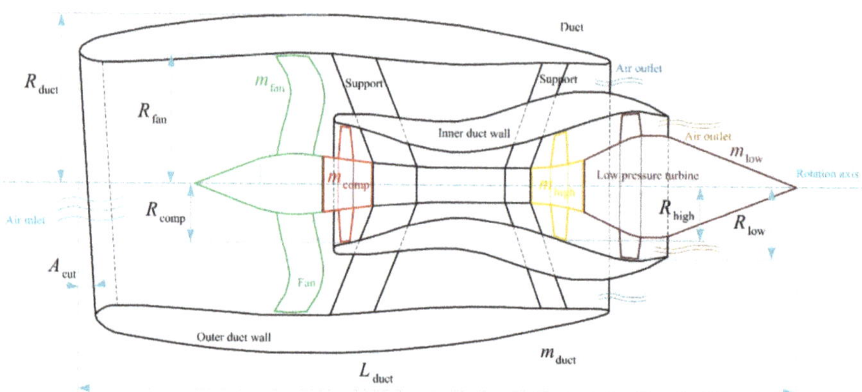

Figure 3. Geometric configuration of a turbofan engine with full factorial design (FFD).

The geometric characteristics of the engine are shown in Table 1, where the four rotors are designed with different airfoil data, including AG 08, AG 25, AH 21, and ARA-D 10%. The outer duct of the entire engine is approximately a spiral body, because the design of A_{cut} makes the upper end of the engine casing entrance slightly longer than the lower end. The number of blades of these rotors is different, which also leads to different base pass times.

Table 1. Main geometric characteristics of the turbofan engine.

Model	m_{fan}	m_{comp}	m_{high}	m_{low}	m_{duct}
Airfoil	AG 08	AG 25	AH 21	ARA-D 10%	—
Outer radius/m	1.485	0.630	0.563	0.806	1.946
Number of blades	12	10	18	16	—

Based on the above geometric design and related parameters, the entire engine model is built as shown in Figure 4, where the total length of the engine is 8.501 m. The engine model consists of the duct, the fan, the compressor, the high-pressure turbine, and the low-pressure turbine. The head direction of the engine points in the positive direction of the x-axis, and the central axis of rotation coincides with the x-axis.

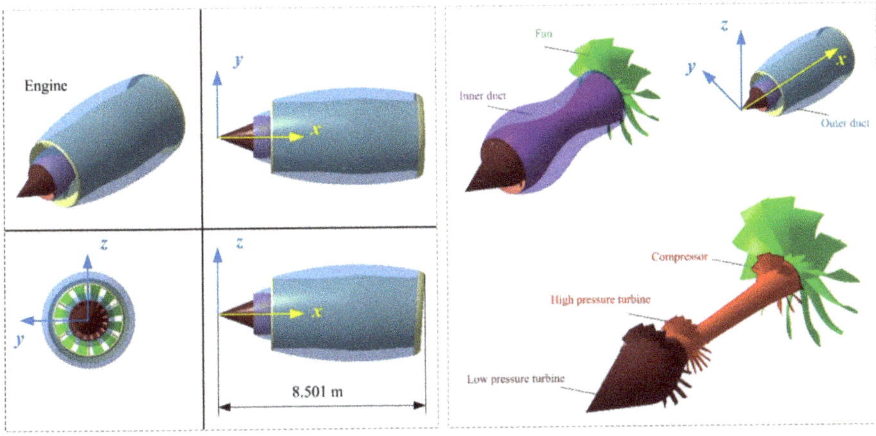

Figure 4. Model building and detail display of the turbofan engine.

The low-speed shaft of the turbofan engine is set to 8000 r/min and the high-speed shaft to 10,000 r/min as shown in Table 2 when the engine is working properly. The duct is set to be stationary, while the fan and the low-pressure turbine rotate around the low-speed shaft, and the compressor and the high-pressure turbine rotate around the high-speed shaft.

Table 2. Motion parameters of the rotors of the turbofan engine.

Model	m_{fan}	m_{comp}	m_{high}	m_{low}
Rotating speed/r/min	8000	10,000	10,000	8000

Fine meshing of this engine and its components is shown in Figure 5, where the duct model is composed of the outer duct, the inner duct, and associated support structures. The size data of each part when determining the high-precision mesh is shown in Table 3, where the global minimum mesh size has a maximum value of 2 mm, and local density increase technology of the grid is used where the geometric size changes dramatically, including the leading and trailing edge regions of each rotor.

Table 3. Grid size for each part of the turbofan engine.

Area	Maximum Value/mm	Area	Maximum Value/mm
Global minimum size	2	Fan blade trailing edge	5
Fan blade leading edge	8	Fan blade	25
Fan hub	35	Blade trailing edge of m_{comp}	4
Blade leading edge of m_{comp}	10	Blade of m_{comp}	25
Hub of m_{comp}	25	Blade trailing edge of m_{high}	5
Blade leading edge of m_{high}	10	Blade of m_{high}	25
Hub of m_{high}	25	Blade trailing edge of m_{low}	6
Blade leading edge of m_{low}	10	Blade of m_{low}	25
Hub of m_{low}	30	Support	50
Inner duct	120	Outer duct	150

Figure 5. Surface mesh distribution of various components of the turbofan engine.

4. Results and Discussion

Figure 6 supports that the electromagnetic scattering effect of the fan surface also changes in a very short time (from $t = 2.083 \times 10^{-4}$ s to $t = 4.167 \times 10^{-4}$ s). In the irradiated area of the radar wave, the upper surface of most of the blades shows a deep red color, indicating that this area is a strong source of electromagnetic scattering. The hub portion presents a lighter red color because the head has a conical design that deflects the radar wave to a nonthreatening direction. The blade root forms a certain occlusion of the joint between the hub and the blade, so a small amount of orange yellow appears here. These results show that DSM could well simulate the rotation of the blade and calculate its electromagnetic scattering characteristics.

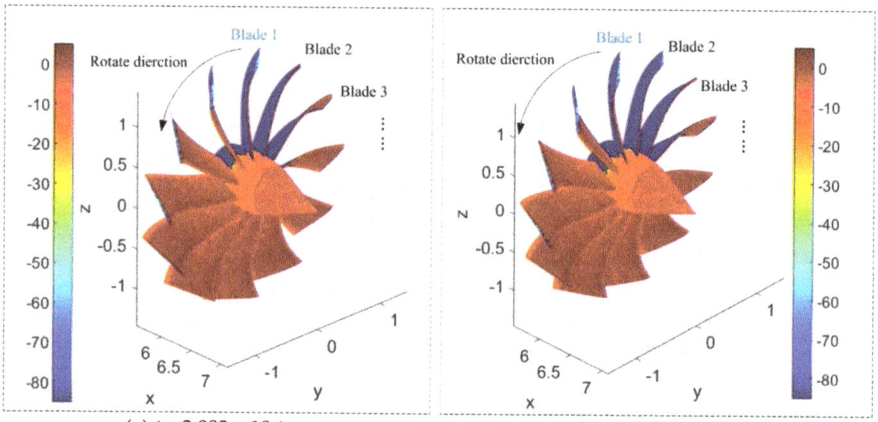

(**a**) $t = 2.083 \times 10^{-4}$ s (**b**) $t = 4.167 \times 10^{-4}$ s

Figure 6. Surface scattering distribution of the fan, $\alpha = 0°$, $\beta = 0°$, and $f_R = 6$ GHz; RCS unit: dBm^2.

Blade rotation is a continuous and long-term process when the turbofan engine is running normally. Therefore, the dynamic electromagnetic scattering characteristics of the various components of the engine and its entirety require a lot of calculation and analysis as shown below.

4.1. *Effect of Individual Rotor*

Figure 7 shows that there is an obvious periodicity in the electromagnetic scattering characteristics of the fan observed in a long time. It could be seen that the period of dynamic electromagnetic scattering of the fan is 6.25×10^{-3} s, which is also the basic pass time because the fan uses a uniform distribution of blade design and the rotating speed remains the same. The maximum RCS of the fan in one observation period is 4.928 dBm^2 appearing at $t = 1.875 \times 10^{-5}$ s, while the minimum size is −17.35 dBm^2 appearing at $t = 5 \times 10{-}4$ s. This shows that the DSM could capture the electromagnetic scattering characteristics of the fan well for a long time, and the rotation of the fan will have a greater impact on this characteristic.

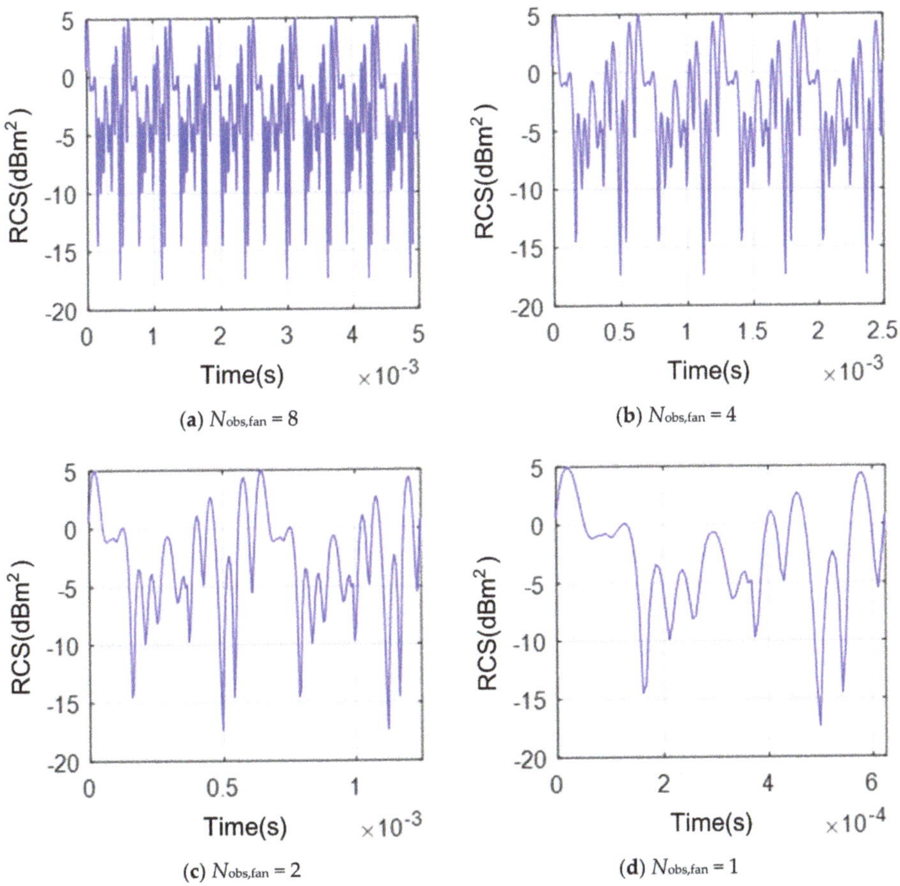

(a) $N_{obs,fan} = 8$ **(b)** $N_{obs,fan} = 4$ **(c)** $N_{obs,fan} = 2$ **(d)** $N_{obs,fan} = 1$

Figure 7. Influence of observation time on dynamic RCS of m_{fan}, $\alpha = 20°$ and $\beta = 0°$.

Figure 8 reveals that the electromagnetic scattering characteristics of the low-pressure turbine also exhibit periodicity within the given observation time. The dynamic RCS period of the low-speed turbine is still 6.25×10^{-4} s because the low-pressure turbine and the fan share a low-speed shaft. The minimum value of the RCS-time curve of the low-pressure turbine is −26.35 dBm^2 at $t = 6.25 \times 10^{-6}$ s, while the maximum value is −5.983 dBm^2 at $t = 5.937 \times 10^{-4}$ s. These results show that the observation time has a non-negligible influence on the dynamic electromagnetic scattering of the low-pressure turbine.

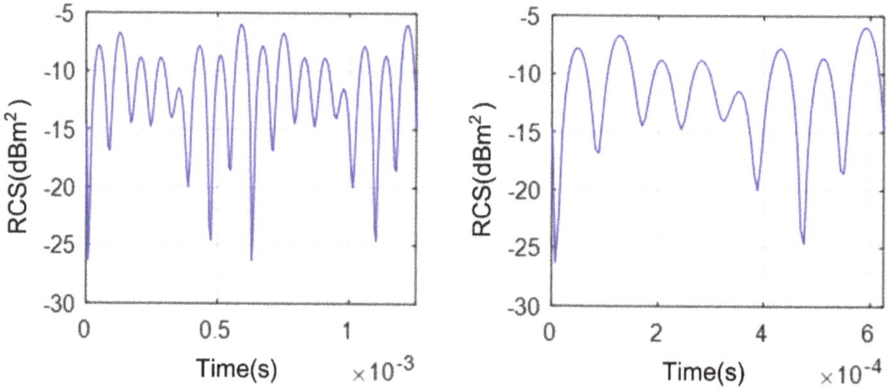

(**a**) $N_{obs,low} = 2$ (**b**) $N_{obs,low} = 1$

Figure 8. Influence of observation time on dynamic radar cross-section (RCS) of m_{low}, $\alpha = 160°$ and $\beta = 0°$.

As the observation time increases, the RCS of the rotor on the low-speed axis of the engine changes periodically, while the degree of undulation is large, which will affect the electromagnetic scattering characteristics of the forward and tail directions of the target.

4.2. Effect of Azimuth Angle

Figure 9 indicates that the dynamic RCS of the fan at different azimuths is very different, including shape, peak, and range of variation. For the RCS-time curve with $\alpha = 0°$, the RCS value is almost always maintained at 1.708 dBm^2, because the fan rotating plane is perpendicular to the radar wave at this time, causing the electromagnetic scattering change caused by the rotation to be too small to be found in the current observation scale. When $\alpha = 10°$, the overall change of RCS-time curve is relatively flat, but a peak of up to 17.04 dBm^2 appears at $t = 2.188 \times 10^{-4}$ s, because the fan rotates 10.5024° at this time and the fan adopts a gradual twist and a variable chord design, this makes the roots of the blades in the upper left part and the leading edges of other blades become strong scattering sources. When α increases from 10° to 50°, the number of fluctuations of the RCS-time curve is obviously increased, and the variation law is also different, because the angle between the normal direction of the rotating surface of the fan and the radar wave is getting bigger and bigger, while the electromagnetic scattering caused by the blades and the hub is getting stronger. The two RCS-time curves of $\alpha = 70°$ and 80° are similar in shape, while the RCS-time curve with $\alpha = 60°$ shows a peak of up to 35.48 dBm^2 at $t = 6.188 \times 10^{-4}$ s, since the fan rotates 35.48° at this time, the upper-right part of the upper surface of the blade makes the main contribution to the peak. It should also be noted that the RCS-time curve at $\alpha = 290°$ shows a peak with a size of 65.86 dBm^2 at $t = 5.125 \times 10^{-4}$ s, because the incident wave is almost perpendicular to the left contour of the front of the hub at this time, the outer end faces of the left blades, the side of the hub, and the root surface of the blades in the illuminated area all have larger RCS. These results indicate that the azimuth has a significant impact on the dynamic RCS of the fan.

Figure 10 manifests that as the azimuth increases, the RCS-time curve of the fan gradually becomes stable overall within the given azimuth range. The RCS-time curve at $\alpha = 320°$ generally shows a wave shape that decreases first and then increases, while the peak value is 8.738 dBm^2 appearing at $t = 1.25 \times 10^{-5}$ s. The minimum value of the RCS-time curve at $\alpha = 330°$ is −16.55 dBm^2 but the peak value of the RCS-time curve at $\alpha = 350°$ is as high as 43.2 dBm^2. When α increases from 320° to 360°, the angle between the normal of the rotating plane of the fan and the radar wave is decreases, which makes the strong scattering source of the fan surface gradually decrease within the same observation scale.

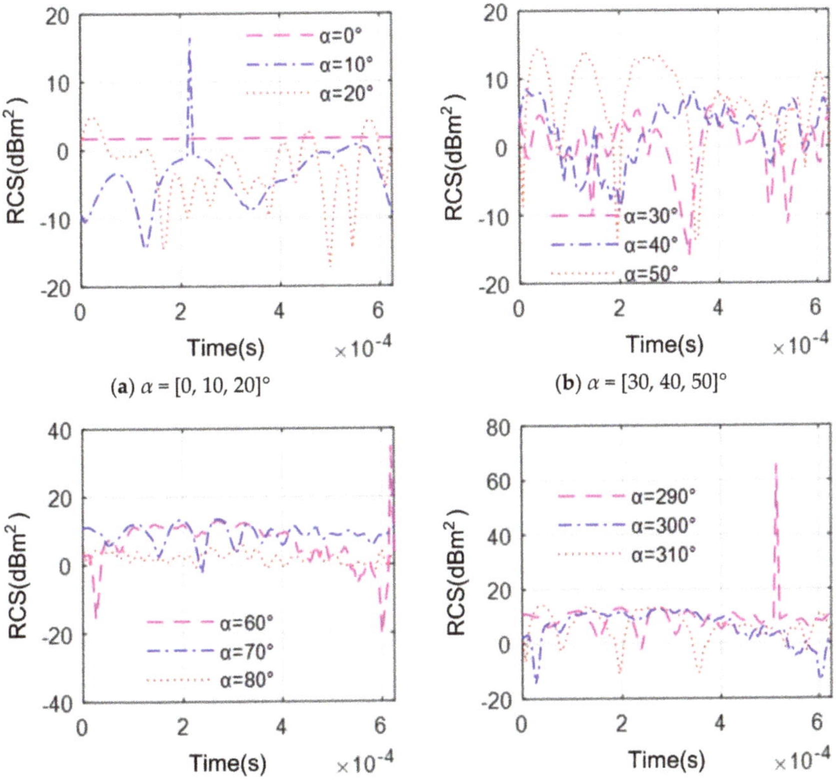

(a) α = [0, 10, 20]°
(b) α = [30, 40, 50]°
(c) α = [60, 70, 80]°
(d) α = [290, 300, 310]°

Figure 9. Influence of azimuth on dynamic RCS of m_{fan}, $N_{obs,fan} = 1$ and $\beta = 0°$.

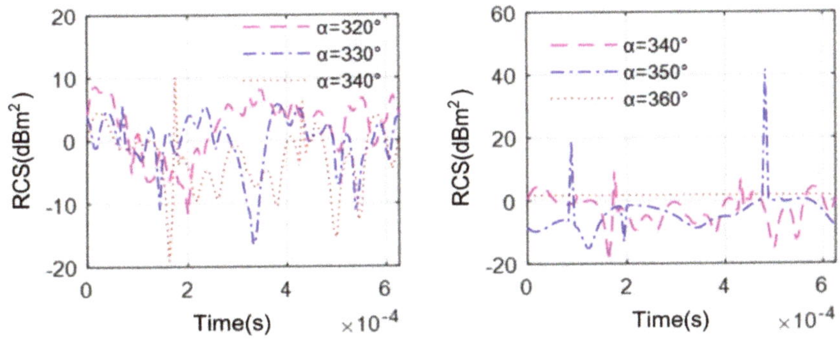

(a) α = [320, 330, 340]°
(b) α = [340, 350, 360]°

Figure 10. Influence of azimuth on dynamic RCS of m_{fan}, $\alpha = 320°$~$360°$, $N_{obs,fan} = 1$, and $\beta = 0°$.

Table 4 presents that azimuth has a very significant effect on the RCS mean of the fan within one observation period. When α increases from −10° to 80°, the RCS mean in one observation period of the fan decreases first and then increases, where the maximum difference is 24.115 dBm^2. The change in RCS within ±10° azimuths of the head indicates that the dynamic electromagnetic scattering of the fan will affect the RCS at important azimuth angles in the head direction of the engine.

Table 4. Influence of azimuth on radar cross-section (RCS) mean of m_{fan}, $N_{obs,fan} = 1$, $\beta = 0°$.

$\alpha/°$	−10	0	10	20	30
RCS mean/dBm^2	23.181	1.709	−0.039	−0.934	1.427
$\alpha/°$	40	50	60	70	80
RCS mean/dBm^2	3.883	9.056	16.284	9.748	2.518

Figure 11 demonstrates that the RCS-time curves of the low-pressure turbine at different azimuths vary widely, including peaks, ranges of changes, and number of fluctuations. For the RCS-time curve at $\alpha = 150°$, the peak value is only −2.155 dBm^2 at $t = 4.594 \times 10^{-4}$ s. When α is increased by 10°, the RCS-time curve changes in a similar pattern, while the RCS-time curve remains substantially straight (RCS equals 6.799 dBm^2) at the current observation scale when α is increased to 180°, for the reason that the normal direction of the rotating plane of the low-pressure turbine is perpendicular to the radar wave, while the size of the hub is larger than that of the blade, which will form strong scattering and cover the RCS changes caused by the rotation of the blades. When α continues to increase to 210°, the shape of its RCS-time curve is similar to that at $\alpha = 150°$, but the peak becomes −2.808 dBm^2 at $t = 3.281 \times 10^{-5}$ s. These results show that the effect of azimuth on the low-pressure turbine is also evident.

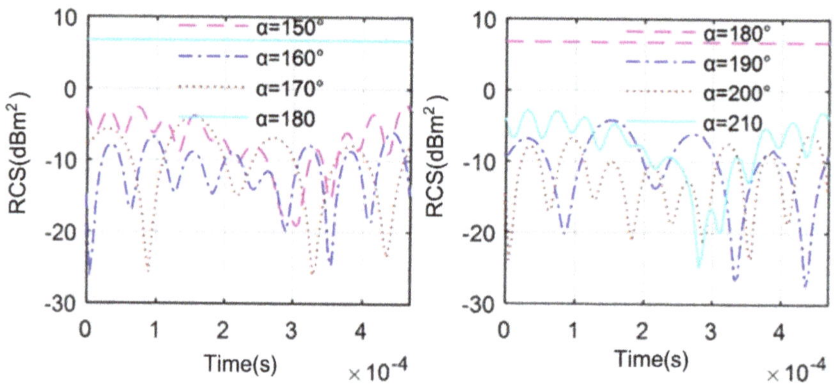

(**a**) α = [150, 160, 170, 180]° (**b**) α = [180, 190, 200, 210]°

Figure 11. Influence of azimuth on dynamic RCS of m_{low}, $N_{obs,low} = 1$, $\alpha = 150°$~$210°$, and $\beta = 0°$.

It could be found that although the outer diameter of the low-pressure turbine is smaller than that of the fan, the RCS generated by the low-pressure turbine in the positive tail direction is larger than the RCS generated by the fan in the positive head direction.

4.3. Effect of Elevation Angle

Figure 12 investigates that the dynamic RCS-time curves of low-pressure turbines at different elevation angles vary widely, including curve shape and minimum and maximum position. When $\alpha = 160°$, the increase in the elevation angle will significantly increase the number of fluctuations in the RCS-time curve. For the RCS curves of $\alpha = 200°$, the curves at $\beta = 0°$ and 20° are similar in shape but the undulation strength is somewhat different, where the former peak is −6.064 dBm^2 at $t = 4.453 \times 10^{-4}$ s and the latter has a peak value of −7.715 dBm^2 at $t = 3.750 \times 10^{-5}$ s. The increase in the elevation angle directly increases the angle between the normal direction of the rotating plane of the low-pressure turbine and the radar wave, which will result in a strong scattering source on the hub and part of the blade surface. These results show that DSM could well demonstrate the effect of elevation angle change on the dynamic RCS of the low-pressure turbine and the change at this time is also not negligible.

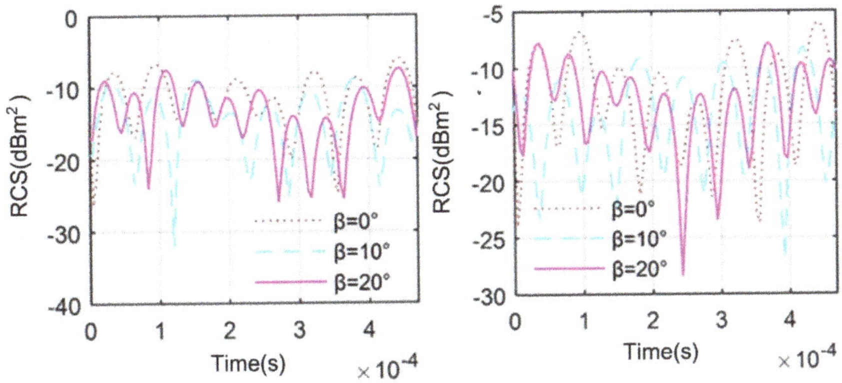

(a) $\alpha = 160°$ (b) $\alpha = 200°$

Figure 12. Influence of elevation angle on dynamic RCS of m_{low}, $N_{\text{obs,low}} = 1$ and $\beta = (0, 10, 20)°$.

Figure 13 presents that the effect of the elevation angle on the dynamic RCS of the low-pressure turbine is more pronounced under the current azimuths, including the aspects of minimum values, peak values, and variation laws. For the RCS-time curve at $\alpha = 150°$, the increase in β significantly increases the fluctuation of the RCS curve, where the minimum values of RCS at $\beta = 0°$, $10°$, and $20°$ are −19.16, −22.25, and −32.06 dBm^2, respectively. When $\alpha = 150°$, the shape of the RCS-time curve at $\beta = 20°$ still appears to change too much, where the maximum value is 3.063 dBm^2 at $t = 1.641 \times 10^{-4}$ s and the minimum value is −30.36 dBm^2 at $t = 9.375 \times 10^{-6}$ s; the reason for the negative peak here is mainly the fact that the trailing edge of the blade in the illuminated area, the upper surface of the right blade, and the head of the propeller hub has a good effect of deflecting radar waves. These results indicate that the engine will have different stealth responses according to the elevation angle under current conditions.

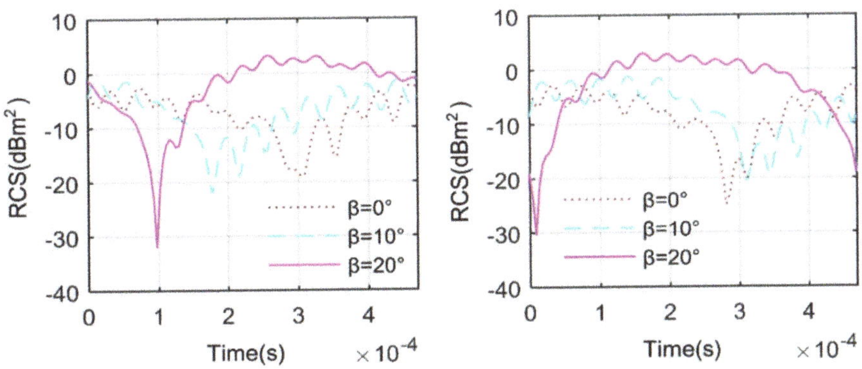

(a) $\alpha = 150°$ (b) $\alpha = 210°$

Figure 13. Influence of elevation angle on dynamic RCS of m_{low}, $N_{\text{obs,low}} = 1$ and $\beta = (0, 10, 20)°$.

Table 5 supports that an increase in the angle of the elevation will have a significant impact on the RCS mean of the low-pressure turbine at the given azimuths. When $\alpha = 150°$ and $210°$, an increase in β will result in an increase in the RCS mean. However, when $\alpha = 160°$ and $200°$, the increase in β will cause the RCS mean to decrease and then increase. This shows that choosing the right azimuth and elevation angle is beneficial to reduce the electromagnetic scattering intensity of the low-pressure turbine.

Table 5. Influence of elevation angle on RCS mean of m_{low}, $N_{obs,low} = 1$, RCS unit: dBm^2.

$\alpha/°$	150	160	200	210
$\beta = 0°$	−6.316	−10.437	−10.862	−6.649
$\beta = 10°$	−5.450	−13.122	−12.813	−5.474
$\beta = 20°$	−0.460	−12.180	−11.966	−0.496

Overall, when approaching the positive tail direction, the RCS index of the low-pressure turbine generally decreases with the increase in β, noting that $\alpha = 180°$ is the positive tail direction.

4.4. Effect of Engine Duct

Figure 14 provides that the addition of engine duct greatly increases the RCS level of the combination with the fan within the given azimuth and observation time. When $\alpha = 0°$, the rotation of the fan still cannot show the change of the RCS of the assembly at the current observation scale, because the radar wave at this time is perpendicular to the normal of the rotation plane of the fan, and the scattering caused by the blade is too weak. When α is gradually increased to 20°, the fluctuation of the RCS-time curve is enhanced, including the number of fluctuations and peaks. The shape of the RCS time curve is similar when $\alpha = 30°$ and 40°, but the latter has more minimum values. In the negative azimuth range of the head direction, the RCS-time curve at $\alpha = 320°$ and 340° varies in a similar range. As the azimuth angle is further restored to the head position, the angle between the radar wave and the rotation plane of the fan gradually approaches zero, which causes the RCS to remain unchanged. Therefore, it could be seen that the RCS-time curve at $\alpha = 350°$ changes slowly overall, except for the two peaks. The RCS of the fan and duct combination at $\alpha = 360°$ is maintained at 13.66 dBm^2, which is much higher than the RCS level of the individual fan (1.708 dBm^2), because the large casing of the duct has more sources of strong scattering, such as the duct lip with a large curvature. These results indicate that the engine's duct is also an important source of strong scattering but not a dynamic source of scattering here.

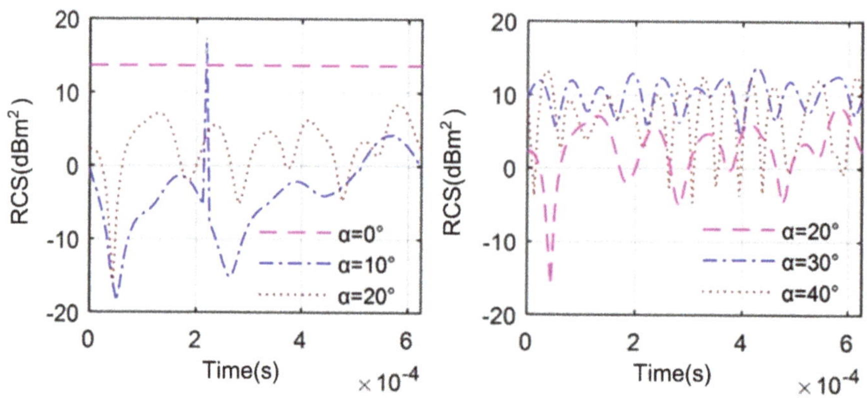

(a) $\alpha = [0, 10, 20]°$. (b) $\alpha = [20, 30, 40]°$

Figure 14. *Cont.*

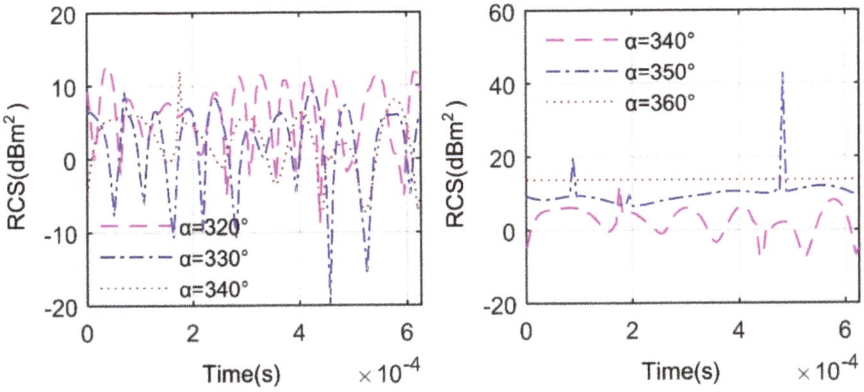

(**c**) α = [320, 330, 340]° (**d**) α = [340, 350, 360]°

Figure 14. The dynamic RCS of $m_{\text{fan}} + m_{\text{duct}}$, $N_{\text{obs,fan}} = 1$, α = (0~40 and 320~360)°, and $\beta = 0°$.

Figure 15 implicates that the RCS curve under the omnidirectional angle of the engine is obvious and cannot be ignored, while the RCS-time curves under different heading azimuths are very different. When $t = 1.875 \times 10^{-4}$ s, the peak value in the omnidirectional angle of the engine is 42.62 dBm^2 at $\alpha = 0°$, while that of the engine at $t = 3.75 \times 10^{-4}$ s is 55.41 dBm^2 at $\alpha = 215.8°$ because the trailing edges and facets of the right blades of the low-pressure turbine is not easy to deflect the incident wave at this time. The RCS peaks at α = 90° and 270° at different times appear larger and stable, which is mainly due to the mediocre design of the huge engine duct. When α increases from 0° to 20°, the dynamic RCS curve of the engine becomes more fluctuating, but the peak and RCS averages decrease, because the strong scattering source at this time is mainly reflected in the duct of the engine, the azimuth angle increases, and the strong scattering source on the lip in the illumination area is transferred to the side of the duct. These results show that the omnidirectional RCS of the engine changes dynamically with time, and the difference of dynamic electromagnetic scattering at different azimuths is obvious.

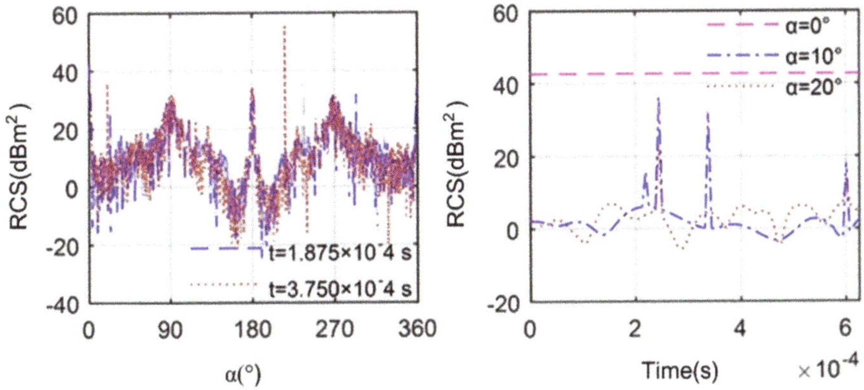

(**a**) t = [1.875, 3.750] × 10^{-4} s (**b**) α = [0, 10, 20]°

Figure 15. The dynamic RCS of m_{engine}, $N_{\text{obs}} = 1$ and $\beta = 0°$.

Figure 16 suggests that the engine duct has an important effect on the electromagnetic scattering of the entire engine in the current observation azimuth and time. For the engine scattering characteristics at $t = 1.875 \times 10^{-4}$ s, it could be found that there are more dark red areas in the lip of the inlet, and more red areas are distributed on the right side of the duct. The left side of the fan and the inner side of the left side of the duct show more orange red. When t is increased to 3.75×10^{-4} s, the light and dark areas of the engine duct surface and its color distribution hardly change, while the color distribution of the fan surface has changed slightly and the least obvious is the blade color of the tail of low-pressure turbine, because the fan, compressor, and inner and outer ducts block more radar waves, making it difficult for them to reach the low-pressure turbine at the tail. These results indicate that the rotors inside the engine will have a dynamic effect on the electromagnetic scattering of the engine while the outer casing occupies more sources of strong scattering.

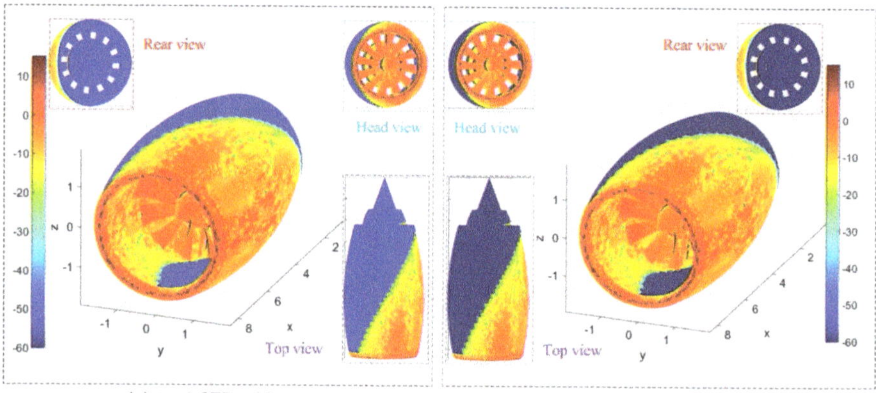

(a) $t = 1.875 \times 10^{-4}$ s **(b)** $t = 3.750 \times 10^{-4}$ s

Figure 16. Surface electromagnetic scattering characteristics of m_{engine}, $\alpha = 20°$ and $\beta = 0°$; RCS unit: dBm^2.

Overall, the dynamic electromagnetic scattering characteristics of the engine are very obvious, and the RCS is different under different azimuth and observation time. Studying the dynamic electromagnetic scattering of the engine is of great significance to the head and tail RCS characteristics of the aircraft.

5. Conclusions

This paper establishes a dynamic scattering method to investigate the electromagnetic scattering characteristics of a biaxial multirotor turbofan engine. Through these studies, the following conclusions could be drawn:

1. The electromagnetic scattering characteristics of each rotor during steady engine operation are periodic, and the period is equal to the base pass time of the rotor at the current rotating speed;
2. The azimuth and elevation angles have a large impact on the dynamic RCS of the engine. The fan has a great influence on the change of the engine head to the RCS, while the low-pressure turbine has a greater influence on the engine tail RCS;
3. The engine duct will greatly increase the RCS level of the entire engine, but will also block the scattering of its internal components. Front and rear rotors are the main influencing factors of engine dynamic RCS.

Future research work can focus on the electromagnetic scattering characteristics of the rotor at different engine speeds, the effect of engine dynamic RCS on aircraft stealth performance, and the change of aircraft attitude on engine RCS.

Author Contributions: Conceptualization and methodology, Z.Z. and J.H.; software, validation, and formal analysis, Z.Z.; investigation and resources, Z.Z. and J.H.; writing and visualization, Z.Z.; funding acquisition, Z.Z. All authors have read and agreed to the published version of the manuscript.

Funding: This work was supported by the project funded by the China Postdoctoral Science Foundation (Grant No. BX20200035).

Acknowledgments: Thanks to Jun Huang for his guidance on this article.

Conflicts of Interest: The authors declare no conflict of interest.

Nomenclature

α	the horizontal angle between radar station-engine connection and the positive x-axis
β	the elevation angle between the radar station and the engine
θ	rotation angle
θ_b	angle between adjacent blades
D	distance between the radar and the target
c	electromagnetic wave propagation speed
V_{tip}	line speed of the blade tip
σ	radar cross-section
ω_l	angular velocity of the low-speed shaft
ω_h	angular velocity of the high-speed shaft
m_{engine}	the model of the engine
M_{engine}	the grid coordinate matrix of the engine model
$\boldsymbol{t_{base}}$	base pass time
N_{fan}	number of fan blades
ε	a custom difference
T_{obs}	total observation time
t	observation time
Subscript	
fan	fan
comp	compressor
high	high-pressure turbine
low	low-pressure turbine

Appendix A

The RCS algorithm test is shown in Figure A1, where the electromagnetic scattering characteristics of the fan model are investigated under different radar wave frequencies, different numbers of grids, and different polarization modes. The RCS-α curves are generally similar at different radar wave frequencies, where RCS average increased from less than 9 to 9.805 dBm^2 as f_R increased from 2 to 12 GHz. When the number of grids is greater than 1.002×10^5, the calculation results of RCS tend to be stable. The fewer meshes, the rougher the model displayed, the faster the calculation speed, but the larger the error. The more the grids, the more accurate the model is described, the calculation speed is reduced but the results are accurate. In order to take into account the calculation speed and grid accuracy, the number of grids and the size of the grid at this time are set as the benchmark. The RCS curves under different polarization methods are almost the same, and the difference is less than 0.051 dBm^2.

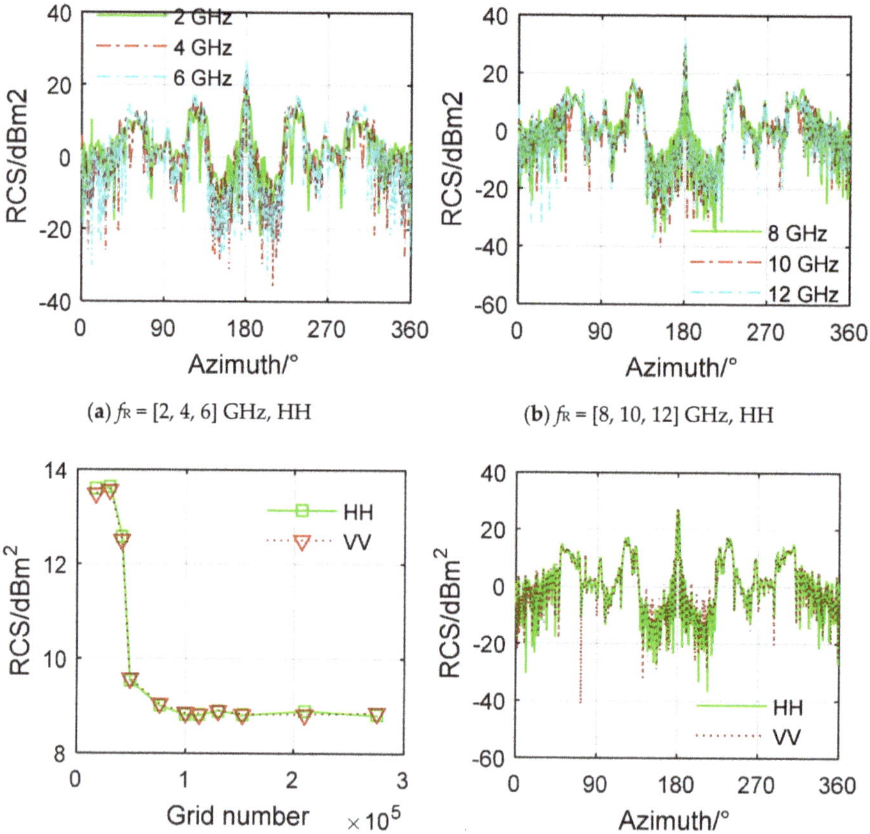

(**a**) f_R = [2, 4, 6] GHz, HH (**b**) f_R = [8, 10, 12] GHz, HH

(**c**) f_R = 6 GHz (**d**) f_R = 6 GHz

Figure A1. RCS algorithm test with m_{fan}, $\alpha = 0{\sim}360°$, $\beta = 0°$, and $t = 1.25 \times 10^{-4}$ s.

References

1. Fedele, L.; Di Vito, L.; Ramundo, F.E. Increasing Efficiency in an Aeronautical Engine through Maintenance Evaluation and Upgrades: Analysis of the Reliability and Performance Improvements under Financial Issues. *Energies* **2020**, *13*, 3059. [CrossRef]
2. Gao, X. Numerical Study of Infrared Radiation and Electromagnetic Scattering Characteristics of Aircraft/Exhaust System. Ph.D. Thesis, Northwestern Polytechnical University, Xi'an, China, 2016.
3. Yang, S.N.; Zhang, Z.X.; Shao, W.R.; Deng, H.W. Numerical Investigation on Electromagnetic Scattering Characteristics for Engine Cavity with Centrocone. *Aeroengine* **2014**, *40*, 48–53.
4. York, M.A.; Hoburg, W.W.; Drela, M. Turbofan engine sizing and tradeoff analysis via signomial programming. *J. Aircr.* **2017**, *55*, 988–1003. [CrossRef]
5. Lu, W.; Huang, G.; Xiang, X.; Wang, J.; Yang, Y. Thermodynamic and Aerodynamic Analysis of an Air-Driven Fan System in Low-Cost High-Bypass-Ratio Turbofan Engine. *Energies* **2019**, *12*, 1917. [CrossRef]
6. Li, Y.F.; Yang, Q.Z.; Gao, X.; Wu, Z.K. Investigation on Radar Cross-Section of S-Shaped Inlets Using IPO and EEC Method. *J. Propuls. Technol.* **2013**, *34*, 577–582.
7. Noh, Y.H.; Park, C.S.; Yook, J.G.; Choi, W.Y.; Jang, Y.H. Analysis for Contribution of Engine Inlet Structure to Monostatic RCS of Aircraft in the VHF Band. In Proceedings of the IEEE International Symposium on Antennas and Propagation & USNC/URSI National Radio Science Meeting, Boston, MA, USA, 8–13 July 2018; pp. 2335–2336.

8. Zhou, Z.Y.; Huang, J.; Wu, N.N. Acoustic and radar integrated stealth design for ducted tail rotor based on comprehensive optimization method. *Aerosp. Sci. Technol.* **2019**, *92*, 244–257. [CrossRef]
9. Chen, J.J.; Huang, J.W.; Xu, J.D.; Pang, C.S.; Li, J.K. Calculation of attitude angle in dynamic radar targets electromagnetic scattering. *J. Proj. Rocket. Missiles Guid.* **2010**, *30*, 183–185.
10. Tasic, M.; Kolundzija, B.; Milosevic, T. Domain decomposition method for scattering from an aircraft with jet engine inlet cavity. In Proceedings of the International Applied Computational Electromagnetics Society Symposium (ACES), Denver, CO, USA, 25–29 March 2018; pp. 1–2.
11. Aiello, G.; Alfonzetti, S.; Rizzo, S.A.; Salerno, N. Multi-Objective Optimization of Thin-Film Silicon Solar Cells with Metallic and Dielectric Nanoparticles. *Energies* **2017**, *10*, 53. [CrossRef]
12. Yang, S.N.; Shao, W.R.; Shang, S.T.; Deng, H.W. Study on Radar Stealth Shaping for Single Expansion Ramp with Spherical 2-D Nozzle. *Aeroengine* **2016**, *42*, 55–62.
13. Schapker, R.L. Frequency effects of electromagnetic scattering from underdense turbulent plasmas. *AIAA J.* **1970**, *8*, 1582–1590. [CrossRef]
14. Wang, R.G.; Cheng, Y.; Chen, W. Calculation of RCS of Intake by Means of Mode Superposition. *J. Air Force Eng. Univ. (Nat. Sci. Ed.)* **2004**, *5*, 21–25.
15. Ling, R.T.; Smith, T.D. Scattering of acoustic and electromagnetic waves by an airfoil. *AIAA J.* **1989**, *27*, 268–273. [CrossRef]
16. Xiao, B.; Zhang, J. Application of SBR-Calculation the RCS of an Arbitrary Shaped Open-Ended Cavity. *J. Appl. Sci.* **1993**, *11*, 328–332.
17. Chen, Q.; Huang, J.; Pan, M.; Lu, F. A Novel Real-Time Mechanism Modeling Approach for Turbofan Engine. *Energies* **2019**, *12*, 3791. [CrossRef]
18. Zhou, Z.Y.; Huang, J.; Yi, M.X. Comprehensive optimization of aerodynamic noise and radar stealth for helicopter rotor based on Pareto solution. *Aerosp. Sci. Technol.* **2018**, *82*, 607–619. [CrossRef]
19. Qin, J.; Pan, M.; Xu, W.; Huang, J. An Output-Based Limit Protection Strategy for Turbofan Engine Propulsion Control with Output Constraints. *Energies* **2019**, *12*, 4043. [CrossRef]
20. Zhou, C.H.; Qian, W.P.; Guo, Y.Q. EM Scattering Model of Dynamic Rigid Target. *J. Sichuan Armed Forces* **2015**, *36*, 138–140.
21. Sobczak, K.; Obidowski, D.; Reorowicz, P.; Marchewka, E. Numerical Investigations of the Savonius Turbine with Deformable Blades. *Energies* **2020**, *13*, 3717. [CrossRef]
22. Jiang, X.W.; Zhao, Q.J. Analysis and evaluation about radar target scattering characteristics of helicopter wing components. *J. Aerosp. Power* **2016**, *31*, 2691–2700.
23. Vijayaraghavan, D.; Keith, T.G. Grid transformation and adaption techniques applied in the analysis of cavitated journal bearings. *J. Tribol.* **1990**, *112*, 52–59. [CrossRef]
24. Song, H.Y.; Liang, L.F.; Zhou, Z.G. The complementary energy principle and generalized variational principles of quasi-static electro-magneto-thermo-elasticity. *J. Harbin Eng. Univ.* **2011**, *32*, 33–37.
25. Eby, M.; Holloway, G. Grid transformation for incorporating the Arctic in a global ocean model. *Clim. Dyn.* **1994**, *10*, 241–247. [CrossRef]
26. Song, H.Y.; Li, H.B.; Liang, L.F.; Zhou, Z.G. The second H-R generalized variational principles of quasi-static electro-magneto-thermo-elasticity. *Struct. Environ. Eng.* **2010**, *37*, 8–16.
27. Zhou, Z.Y.; Huang, J. Target head direction far field radar cross section reduction based on narrow surface element method. *Optik* **2020**, *200*, 163230. [CrossRef]
28. Chen, H.; Wang, Z.L.; Li, J.; Zhang, M.H. Launch vehicle RCS simulation and computing method based on component decomposition. *J. Spacecr. TT C Technol.* **2014**, *33*, 198–201.
29. Tomita, H. A stretched icosahedral grid by a new grid transformation. *J. Meteorol. Soc. Jpn. Ser. II* **2008**, *86*, 107–119. [CrossRef]
30. Singh, S.K.; Khanna, K.; Bose, R.; Panigrahi, B.K.; Joshi, A. Joint-transformation-based detection of false data injection attacks in smart grid. *IEEE Trans. Ind. Inform.* **2017**, *14*, 89–97. [CrossRef]
31. Chen, B.; Tong, C.M.; Li, X.M.; Liu, K.Y. Aircraft Target Recognition Based on HRRP. *J. Microw.* **2017**, *33*, 72–77.
32. Qi, Y.T.; Zhang, X.Y.; Lin, G.; Li, J.Z. A simulation method of aircraft dynamic RCS. *Chin. J. Radio Sci.* **2018**, *34*, 97–103.

33. Daroukh, M.; Moreau, S.; Gourdain, N.; Boussuge, J.F.; Sensiau, C. Tonal Noise Prediction of a Modern Turbofan Engine With Large Upstream and Downstream Distortion. *J. Turbomach.* **2019**, *141*, 021010. [CrossRef]
34. Lv, Z.Y.; Wei, P. Application of small scale grid tomography velocity modeling method in timedepth conversion of low-amplitude structural region. *Prog. Geophys.* **2020**, *1*, 1–9.
35. Pan, R.; Song, Z.; Liu, B. Optimization Design and Analysis of Supersonic Tandem Rotor Blades. *Energies* **2020**, *13*, 3228. [CrossRef]

energies

Article

Development of an Efficient Thermal Electric Skipping Strategy for the Management of a Series/Parallel Hybrid Powertrain †

Vincenzo De Bellis [1], Enrica Malfi [1,*] and Jean-Marc Zaccardi [2]

1 Dipartimento Di Ingegneria Industriale, Università Degli Studi Di Napoli Federico II, 80125 Napoli, Italy; vincenzo.debellis@unina.it
2 IFP Energies Nouvelles, Institut Carnot IFPEN Transports Energie, Rond-Point De L'échangeur de Solaize, BP 3, 69360 Solaize, France; jean-marc.zaccardi@ifpen.fr
* Correspondence: enrica.malfi@unina.it; Tel.: +39-081-768-3264
† This paper is an extended version of our paper published in 2020 WCX SAE World Congress Experience Conference, Detroit, MI, USA, 16–18 June 2020.

Abstract: In recent years, the development of hybrid powertrain allowed to substantially reduce the CO_2 and pollutant emissions of vehicles. The optimal management of such power units represents a challenging task since more degrees of freedom are available compared to a conventional pure-thermal engine powertrain. The a priori knowledge of the driving mission allows identifying the actual optimal control strategy at the expense of a quite relevant computational effort. This is realized by the off-line optimization strategies, such as Pontryagin minimum principle—PMP—or dynamic programming. On the other hand, for an on-vehicle application, the driving mission is unknown, and a certain performance degradation must be expected, depending on the degree of simplification and the computational burden of the adopted control strategy. This work is focused on the development of a simplified control strategy, labeled as efficient thermal electric skipping strategy—ETESS, which presents performance similar to off-line strategies, but with a much-reduced computational effort. This is based on the alternative vehicle driving by either thermal engine or electric unit (no power-split between the power units). The ETESS is tested in a "backward-facing" vehicle simulator referring to a segment C car, fitted with a hybrid series-parallel powertrain. The reliability of the method is verified along different driving cycles, sizing, and efficiency of the power unit components and assessed with conventional control strategies. The outcomes put into evidence that ETESS gives fuel consumption close to PMP strategy, with the advantage of a drastically reduced computational time. The ETESS is extended to an online implementation by introducing an adaptative factor, resulting in performance similar to the well-assessed equivalent consumption minimization strategy, preserving the computational effort.

Keywords: hybrid powertrain; optimization strategy; computational efficiency; energy management; fuel economy

Citation: De Bellis, V.; Malfi, E.; Zaccardi, J.-M. Development of an Efficient Thermal Electric Skipping Strategy for the Management of a Series/Parallel Hybrid Powertrain . *Energies* **2021**, *14*, 889. https://doi.org/10.3390/en14040889

Academic Editor: T M Indra Mahlia
Received: 4 January 2021
Accepted: 2 February 2021
Published: 8 February 2021

1. Introduction

Nowadays, the main purpose for vehicle manufacturers is the reduction of CO_2 and pollutant emissions. Since hybrid electric vehicles (HEVs) have shown a high potential to pursue this aim, if compared to conventional vehicles, their development is continuously improving [1]. The way forward to achieve this outcome is the realization of a more sophisticated control strategy of the powertrain. A hybrid powertrain consists of a thermal unit (internal combustion engine—ICE), coupled to one or more electric units in series and/or parallel, connected to an energy storage device, generally a battery [2]. Regardless of the hybrid architecture, a key role is played by the control strategy. Its task, once fixed, the power demand at the vehicle wheels is to determine the optimal power to be delivered/absorbed by the available units. As known, the focus for a control strategy is the minimization of the fuel consumption along a path, instead of decreasing, at each instant of time, the fuel flow rate.

Hybrid powertrains most frequently exhibit a disposition in parallel to the ICE and of an electric motor (EM). In such configurations, at each time, the control problem provides for the identification of the power-split (PS) between the units. Aiming to the more adaptable and efficient management of the phases to charge the battery, a second electric unit (electric generator—EG), linked to the thermal engine in series, is sometimes installed. Several optimization logics have been developed for HEV in order to maximize the fuel economy.

One of the most widespread approaches is the dynamic programming (DP) method [3] that solves the problem numerically to identify optimal global behavior according to the complete speed profile and topology of a driving scenario [4]. This procedure, in addition to being highly time-demanding, cannot be directly employed in a real-time application since it needs future event information. The DP, along with other methodologies, because of "a priori" knowledge of future data, is qualified as a global optimization strategy (GOS). This strategy can be effectively utilized to develop the design of a new powertrain architecture, and it is able to provide useful directions to define heuristic control strategies [5]. To get around the problem of knowing the future events, the stochastic DP algorithm was developed by determining, over different driving cycles, the driver power demand sequence on the basis of the Markov chain [6]. Nevertheless, the abovementioned DP variant is however affected by the computational matter for real-time implementation.

A different approach consists of the use of conventional analytical optimization methods to solve the problem of HEV energy management, e.g., the Pontryagin minimum principle (PMP) [7]. It relies on the instantaneous minimization of the Hamiltonian function when determined the costate optimal trajectory. This method needs to know the driving mission to achieve the energy balance condition for the battery between the start and end of the driving mission, which is a regulatory requirement for HEV certification. Further, difficulties in the PMP application arise when including state constraints in the problem definition. A possible solution is to combine the PMP with a penalty function approach [8]. The aim is to increase the Hamiltonian value whenever the optimal trajectory violates its constraints. An effective penalty approach through an implicit Hamiltonian minimization is employed in [9] with several states and inputs under mixed input–state constraint.

The application of real-time local optimal solution leads to long optimization time and computational complexity [10]. Because of these issues, to improve the efficiency of the optimization process, different solutions have been proposed, such as the application of an approximate PMP algorithm [11]. Through the introduction of a simple convex approximation to the local Hamiltonian, this strategy, before deciding on the optimal control for the powertrain, only requires the calculation and comparison of five candidate Hamiltonians. Another crucial matter of these approaches is related to the operating domain discretization of the powertrain components, relying on the conflicting requirements of fine control and computational effort.

The above-described strategies can be upgraded to an online version, overcoming the problem due to the lack of information on future events. One of the most common methodologies is the equivalent consumption minimization strategy (ECMS) [12], considered as an extension of the PMP [13]. This approach aims to minimize online equivalent fuel consumption, also taking into account a contribution associated with the battery power consumption via an equivalence factor, s_0. To realize online applications, the adequacy is achieved by an adaptive s_0, modified by a fuzzy PI controller [14], or by a correction term associated with the battery state of charge (SoC) [15]. Another option for the correction is the use of a 2-dimensional look-up table, derived from an equivalence factor optimization and then applied for real-time adjustments [16]. Once tuned, these approaches showed suboptimal performance, although quite similar to off-line approaches [17–19]. Once more, if it is required a dense discretization to explore the performance maps of powertrain subcomponents, the computational time represents a remarkable issue.

In the light of the above evidence, the main purpose of this work is to develop a simplified control strategy, simultaneously featured by performance similar to PMP/ECMS one, but extremely efficient from a computational point of view compared to the above approaches. As highlighted by the analysis of the current literature [9–11], the computational effort still remains an open point in the development of optimal control strategies, especially if those strategies are intended for real-time implementation. The computational efficiency is the most relevant aspect addressed in the development of the proposed strategy.

The power-split principle is not utilized, substituted by an alternate utilization of thermal and electric units named efficient thermal electric skipping strategy (ETESS). The choice between the traction modality, at each time, relies on the evaluation of an equivalent fuel rate considering pure electric driving, compared to actual fuel rate considering a pure ICE driving. The ETESS is employed in an "in-house developed" simulation platform and tested on a reference C-class vehicle, considering different powertrain variants and along different driving cycles. In the following, the tested HEV architecture and features are described, then the ETESS is detailed. Lastly, the outcomes of the proposed control strategy are processed and compared to PMP and ECMS methodologies, in off-line and online variants, respectively.

2. HEV Architecture

The powertrain architecture of the tested C-class vehicle is a combined series/parallel hybrid power-unit, as represented in Figure 1. The powertrain is featured by an ultra-efficient ICE (labeled as engine 1), two electric motor/generator units (EM and EG), a battery (Ba), three clutches (Cl_{1-3}) and two gearboxes (GB_{1-2}). The main data concerning the vehicle and the powertrain components are collected in Table 1 [20,21]. Due to the presence of three clutches, the powertrain can flexibly switch between series and parallel modes, excluding the mechanical connection with the components which are not used, to minimize the losses. In the series modality, the EM moves the vehicle, while in parallel mode, both thermal engine and EM are used to fulfill the power demand at the wheels. In this last case, the most common optimal control strategies involve a power-split between EM and ICE. In both series and parallel modes, the battery can be charged through the EG. The regenerative braking is made by EM.

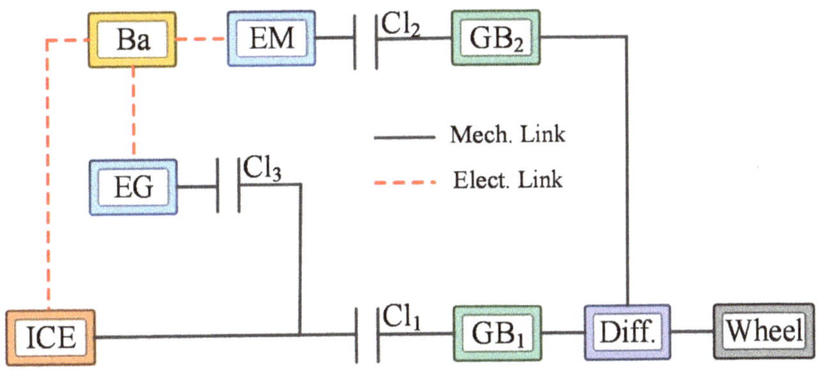

Figure 1. Powertrain schematic of the tested hybrid electric vehicles (HEV).

This work deals with a prototype vehicle, equipped by a very efficient ICE (engine 1), under development [22], and whose features are listed in Table 2. The main feature of engine 1 is to operate in ultra-lean conditions, resulting in very high efficiencies over the entire operating domain. The ultra-lean mixture is realized thanks to a two-stage boosting system composed of a variable geometry turbocharger and an E-compressor. The battery supplies the energy to move the E-compressor, as schematically represented in Figure 1 by a red dashed link which connects Ba and ICE icons.

Table 1. Main characteristics of the tested HEV.

Vehicle	
Mass, kg	1730
Car aero drag, m^2	0.775
Tire rolling resistance coef., -	0.008
Wheel diameter, m	0.723
Axle ratio, -	4.4
Axle inertia, kgm^2	1.5
Electric Generator	
Max power, kW	55
Max torque, Nm	165
Inertia, kgm^2	0.10
Electric Motor	
Max power, kW	50
Max torque, Nm	240
Inertia, kgm^2	0.10
Battery	
Internal resistance, ohm	0.375
Voltage, volt	400.0
Energy density, Wh/kg	170.0
Usable battery sizing, kWh	0.50
SoC limits, -	0.2–0.9
Gear-Box$_1$	
Gear 1 ratio, -	2.72
Gear 2 ratio, -	1.64
Gear 3 ratio, -	0.99
Gear 4 ratio, -	0.60
Gear-Box$_2$	
Gear 1 ratio, -	2.67
Gear 2 ratio, -	1.03

Table 2. Main characteristics of the engines 1 and 2.

Engine Main Specifics	Engine 1	Engine 2
Displacement, cm^3	1633.1	875.4
Max power, kW	125	62.6
Minimum BSFC, g/kWh	182	240
Inertia, kgm^2	0.35	0.29

The model-estimated brake specific fuel consumption (BSFC) map of engine 1 is reported in Figure 2a. This last also shows an intermediate black dashed line that corresponds to a smaller engine characterized by a halved rated torque. The map of the power consumption of the E-Compressor is shown in Figure 2b. The efficiency maps of EM and EG were generated with an electric motor map creation tool in Simcenter Amesim [23]. These maps, depicted in Figure 3a,b, are representative of typical synchronous electric motors under 400 V. The efficiency maps and the maximum torque are modeled assuming a perfect symmetry of the performance for motor/brake operations. This means that the maps in Figure 3a,b are also representative of efficiency with a negative torque, and the maximum absorbed torque is equal to the maximum delivered torque.

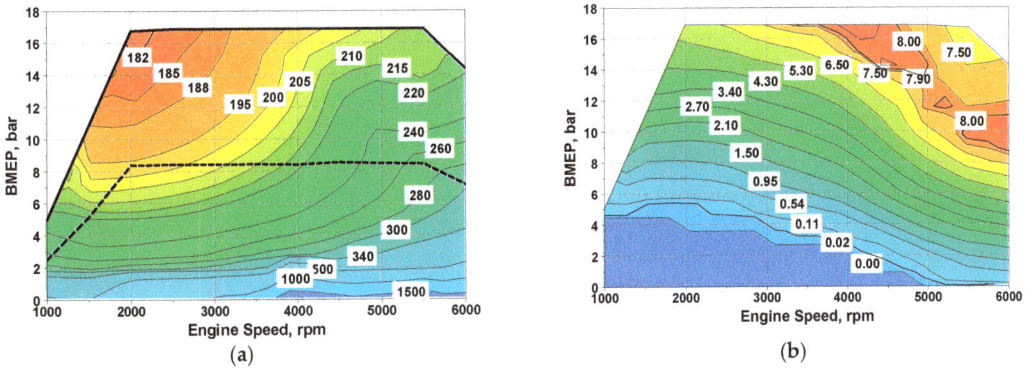

Figure 2. Maps of thermal engine 1 brake specific fuel consumption (BSFC), g/kWh, (**a**) and E-compressor power consumption, kW, (**b**).

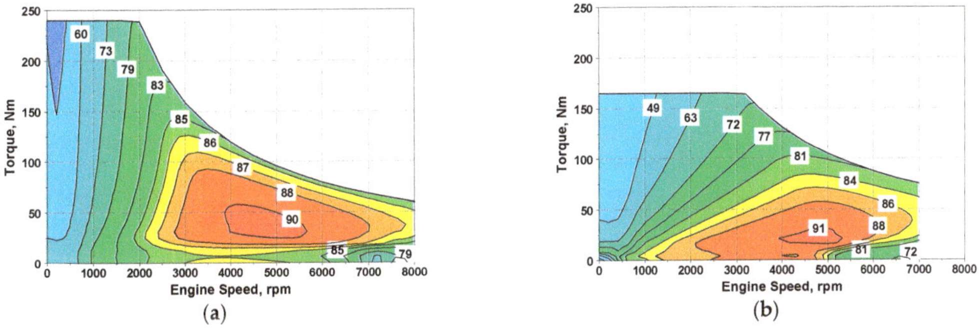

Figure 3. Efficiency maps (-) of electric motor (EM) (**a**) and electric generator (EG) (**b**).

Aiming at testing the robustness of the proposed control strategy, as said, some analyses are performed with a more conventional turbocharged downsized stoichiometric engine, labeled as engine 2 [24], whose BSFC map is shown in Figure 4.

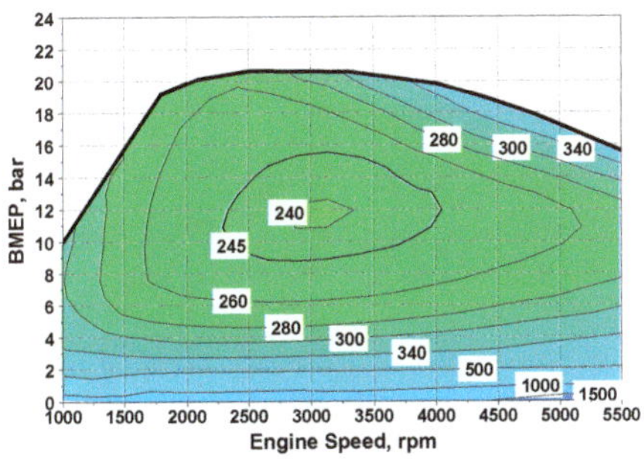

Figure 4. Thermal engine 2 BSFC map (g/kWh).

3. Vehicle and Powertrain Modeling

The simulation platform is an "in-house developed" software implemented in the Fortran language (UniNa vehicle simulation, UNVS) [21]. In this framework, the vehicle is characterized by the data listed in Table 1 (mass, aerodynamic coefficient, etc.), while each component of the powertrain and of the vehicle is defined by a lumped-parameter approach. The control is handled by a "backward-facing" (quasi-static) method [25]. The tractive demand at the wheels takes into account the inertial forces (associated with the vehicle and rotating parts), resistances (aerodynamic and rolling load) and road grade. Finally, the thermal unit and the electric motors are described by a quasi-steady map-based approach. Particularly, for the ICE, the BSFC map is implemented, storing the BSFC levels as a function of the engine rotational speed and brake mean effective pressure (BMEP), while, for the electric units, the efficiency maps are considered, the function of rotational speed and delivered/absorbed torque. For both thermal and electric units, the maximum and minimum torque curves are assigned. A linear interpolation approach is employed to extract BSFC, efficiency, maximum and minimum torque values from ICE and electric unit performance maps. A simplified SoC model is employed to describe the battery behavior, where Joule losses are introduced by a constant internal resistance [13]. The gearboxes are characterized by constant efficiencies of 0.97. The fuel consumption of the ICE at zero or negative load is estimated by a torque-dependent linear extrapolation method, by following [26]. The ICE thermal transient is not modeled, resulting in a null fuel consumption penalization at cold start. The reliability of the physics behind the adopted simulation platform has been checked in a previous work through the assessment with commercial software, as detailed in [21]. Despite the abovementioned simplifications, the adopted approach can be considered accurate enough to be employed for the illustration of the potential of the ETESS.

4. State of Art for Hybrid Powertrain Management Strategy

Any control strategy for vehicle powertrain aims to minimize predetermined quantities such as the consumed fuel or the pollutant emissions along a driving mission, fulfilling some constraints, e.g., complying with maximum or minimum engine torque or rotational speed, etc. A simplified procedure, largely applied, requires minimizing a combination of the abovementioned parameters resulting in the following mathematical formulation of the problem:

$$\begin{array}{l} \underset{u(t)}{\operatorname{argmin}} J[x(t),t] \\ u(t) \in U \\ x(t) \in X \end{array} \tag{1}$$

where J represents the performance index to minimize, x indicates the generic state variable, and u is the generic control variable, while X and U are the related ranges of variation. J is the integral of a cost function L from t_0 to t plus the difference between the current and the initial state variable, through the penalization factor b.

$$J[x(t),t] = \int_{t_0}^{t} L[x(t),u(t),t]dt + \beta(x(t_0) - x(t)) \tag{2}$$

Since the widespread hypothesis considers the consumed fuel along the driving cycle as the parameter to minimize, the only state variable given is the battery SoC, while the power-split between the thermal engine and electric units ($u = P_{el}/P_{dem}$) represents the control variable. Under those simplifications, the cost function is arranged as:

$$J[x(t),t] = \int_{t_0}^{t} \dot{m}_f[u(t),t]dt + \beta(\mathrm{SoC}(t_0) - \mathrm{SoC}(t)) \tag{3}$$

On the right of Equation (3), the second term is a global constraint for the considered state variable.

4.1. Pontryagin Minimum Principle

For the PMP strategy, at each time, the optimal solution is found through the minimization of the Hamiltonian function:

$$H[u(t),\mathrm{SoC}(t),t,\lambda(t)] = \dot{m}_f[u(t),t] + \lambda(t)\dot{\mathrm{SoC}}[u(t),\mathrm{SoC}(t),t] \quad (4)$$

where $\lambda(t)$ is the costate, and its dynamic equation is given by:

$$\dot{\lambda}(t) = -\frac{H[u(t),\mathrm{SoC}(t),\lambda(t),t]}{\partial \mathrm{SoC}} = -\lambda(t)\frac{\partial \dot{\mathrm{SoC}}[u(t),\mathrm{SoC}(t),t]}{\partial \mathrm{SoC}} \quad (5)$$

According to the prevailing assumption, for which the SoC time derivative is not dependent on its current level [13], the costate is constant over time, whereas the optimal costate, identified as λ^*, needs only to fulfill the energy balance for the battery between the start and the end of the driving cycle:

$$\mathrm{SoC}(t_0) = \mathrm{SoC}\left(t_f\right) \quad (6)$$

λ^* can be evaluated only by defining "a priori" the vehicle driving mission, depending on the knowledge of future information. If the Hamiltonian cannot be given as an explicit function of the control variable, to solve the problem, a discretization of the control variable domain at each simulation step is mandatory. By varying the grid sizing, the problem's solution may exhibit variations, resulting in a quite different cost function minimum and control variable trajectory. Using finer grids leads to better outcomes, but the computational time could become a problem to be reckoned with.

4.2. Equivalent Consumption Minimization Strategy

The ECMS may be deemed as an online variant of the PMP [13]. This method requires to minimize an equivalent fuel rate at each time, that is, the sum of the actual fuel rate and of a contribution arising from the battery electrical power by using an equivalence factor, as stated here below:

$$\dot{m}_{eq}[u(t),t] = \dot{m}_f[u(t),t] + s_0\frac{P_{bat}[u(t),t]}{LHV} \quad (7)$$

LHV is the lower heating value of the fuel, P_{bat} represents the power released or absorbed by the battery, and s_0 an equivalence factor. A piecewise linear type of description concerning s_0 (different for battery charge and discharge phases) demonstrates realizing very close to the optimal powertrain management but requires to be adjusted depending on the vehicle characteristics and driving mission [18]. Several methodologies have been developed aiming at achieving an adaptative adjustment of the equivalence factor [14,15,18,19]. Some of these are developed according to the outputs deriving from off-line optimization strategies [15,16]. The impact on performance by applying a constant value for s_0 was evaluated in [18] by determining results very close to the optimality.

Between the available alternatives, a very robust methodology presents an equivalence factor correction based on the difference between the current SoC level and a target value [14]. The correction is computed by a PID controller, expressed as:

$$s_{corr}(t) - s_0 = K_p\Delta\mathrm{SoC} + K_i\int_0^t \Delta\mathrm{SoC}d\tau + K_d\frac{d}{dt}\Delta\mathrm{SoC} \quad (8)$$

$$\Delta\mathrm{SoC} = \mathrm{SoC}(t) - \mathrm{SoC}_{target} \quad (9)$$

where SoC_{target} is a reference SoC level, and ΔSoC is the error between the current SoC and the above reference. The first term on the right hand of Equation (8) represents a proportional correction term, whereas the second and the third ones are an integral and a derivative correction, respectively. Considering the online optimization described in the next sections, this widespread method is chosen to achieve the strategy adaptivity.

5. Description of the Efficient Thermal Electric Skipping Strategy

As opposed to the power-split concept, the basic idea of ETESS is to alternatively employ the electric units and thermal engine to fulfill the power demand at the vehicle wheels, P_{dem}. The choice between the power units is realized, at each time, comparing the actual fuel consumption of the thermal engine that operates to fully satisfy the power demand, $\dot{m}_{f,th}$, and equivalent fuel consumption, $\dot{m}_{f,el}$, related to a vehicle pure electric driving. While the first fuel rate, $\dot{m}_{f,th}$ is straightforwardly calculable, the definition of the second one is the most critical issue for the strategy implementation. The value of $\dot{m}_{f,th}$, for each available gear ratio, n_{GB1}, is based on the power demand, P_{dem}, and on the losses in the GB_1 and in the differential (see power flux in Figure 5a), resulting in the following expression:

$$\dot{m}_{f,th}(n_{GB_1}) = \frac{P_{dem} \cdot BSFC(n_{GB_1})}{\eta_{GB_1}(n_{GB_1})\eta_{diff}} \tag{10}$$

where η_{GB1} is the efficiency of GB_1 and BSFC is the actual fuel consumption of the engine that operates with the load and speed enforced by the vehicle speed and by P_{dem}.

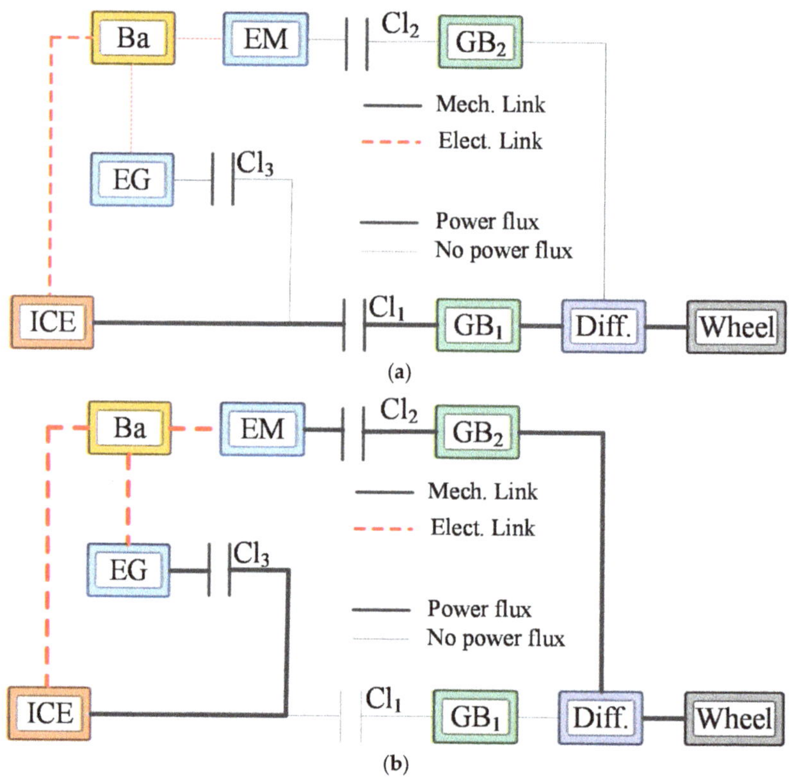

Figure 5. Power flux in pure thermal (**a**) and pure electric driving (**b**).

The idea to identify the fuel rate $\dot{m}_{f,el}$ is that in a pure series-electric driving, the power delivered by the EM is produced by the thermal engine in an undefined time, and working in its optimal operating point, featured by a $BSFC_{min}$. In a pure series driving, the power flux from the thermal engine to the wheel entails some losses in the EG, in the EM, in the battery, in the GB_2, and in the differential, and it can be quantified by the efficiencies of each component (see power flux in Figure 5b). In a pure electric mode, the equivalent fuel consumption is thus defined by multiplying P_{dem} and an "adapted" $BSFC_{min}$, corrected by the abovementioned efficiencies to consider the losses from the ICE to the wheels.

$$\dot{m}_{f,el} = c_0 \cdot \frac{P_{dem} \cdot BSFC_{min}}{\eta_{GB_2}\, \eta_{EG,charge}\, \eta_{EM}\, \eta_{diff}} \tag{11}$$

where η_{GB2}, $\eta_{EG,charge}$, η_{EM} and η_{diff} are the efficiencies of GB_2, EG, EM, and differential, respectively, and c_0 is a tuning constant. Note that $\eta_{EG,charge}$ represents the EG efficiency computed in the torque-speed couple where the battery charge occurs, which in turn depends on the engine operating point of minimum BSFC. The tuning constant c_0 is introduced to achieve the energy balance for the battery, as expressed by Equation (6). The choice between either pure thermal or electric driving is carried out based on the inequality below:

$$\begin{cases} \dot{m}_{f,el} < \dot{m}_{f,th} & \Rightarrow \text{ pure electric mode} \\ \dot{m}_{f,el} > \dot{m}_{f,th} & \Rightarrow \text{ pure thermal mode} \end{cases} \tag{12}$$

In this comparison, the gear ratio which involves the lowest fuel consumption is chosen for the definition of $\dot{m}_{f,th}$. If the maximum power delivered by the thermal engine for a certain gear ratio, $P_{ICE,max}$ (n_{GB1}), is lower than the power demand, the fuel rate is corrected as:

$$\dot{m}_{f,th}(n_{GB_1}) = P_{ICE,max}(n_{GB_1}) \cdot BSFC(n_{GB_1}) + \Delta\dot{m}_{f,el}(n_{GB_1}) \tag{13}$$

$$\Delta\dot{m}_{f,el}(n_{GB_1}) = c_0 \cdot \frac{\left(P_{dem} - P_{ICE,max}(n_{GB_1})\eta_{GB_1}(n_{GB_1})\eta_{diff}\right) \cdot BSFC_{min}}{\eta_{GB_2}\, \eta_{EG,charge}\, \eta_{EM}\, \eta_{diff}} \tag{14}$$

where the first term of Equation (13) is the fuel rate when the engine operates at its maximum power and the second term, $\Delta\dot{m}_{f,el}$, represents an equivalent fuel consumption needed to fulfill the power demand with the support of the electric motor. This is expressed by Equation (14).

A simplified explanation of the ETESS principle can be associated with a specialization of the ECMS, where the only acceptable values for the power-split are either 0 or 1. Even if introducing this simplification means to realize a certain fuel economy penalization, a drastic decrease in the computational effort is expected. This tradeoff between computational effort and fuel economy will be illustrated in Section 7.

Returning to the ETESS logic description, once the wheel power demand becomes negative, regenerative braking is realized by the EM.

A flowchart summarizes the ETESS logics in Figure 6. This underlines the choice between a pure electric or thermal driving, based on the fuel rates $\dot{m}_{f,th}$ and $\dot{m}_{f,el}$, and the activation of a parallel mode only when the ICE or the EM cannot fulfill by itself the power demand. Depending on this methodology, the battery charge is actuated, especially when the vehicle brakes rather than in a phase featured by positive power demand. Thus, it is possible to reduce as much as possible the energy flux from the ICE to the battery (throughout the electric units) with the aim to minimize the correlated unavoidable mechanical and electrical losses.

Along a driving cycle, if the torque limits for each unit (thermal and electric) are not overcome, the only energy available to perform a pure electric driving is the one recuperated by the regenerative braking, while the thermal engine, once switched on, will provide the power strictly required to drive the vehicle.

According to the simple inequality of Equation (12), choosing between pure electric or thermal driving is straightforward, without requiring a discrete map exploration. To evaluate $\dot{m}_{f,th}$, the engine operating point is univocally identified by the vehicle speed, by the tractive power demand, and by the losses along the driveline from the wheels to the engine. Likewise, also the fuel rate $\dot{m}_{f,el}$ for a pure electric driving is univocally defined by the vehicle speed, by the traction power demand, and by the losses along the driveline, and furthermore by the dissipations in the electric units.

Finally, it is worth highlighting the versatility of the proposed control strategy, also considering its suitability to any hybrid powertrain fitted with a gearbox. The application here reported will represent just an example for a quite complex test case.

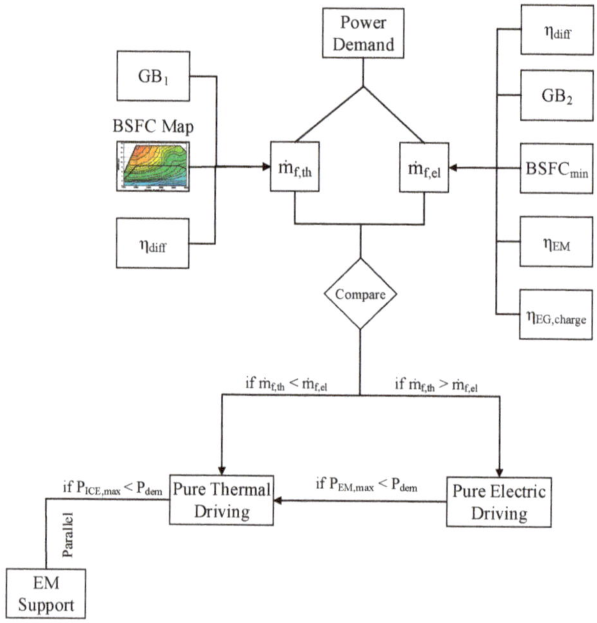

Figure 6. Flowchart that schematizes the ETESS logics.

6. Issue for the Strategy Implementation in the Vehicle Simulator

For the PMP and ECMS implementations, a grid is defined for the exploration of the EM and EG efficiency maps, composed of 29 and 19 torque levels for each rotational speed, respectively. This choice represents a reasonable compromise between computational effort and the degree of accuracy of the solution, as proved by the parametric analysis discussed in the next section. No gridding is needed for access to the engine BSFC map since the engine load level is univocally identified by the power demand at the vehicle wheels and by the torque levels of the electric units, which in turn are iteratively explored by the above grids.

For the ETESS implementation, the power request for both EM, EG and ICE is univocally determined (once assigned the gear number), and for this reason, no gridding is required. This characteristic is expected to drastically improve its simulation time in comparison with PMP, ECMS or GOS in general. However, the ETESS undergoes a drawback analogous to the PMP one, i.e., the requirement of a priori knowledge of the speed-profile aimed at selecting the value of c_0. Anyway, it can be easily extended to a real-time implementation introducing an adaptive correction for c_0, similarly to the ECMS method.

Whatever is the employed strategy, either PMP, ECMS or ETESS, a minimum dwell time of 1 s for the vehicle state (gear number or driving mode—either series or parallel) is imposed, and a fuel consumption penalization of 0.5% is introduced in case of state variation. In this track, a battery power consumption is assumed when the engine is turned on (4.8 kW along 1 s). As known, the intensity of the above penalizations, on one side, allows to reduce control instabilities, but, on the other side, affects the overall fuel consumption over a driving mission. The selected intensities of the above penalizations aim to result in an as low as possible fuel consumption but would require some adjustments if the strategies are applied in more advanced simulation platforms (dynamics forward-facing models) or on real on-vehicle testing.

7. Discussion of ETESS Potential

Preliminarily, the assessment in off-line simulations between the ETESS and the PMP is described for the tested HEV by considering different powertrain designs and driving cycles. For each tested case, the constants λ^* and c_0 are case-by-case tuned, aiming at obtaining the battery energy balance between the cycle start and end (Equation (6)). In Table 3, all the examined configurations are reported, pointing out the driving cycles and the powertrain features. The driving cycles considered are six. Specifically, cases from #1 to #7, cases #10 and #11 are referred to common speed missions (standardized WLTC and Artemis variants), where cases #8 and #9 correspond to real driving emissions (RDE) compliant cycles provided by the European Commission's Joint Research Center. For those two cases, the speed and altitude profiles are depicted in Figure 7, whereas in Table 4, their main data are listed (additional information on these two RDE-compliant cycles is available in the annex of [27]). The choice to include the analyses along RDE-compliant cycles is not estimating the CO_2 emissions, for which those cycles were defined, but to verify the ETESS potential also in real driving conditions.

Table 3. Simulation plan.

Case #	Driving Cycle	ICE	EM	EG	Ba
1	WLTC	$Base_1$	Base	Base	Base
2	WLTC	Red_1	Base	Base	Base
3	WLTC	$Base_1$	Red	Red	Base
4	WLTC	Red_1	Red	Red	Base
5	Artemis Motorway	$Base_1$	Base	Base	Base
6	Artemis Road	$Base_1$	Base	Base	Base
7	Artemis Urban	$Base_1$	Base	Base	Base
8	RDE1	$Base_1$	Base	Base	Big
9	RDE2	$Base_1$	Base	Base	Big
10	WLTC	$Base_2$	Base	Base	Base
11	WLTC	$Base_2$	Red	Red	Base

Table 4. RDE-compliant cycles main data. Reproduced from [27], Publications Office of the European Union: 2019.

Driving Cycle Main Characteristics	RDE1	RDE2
Length, m	93,939	78,853
Duration, s	6693	5599
Mean speed, km/h	56.2	56.3
Max speed, km/h	126	129
Mean accel., m/s^2	0.39	0.41
Max accel., m/s^2	3.33	5.04
Mean decel., m/s^2	−0.42	−0.43
Max decel., m/s^2	−3.14	−3.38

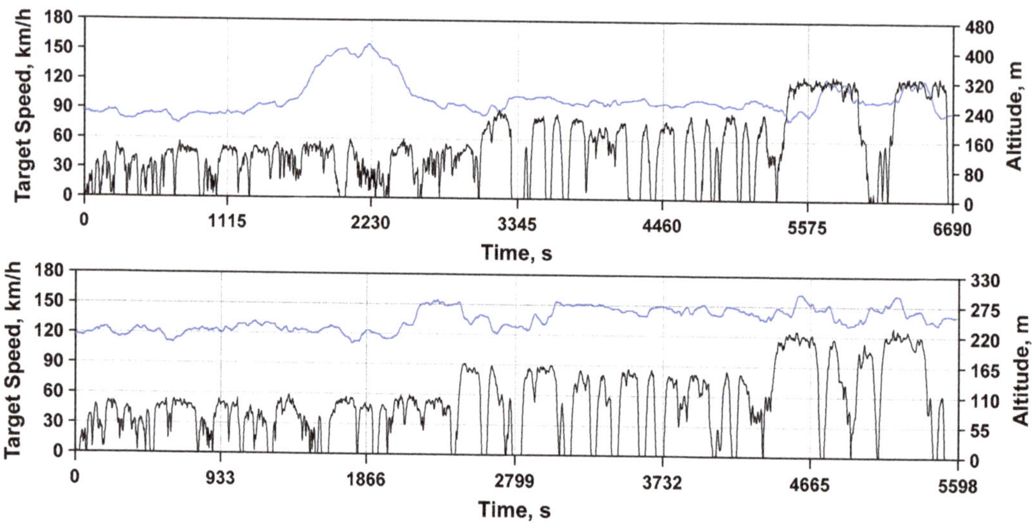

Figure 7. RDE-compliant cycles target speed and altitude profiles.

The variants of the powertrain are synthetically described in Table 3, where "Base" is referred to the baseline performance, while "Red" refers to motor or engine with reduced maximum performance, keeping the same rotational speed range. Considering the ICE column, the subscript 1 or 2 indicates the thermal engine used between the two ones introduced in Section 2. The "Red" configuration of the thermal engine entails maximum performance halved compared to the baseline. The "Red" variant of electric units presents a power limit reduction of 80% with respect to the reference. For both thermal and electric units, BSFC and efficiency iso-contours are not adapted as against the baseline engine ones, shown in Figures 2 and 3. In other words, the power units are assumed to work with reduced limits but without modifying their efficiency. The last column, describing the battery size, indicates for cases #8 and #9 a doubled capacity ("Big"). Without a resizing of the battery, the RDE-compliant cycles could not be performed without fully discharging the battery, hence without respecting the constraints of the optimization problem expressed by Equation (1).

For the sake of brevity, in the following detailed results will be presented for cases #1, #4, #10 and #11. Starting from case #1, as can be observed in Figure 8, ETESS and PMP provide almost superimposed results of EM (Figure 8c) and ICE (Figure 8b) powers, which reflects on the trends of fuel rate (Figure 8f), SoC (Figure 8e) and selected gear number (Figure 8g). The power-split trend in Figure 8h (upper side) points out that the PMP switches between 0 and 1, even if a modulation between those extreme levels is potentially available. The ETESS profile is superimposed in most parts of the cycle. The bottom part of Figure 8h shows the ratio between the power of EG and ICE. It can be observed that the PMP chooses to realize a very limited battery charging through the EG, which determines an SoC profile similar to the ETESS one, for which the battery charging by the ICE is disabled. Although not explicitly shown in the presented results, it is worth noting that, when the power-split is equal to 1, the series mode is activated and the clutch number 1 (see Figure 1) is open, while the opposite occurs if the power-split is lower than 1 (parallel mode).

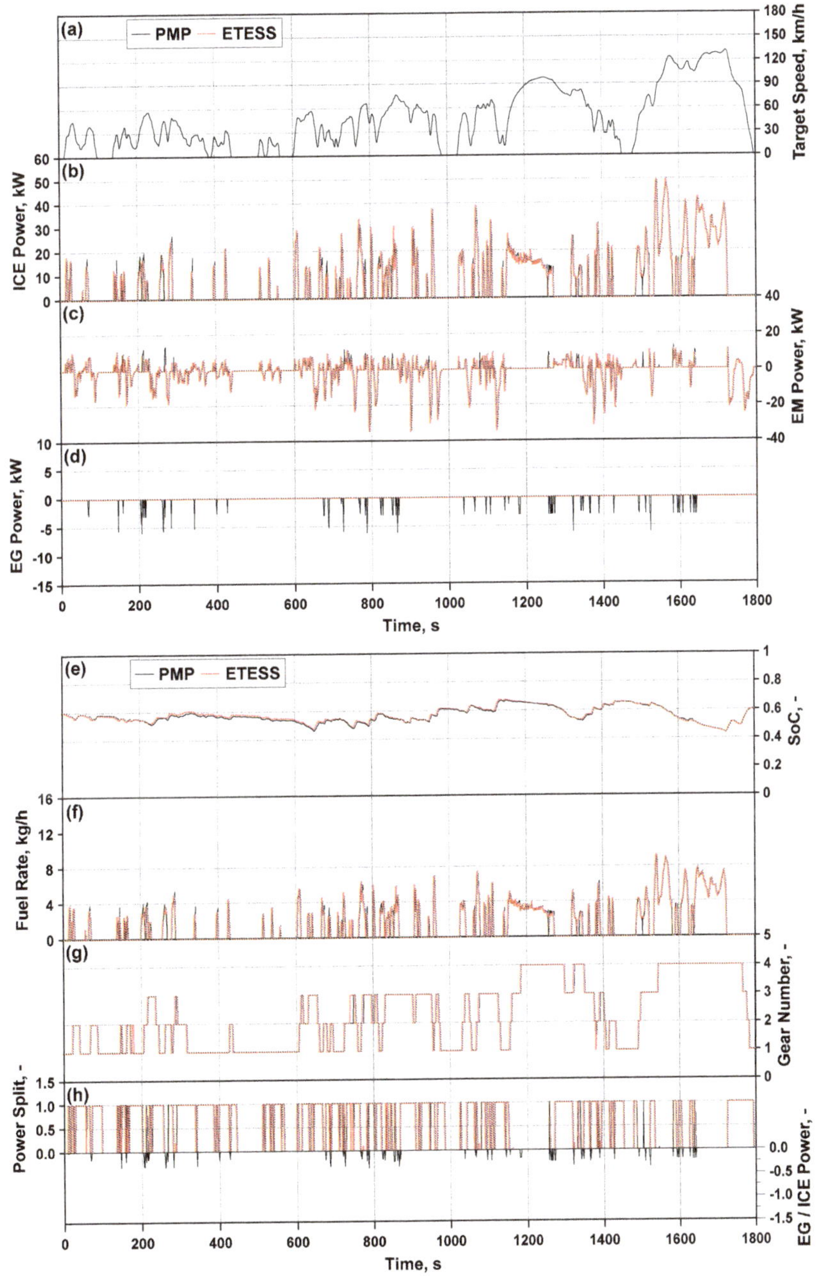

Figure 8. Pontryagin minimum principle (PMP)-ETESS comparisons of ICE power (**b**), EM power (**c**), EG power (**d**), SoC (**e**), fuel rate (**f**), gearboxes (GB_1) number (**g**) and vehicle mode (**h**) along the WLTC (vehicle speed—(**a**))—case #1.

As stated above, in case #4, the powertrain characteristics are changed due to a reduction of the maximum ICE torque that corresponds to the BMEP dashed line in Figure 2, and by decreasing of 80% the maximum and minimum torque for the electric motors. The assessments in Figure 9 depict greater differences between the PMP and ETESS, compared to case #1.

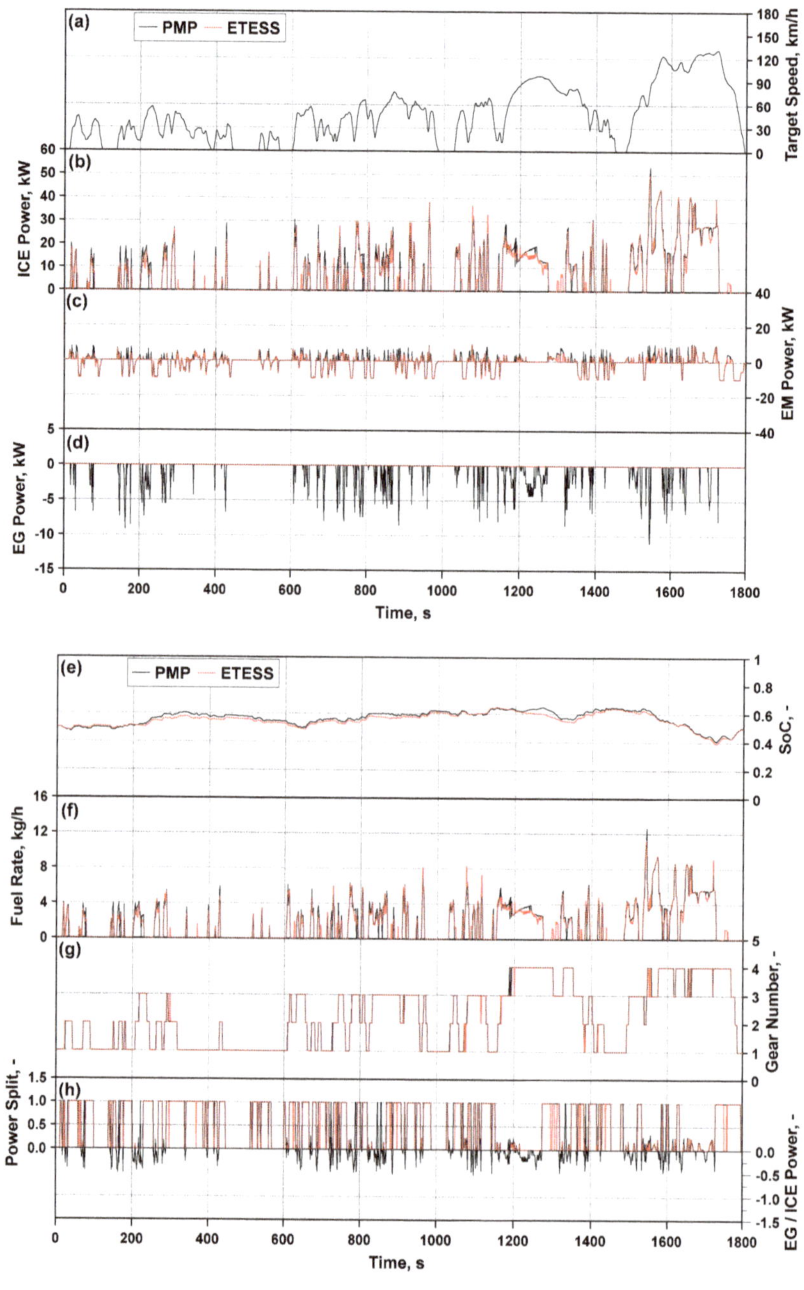

Figure 9. PMP-ETESS comparisons of ICE power (**b**), EM power (**c**), EG power (**d**), SoC (**e**), fuel rate (**f**), GB_1 number (**g**) and vehicle mode (**h**) along the WLTC (vehicle speed—(**a**))—case #4.

For the ETESS, because of the impossibility of an EG battery charging, pure ICE driving is very often chosen (Figure 9b). On the contrary, for the PMP, in most parts of the driving cycle with positive power demand, the EG battery charging is activated (Figure 9d). This allows employing the EM more frequently compared to ETESS (Figure 9c). As well as case #1, an alternate utilization of ICE and EM is prevalently chosen by both PMP and ETESS (power-split either 0 or 1 in Figure 9h). However, in case #4, a power-split is applied by both control strategies, especially when a higher gear ratio is chosen. In those circumstances, the ICE attains the maximum power limits and support from the EM are needed to fulfill the power demand. To this aim, it can be observed, as an example, the trends around 1700 s. At this time, the longest available gear number is selected (Figure 9g); consequently, the thermal engine operates at a reduced speed, its output limit is attained, and a certain contribution of EM is mandatory.

Cases #10 and #11 differ from cases #1 and #2 because of the thermal unit. The thermal engine 2 is characterized by a displacement and a maximum power about halved if compared to thermal engine 1. Its BSFC map, in Figure 3, shows higher values of consumption all over the operating domain. For the PMP, the worst performance of the thermal unit reflects in less use of battery charging through the EG, as shown in Figures 10d and 11d. Once again, an alternate utilization of pure ICE or EM driving is mainly preferred, as shown in Figures 10h and 11h. Those choices make the ETESS control very similar to the PMP, as highlighted by the power profiles of ICE and EM, in Figure 10b,c and Figure 11b,c. Some differences emerge in the high-speed portion of the driving cycle (between 1550 s and 1700 s), where sometimes a hybrid ICE/EM driving is selected by the PMP, while this does not occur for the ETESS.

From an overall analysis of the instantaneous profiles for the two confronted control strategies, it turns out their significant coherence. A global comparison between ETESS and PMP is made by the bar charts shown in Figure 12. The bars correspond to the consumed mass of fuel per kilometer for all the considered cases, and, over each couple of bars, is reported the fuel consumption percent difference (assuming as a reference the PMP level). The ETESS behaves similarly to PMP, with an average fuel consumption increase of about 0.4% and, in most cases, below 0.5%. Limiting the analysis to the cases with engine 1, greater differences emerge when the electric unit sizing is reduced (cases #3 and #4) or when the electric driving is limited (cases #7). Following in the ETESS/PMP comparison assessment, very similar fuel consumptions occur for the base powertrain over substantially different driving cycles (Cases #1, #5, #6, #7, #8). This result appears relevant considering that those cycles differ in terms of both power demand, vehicle medium and maximum speed, vehicle medium and maximum acceleration/deceleration, duration, and length. The performance of ETESS slightly worsens for the less efficient engine and a sufficient sizing of the electric units, looking at the comparison between cases #1 and #10. In this case, the PMP strategy allows achieving a lower fuel consumption, however, with a difference with the ETESS smaller than 1%. In case #2, the ETESS even performs better than PMP. This apparent incongruence is explained by a parametric analysis on the gridding of the maps of EM and EG, whose results are reported in Figure 13. This last shows the kilometric fuel consumption for different gridding of the torque levels for EM and EG (whose number of breakpoints are represented in the figure by the notation $n_{EM} \times n_{EG}$) and the simulation time normalized by the physical time. The setting with a grid of 29 × 19 is assumed as a reference for the definition of the fuel consumption percent difference reported on each bar. For the sake of completeness, the values related to the ETESS are shown, as well. Figure 13 underlines that refining the grid, the fuel consumption slightly reduces, but the computational time increases exponentially, as expected. The lowest fuel consumption level is reached by the PMP with the finest grid (79 × 69 points) among the ones considered but with a simulation time about 10 times longer than the reference setting (29 × 19). Compared to the ETESS, the computational time is about three orders of magnitude higher, with only a slight increased fuel consumption. Among the tested gridding, the overall difference between the best and worst PMP cases is about 1.3%, proving the relevance of this aspect for the identification of the optimal strategy and fuel consumption. Analogous sensitivity analyses are repeated for all the other cases of Table 3, from which emerge that the differences between the gridding settings are less evident. As an example, the results of the analyses are shown for case #1 in Figure 14. In this case, the percent difference between the extreme PMP reduced to about 0.1%. It can be concluded that the most critical case from the gridding sensitivity viewpoint is the one where the operating limits of the thermal engine are more frequently reached (case #2 presents the depowered engine 1 variant). This is basically due to the choice of gridding the electric unit maps and not the map of the ICE. A different choice would determine a similar issue for the cases in which the operating limits of the electric units were reached.

Coming back to the comparison between the considered cases of Table 3, the bar chart in Figure 15 depicts the duration when a hybrid thermal/electric driving is chosen (PS greater than 0 and lower than 1) normalized by the total cycle duration. It can be noted that from cases #1 to #9, characterized by a more efficient ICE, a power-split is applied both from PMP and ETESS with a comparable extent. The time in PS becomes relevant only when the ICE power limit is reduced (cases #2 and 4#) and support of the EM is needed to fulfill the power demand.

A different behavior appears for cases #10 and #11, which differs from cases #1 and #3 because of the less efficient thermal unit. In those cases, the ETESS still involves an almost null time in PS mode, while the PMP determines a hybrid driving for the high-speed cycle portion, as already discussed for the instantaneous results.

The time percentage in charging mode is shown in Figure 16 for the PMP strategy. It is worth recalling that the same plot for the ETESS would have been meaningless, not being allowed battery charging through the EG. Figure 16 underlines that by reducing the size either of the thermal unit (case #2), of the electric ones (case #3) or of both (case #4), the time percentage in charging mode becomes bigger. The longer the time in charging mode, the longer is the activation of the electric driving in comparison with the ETESS, as shown, for instance, in the EM power profile in Figure 9c. Figure 16 also highlights that, for the same powertrain, more burdensome cycles require more time in charging mode (comparison between cases #1, #5, #6, #7, #8). Another outcome arises by observing differences between the analyses performed along the same driving cycle but with different ICEs (couple #1–3 and #10–11). The battery charging is more frequent if a more efficient engine is employed. Otherwise, the battery charging results to not be convenient from a global viewpoint, and EG is almost not used, as shown in Figures 10d and 11d.

It is worth underlining that concerning the off-line simulations discussed above, the ETESS gives fuel consumption performance very close to PMP ones in all tested cases, but with a shorter computational time (about two orders of magnitude lower). The ETESS executes three orders of magnitude faster than the physical system, demonstrating the potential for real-time implementation on the vehicle.

The second part of the numerical activity concerns the verification of the robustness of the ETESS online version for different vehicle variants and driving cycles (all the cases listed in Table 3), in comparison to a well-assessed online methodology such as the ECMS. The ETESS online variant is verified, establishing an adaptative correction for c_0 in Equation (11), achieved by using a PID controller. The latter minimizes the error between the current SoC and a predefined target of 0.55. The correction is applied to start from an initial value that is equal, for each case, to the one identified for the corresponding off-line simulation. A similar approach is adopted for the ECMS analyses, where the control adaptivity is realized by a PID controller acting on s_0 in Equation (7).

The online simulations consist of six simulations for each case, having different initial SoC but the same final target. The outcomes of these analyses are arranged, in accordance with the WLTP procedure [28,29], to obtain a corrected kilometric fuel consumption. The findings are reported in the bar chart of Figure 17. The ETESS returns values comparable to the ECMS ones, proving the methodology robustness. A fuel consumption penalization of about 0.1% arises on average. ETESS provides higher fuel consumption than ECMS only in case #4, while it provides even lower levels in cases #2 and #5. Considering the online ETESS variant, if compared to the ECMS, the benefits in the simulation time are confirmed, to a similar extent.

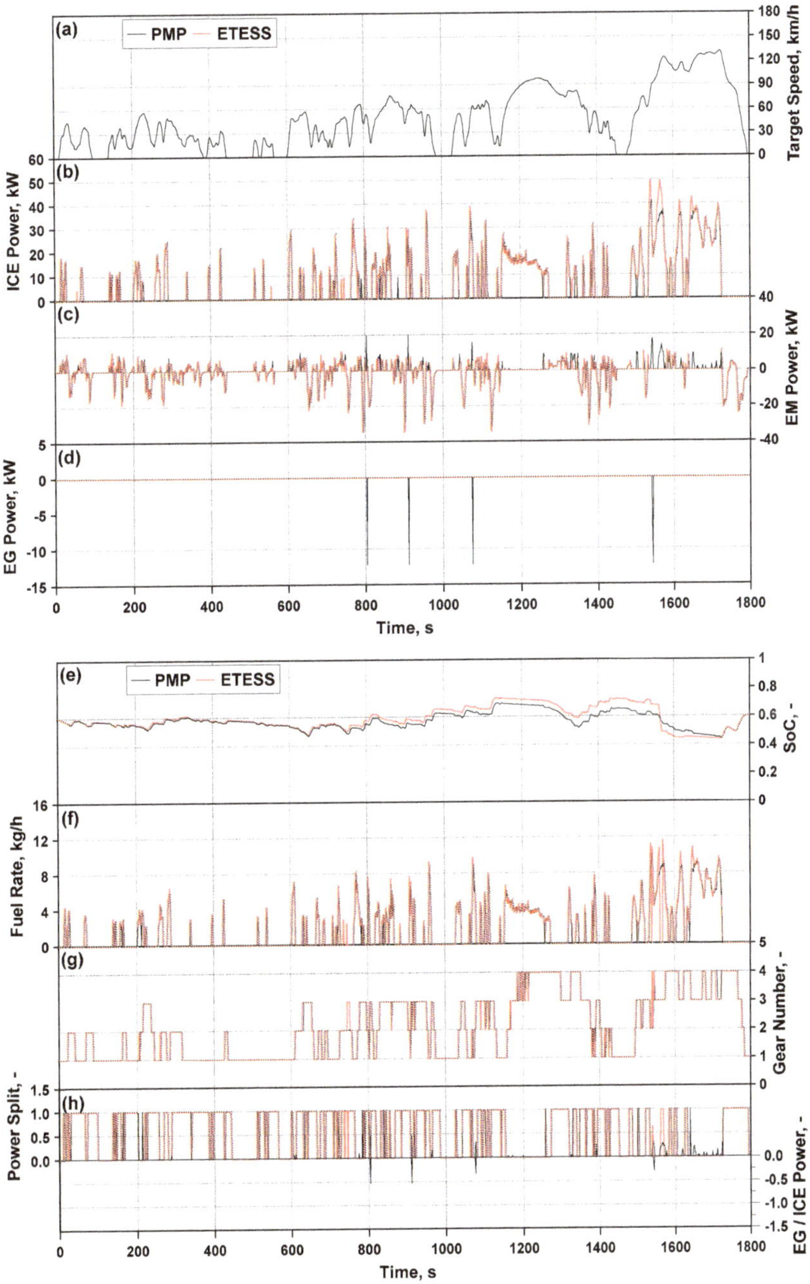

Figure 10. PMP-ETESS comparisons of ICE power (**b**), EM power (**c**), EG power (**d**), SoC (**e**), fuel rate (**f**), GB_1 number (**g**) and vehicle mode (**h**) along the WLTC (vehicle speed—(**a**))—case #10.

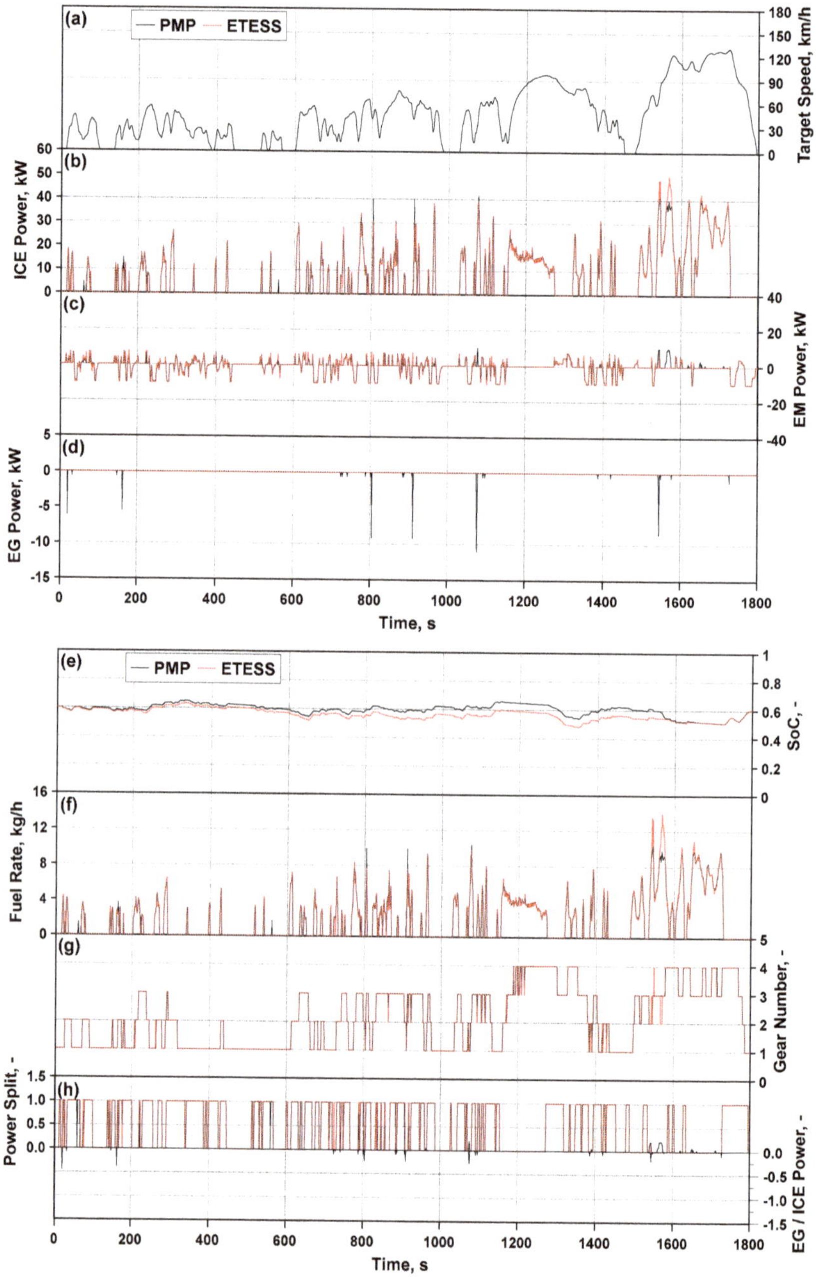

Figure 11. PMP-ETESS comparisons of ICE power (**b**), EM power (**c**), EG power (**d**), SoC (**e**), fuel rate (**f**), GB_1 number (**g**) and vehicle mode (**h**) along the WLTC (vehicle speed—(**a**))—case #11.

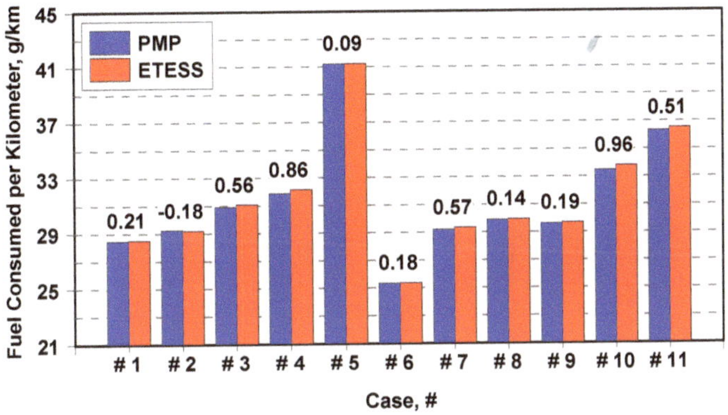

Figure 12. Assessment of kilometric consumed fuel in the cases of Table 3 between off-line ETESS and PMP.

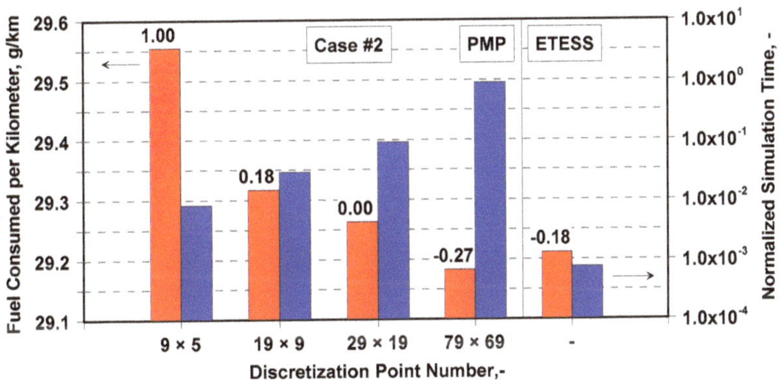

Figure 13. Assessment of kilometric consumed fuel and normalized simulation time in case #2 of Table 3, between off-line PMP, for different map grid discretization, and ETESS.

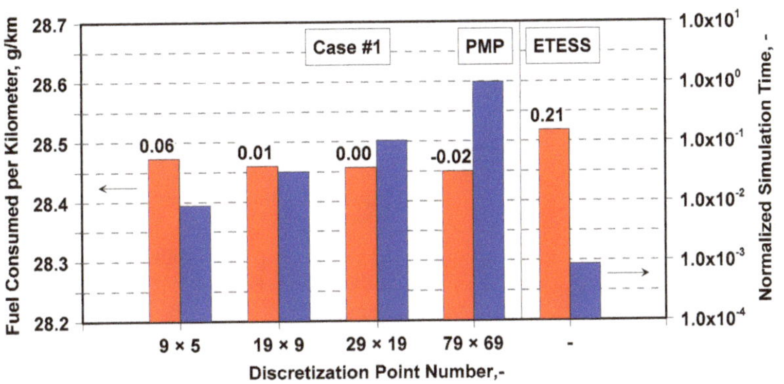

Figure 14. Assessment of kilometric consumed fuel and normalized simulation time in case #1 of Table 3, between off-line PMP, for different map grid discretization, and ETESS.

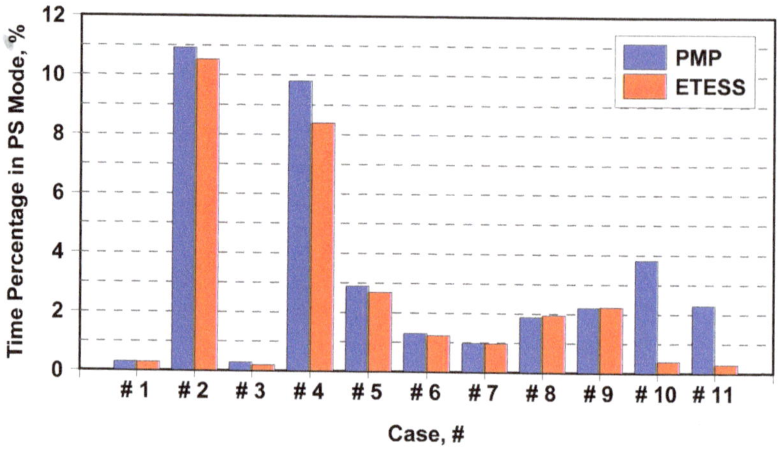

Figure 15. Assessment of time percentage in power-split mode in the cases of Table 3 between off-line ETESS and PMP.

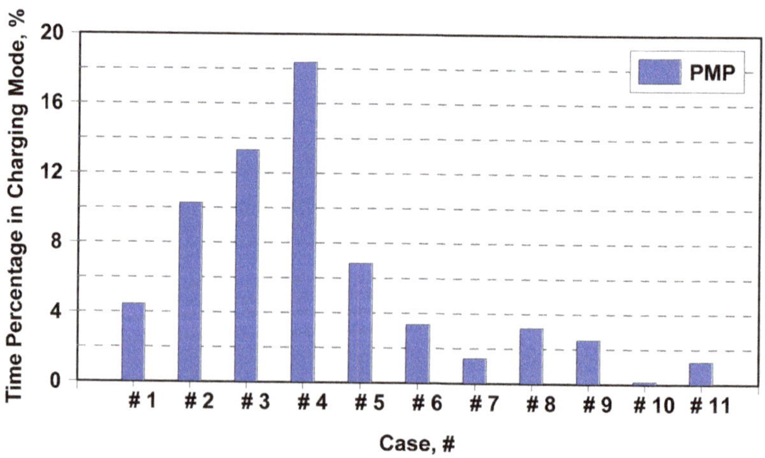

Figure 16. Time percentage in charging mode for PMP in the cases of Table 3.

As a final consideration, the slightly different performance of ETESS in comparison to well-assessed methods (PMP and ECMS) appears acceptable because of its computational efficiency. The near-optimal outcomes essentially arise from the theoretical background of the ECMS approach. The further evidence that emerged by the proposed analyses is that a fine exploration of whatever power-split is not compulsory since its evaluation in two significant values (either 0 or 1) is enough to achieve results near to optimality. Further, the ETESS computational efficiency proves the potential for real-time implementation and for the handling of situational information, which is required for the control of connected vehicles. The strategy provides certain robustness, not only for different vehicle configurations but also for different driving missions.

The further progress of this activity will concern the verification of the consistency of the ETESS considering more complex modeling of some powertrain subcomponents (for example, variable efficiencies of battery and gearbox) and in a dynamic "forward-facing" simulation. Moreover, the extension of the ETESS approach to the control of powertrains fitted with epicyclic gearing as power-split devices will be investigated.

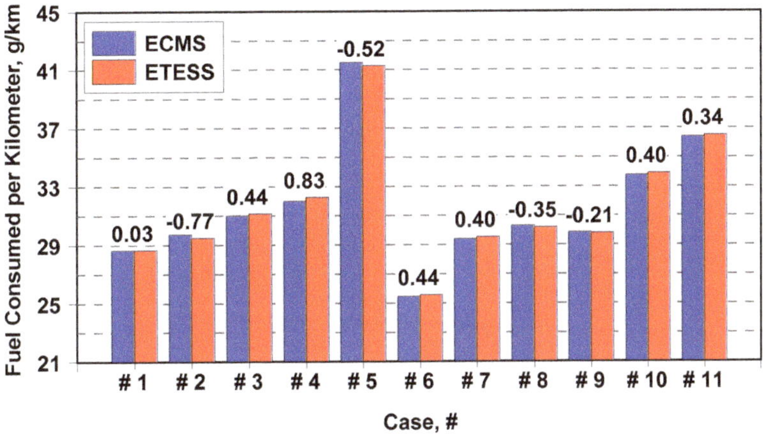

Figure 17. Assessment between online ETESS and ECMS of kilometric consumed fuel and percent difference in the cases of Table 3.

8. Conclusions

This work presents an efficient control strategy for hybrid powertrains, named ETESS. The basic concept of the ETESS is to alternatively utilize thermal and electric units to fulfill the vehicle power demand. ETESS is numerically tested by an "in-house" coded model, with reference to a C-class vehicle.

The methodology is verified for different powertrain variants in terms of power units sizing and efficiency, and along various driving cycles, both regulatory and RDE-compliant cycles. For each investigated case, the ETESS is assessed with the well-known PMP approach in off-line analyses. As known, the PMP, conversely to ETESS, relies on the power-split concept, involving the possibility of a combined driving by both thermal and electric units. Off-line vehicle simulations highlight that the proposed approach performs similarly to PMP, resulting in higher fuel consumption of about 0.4% on average. Major fuel consumption differences appear in the tests with a less efficient thermal unit, for which a power-split sometimes appears preferable along the driving cycle, and the ETESS logic fails.

A parametric analysis in a representative test case shows that the map gridding affects PMP capability to identify optimal control, with an impact on the computational effort. Indeed, the fuel consumption difference between the most refined and most coarse map gridding is about 1.3% at the expense of a variation of two orders of magnitude in the computational time. Those issues do not emerge for the ETESS approach.

With reference to the most efficient engine here considered, the need for a power-split between thermal and electric units during the cycle emerges primarily if the thermal engine presents limited power/torque performances, both following PMP and ETESS strategies. If the less efficient thermal unit is adopted, greater differences appear in the comparison between PMP and ETESS, where the first one more frequently resorts to a hybrid driving.

Simulation results adopting PMP underlines that the time spent in charging mode (due to the ICE power supply) becomes greater, reducing the size of either the thermal unit, the electric ones or both. Moreover, more burdensome cycles require longer time in charging mode. On the other hand, whatever are the engine characteristics and the driving mission, the ETESS strategy does not involve the possibility of battery charging, except for the regenerative braking.

The robustness of the ETESS is tested in online simulations, with the aim to confirm the possibility of being employed in a real-time vehicle application. In this framework, an assessment with the ECMS approach is carried out. The online calculations underline that the ETESS behaves similarly to ECMS in terms of fuel consumption, with a very reduced difference (fuel consumption increase of 0.1% on average), and in all considered cases below 1%. Once again, the main benefit is a hugely reduced calculation effort.

As a future development of this study, the ETESS reliability will be tested in more complex simulation frameworks (in a forward-facing simulator, also accounting for the dynamics of the powertrain components) and for different powertrain architectures (for instance, powertrain fitted with epicyclic gearing as power-split device). Moreover, the robustness of the ETESS online variant for different driving missions will be further investigated to avoid the need for an off-line identification of the initial c_0 value.

Author Contributions: Conceptualization, V.D.B.; methodology, V.D.B.; software, E.M.; validation, E.M., Data Curation, E.M., J.-M.Z.; writing—original draft preparation, V.D.B., E.M., J.-M.Z.; writing—review and editing, V.D.B., E.M., J.-M.Z.; visualization, E.M.; supervision, V.D.B.; project administration, V.D.B., J.-M.Z.; funding acquisition, V.D.B., J.-M.Z. All authors have read and agreed to the published version of the manuscript.

Funding: This project has received funding from the European Union's Horizon 2020 research and innovation program under grant agreement No. 724084.

Institutional Review Board Statement: Not applicable.

Informed Consent Statement: Not applicable.

Data Availability Statement: Data sharing not applicable.

Acknowledgments: The authors would like to thank J-C. Dabadie, G. Alix, A. Abdelli, J. Melgar Sossa and F. Le Berr (all IFPEN) for their support and inputs for vehicle simulations, V. Valverde (Joint Research Cent) for providing the two RDE-compliant driving cycles, as well as C. Libert (Renault SA) for the support in the vehicle characteristics definition.

Conflicts of Interest: The authors declare no conflict of interest.

Abbreviations

Notations

c_0	Tuning constant
f	Function
H	Hamiltonian
J	Performance index
K	Constant
L	Cost function
m	Mass
n	Gear ratio
P	Power
pth	Tolerance of the hyperbolic tangent function
s_0	Equivalence factor
s_{corr}	Equivalence factor correction
t	Time
u	Control variable, power-split
U	Variation range of the control variable
x	State variable
X	Variation range of the state variable
Greeks	
b	Penalization factor
h	Efficiency
l	Costate
Acronyms	
Cl	Clutch
Ba	Battery
BMEP	Brake mean effective pressure
BSFC	Brake specific fuel consumption
DP	Dynamic programming

ECMS	Equivalent consumption minimization strategy
ETESS	Efficient thermal electric skipping strategy
EM	Electric motor
EG	Electric generator
GB	Gear-boxes
GOS	Global optimization strategy
HEV	Hybrid electric vehicle
ICE	Internal combustion engine
LHV	Lower heating value
NEDC	New European driving cycle
PI	Proportional-integrative
PID	Proportional-integrative-derivative
PMP	Pontryagin minimum principle
PS	Power-split
RDE	Real driving emission
SI	Spark ignition
SoC	State of charge
UNVS	UniNa vehicle simulation
UniNa	University of Naples
WLTC	Worldwide harmonized light-duty vehicles test cycle
WLTP	Worldwide harmonized light-duty vehicles test procedure
Subscripts	
0	Initial
batt	Battery
d	Derivative
dem	Demand
diff	Differential
el	Electric
eq	Equivalent
f	Final, fuel
i	integrative
max	Maximum
min	Minimum
p	Proportional
th	Thermal
Superscripts	
.	Temporal derivative
*	Optimal

References

1. Morita, K. Automotive power source in 21st century. *JSAE Rev.* **2003**, *24*, 3–7. [CrossRef]
2. Guzzella, L.; Sciarretta, A. Electric and Hybrid-Electric Propulsion Systems. In *Vehicle Propulsion Systems: Introduction to Modeling and Optimization*, 3rd ed.; Springer: Berlin, Germany, 2013. [CrossRef]
3. Patil, R.M.; Filipi, Z.; Fathy, H.K. Comparison of Supervisory Control Strategies for Series Plug-In Hybrid Electric Vehicle Powertrains Through Dynamic Programming. *IEEE Trans. Control Syst. Technol.* **2014**, *22*, 502–509. [CrossRef]
4. Brahma, A.; Guezennec, Y.; Rizzoni, G. Optimal energy management in series hybrid electric vehicles. In Proceedings of the 2000 American Control Conference, Chicago, IL, USA, 28–30 June 2000; Volume 1, pp. 60–64. [CrossRef]
5. Lin, C.C.; Peng, H.; Grizzle, J.W.; Liu, J.; Busdiecker, M. Control System Development for an Advanced-Technology Medium-Duty Hybrid Electric Truck. *SAE Tech. Pap.* **2003**. [CrossRef]
6. Moura, S.J.; Fathy, H.K.; Callaway, D.S.; Stein, J.L. A Stochastic Optimal Control Approach for Power Management in Plug-In Hybrid Electric Vehicles. *IEEE Trans. Control Syst. Technol.* **2011**, *19*, 545–555. [CrossRef]
7. Kim, N.; Cha, S.; Peng, H. Optimal control of hybrid electric vehicles based on Pontryagin's minimum principle. *IEEE Trans. Control Syst. Technol.* **2011**, *19*, 1279–1287. [CrossRef]
8. Kareemulla, T.; Delprat, S.; Czelecz, L. State constrained hybrid vehicle optimal energy management: An interior penalty approach. *IFAC-PapersOnLine* **2017**, *50*, 10040–10045. [CrossRef]
9. Sánchez, M.; Delprat, S.; Hofman, T. Energy management of hybrid vehicles with state constraints: A penalty and implicit Hamiltonian minimization approach. *Appl. Energy* **2020**, *260*, 114149. [CrossRef]

10. Yeqin, W.; Zhen, W.; Yuyan, W.; Aoyun, X.; Chang, G.; Zhongyi, T. Research on energy optimization control strategy of the hybrid electric vehicle based on Pontryagin's minimum principle. *Comput. Electr. Eng.* **2018**, *72*, 203–213. [CrossRef]
11. Chong, H.; Minggao, O.; Liangfei, X.; Hewu, W. Approximate Pontryagin's minimum principle applied to the energy management of plug-in hybrid electric vehicles. *Appl. Energy* **2014**, *115*, 174–198. [CrossRef]
12. Paganelli, G.; Delprat, S.; Guerra, T.M.; Rimaux, J.; Santin, J.J. Equivalent consumption minimization strategy for parallel hybrid powertrains. In Proceedings of the Vehicular Technology Conference IEEE 55th Vehicular Technology Conference VTC Spring 2002, Birmingham, AL, USA, 6–9 May 2002. [CrossRef]
13. Sciarretta, A.; Guzzella, L. Control of hybrid electric vehicles. *IEEE Control Syst. Mag.* **2007**, *27*, 60–70. [CrossRef]
14. Zhang, F.; Xi, J.Q.; Langari, R. An adaptive equivalent consumption minimization strategy for parallel hybrid electric vehicle based on Fuzzy PI. In Proceedings of the 2016 IEEE Intelligent Vehicles Symposium (IV), Gothenburg, Sweden, 19–22 June 2016. [CrossRef]
15. Pei, D.; Leamy, M., J. Dynamic Programming-Informed Equivalent Cost Minimization Control Strategies for Hybrid-Electric Vehicles. *J. Dyn. Syst. Meas. Control* **2013**, *135*, 051013. [CrossRef]
16. Chao, Y.; Siyu, D.; Liang, L.; Sixong, Y.; Yiyong, Y.; Yue, Z. Adaptive real-time optimal energy management strategy based on equivalent factors optimization for plug-in hybrid electric vehicle. *Appl. Energy* **2017**, *203*, 883–896. [CrossRef]
17. Serrao, L.; Onori, S.; Rizzoni, G. A comparative analysis of energy management strategies for hybrid electric vehicles. *J. Dyn. Syst. Meas. Control* **2011**, *133*, 031012. [CrossRef]
18. Musardo, C.; Rizzoni, G.; Guezennec, Y.; Staccia, B. A-ECMS: An Adaptive Algorithm for Hybrid Electric Vehicle Energy Management. *Eur. J. Control* **2013**, *11*, 509–524. [CrossRef]
19. Chao, S.; Fengchun, S.; Hongwen, H. Investigating adaptive-ECMS with velocity forecast ability for hybrid electric vehicles. *Appl. Energy* **2017**, *185*, 1644–1653. [CrossRef]
20. Tufano, D.; De Bellis, V.; Malfi, E. Development of an on-line energy management strategy for hybrid electric vehicle. *Energy Procedia* **2018**, *148*, 106–113. [CrossRef]
21. De Bellis, V.; Malfi, E.; Tufano, D.; Bozza, F. Efficient Thermal Electric Skipping Strategy applied to the Control of Series/Parallel Hybrid Powertrain. *SAE Tech. Pap.* **2020**. [CrossRef]
22. Bozza, F.; Tufano, D.; Malfi, E.; Teodosio, L.; Libert, C.; De Bellis, V. Performance and Emissions of an Advanced Multi-Cylinder SI Engine Operating in Ultra-Lean Conditions. *SAE Tech. Pap.* **2019**. [CrossRef]
23. Dabadie, J.; Sciaretta, A.; Font, G.; Le Berr, F. Automatic Generation of Online Optimal Energy Management Strategies for Hybrid Powertrain Simulation. *SAE Tech. Pap.* **2017**. [CrossRef]
24. Bozza, F.; De Bellis, V.; Teodosio, L.; Tufano, D.; Malfi, E. Techniques for CO_2 Emission Reduction over a WLTC. A Numerical Comparison of Increased Compression Ratio, Cooled EGR and Water Injection. *SAE Tech. Pap.* **2018**. [CrossRef]
25. Wipke, K.B.; Cuddy, M.R.; Burch, S.D. ADVISOR 2.1: A user-friendly advanced powertrain simulation using a combined backward/forward approach. *IEEE Trans. Veh. Technol.* **1999**, *48*, 1751–1761. [CrossRef]
26. Dekraker, P.; Barba, D.; Moskalik, A.; Butters, K. Constructing Engine Maps for Full Vehicle Simulation Modeling. *SAE Tech. Pap.* **2018**. [CrossRef]
27. Valverde, V.; Clairotte, M.; Bonnel, P.; Giechaskiel, B.; Carriero, M.; Otura Garcia, M.; Gruening, C.; Fontaras, G.; Pavlovic, J.; Martini, G.; et al. *Joint Research Centre 2018 Light-Duty Vehicles Emissions Testing—Contribution to the EU Market. Surveillance: Testing Protocols and Vehicle Emissions Performance EUR29897 EN*; JRC117625; Publications Office of the European Union: Luxembourg, 2019; ISBN 978-92-76-12333-0. [CrossRef]
28. Mock, P.; Kühlwein, J.; Tietge, U.; Franco, V.; Bandivadekar, A.; German, J. The WLTP: How a New Test Procedure for Cars will Affect Fuel Consumption Values in the EU. Working Paper 2014-9. Available online: https://theicct.org/publications/wltp-how-new-test-procedure-cars-will-affect-fuel-consumption-values-eu (accessed on 17 April 2020).
29. Commission Regulation EU. 2018/1832 of 5 November 2018, Official Journal of the European Union L 301/1. Available online: https://eur-lex.europa.eu/legal-content/EN/TXT/PDF/?uri=CELEX:32018R1832&from=FR (accessed on 2 November 2019).

energies

MDPI

Article

Experimental and Simulation of Diesel Engine Fueled with Biodiesel with Variations in Heat Loss Model

Daniel Romeo Kamta Legue [1,2,*], Zacharie Merlin Ayissi [2,3,4,*], Mahamat Hassane Babikir [1], Marcel Obounou [1,2] and Henri Paul Ekobena Fouda [1]

1 Department of Physics, Energy—Electrical and Electronic Systems, University of Yaounde 1, Yaoundé P.O. Box 812, Cameroon; hassanemahamat6@gmail.com (M.H.B.); marcelobounou@yahoo.fr (M.O.); hekobena@gmail.com (H.P.E.F.)
2 Laboratory E3M, University of Douala, Douala P.O. Box 2701, Cameroon
3 Nationale Higher Polytechnique School, University of Douala, Douala P.O. Box 2701, Cameroon
4 Department of Automotive Engineering and Mécatronics, ENSPD, University of Douala, Douala P.O. Box 2701, Cameroon
* Correspondence: campusromo@yahoo.fr (D.R.K.L.); merlin.ayissi@gmail.com (Z.M.A.)

Abstract: This study presents an experimental investigation and thermodynamic 0D modeling of the combustion of a compression-ignition engine, fueled by an alternative fuel based on neem biodiesel (B100) as well as conventional diesel (D100). The study highlights the effects of the engine load at 50%, 75% and 100% and the influence of the heat loss models proposed by Woschni, Eichelberg and Hohenberg on the variation in the cylinder pressure. The study shows that the heat loss through the cylinder wall is more pronounced during diffusion combustion regardless of the nature of the fuels tested and the load range required. The cylinder pressures when using B100 estimated at 89 bars are relatively higher than when using D100, about 3.3% greater under the same experimental conditions. It is also observed that the problem of the high pressure associated with the use of biodiesels in engines can be solved by optimizing the ignition delay. The net heat release rate remains roughly the same when using D100 and B100 at 100% load. At low loads, the D100 heat release rate is higher than B100. The investigation shows how wall heat losses are more pronounced in the diffusion combustion phase, relative to the premix phase, by presenting variations in the curves.

Keywords: biodiesel; load; heat loss; cylinder pressure; heat release rate; python

Citation: Legue, D.R.K.; Ayissi, Z.M.; Babikir, M.H.; Obounou, M.; Ekobena Fouda, H.P. Experimental and Simulation of Diesel Engine Fueled with Biodiesel with Variations in Heat Loss Model. *Energies* **2021**, *14*, 1622. https://doi.org/10.3390/en14061622

Academic Editor: Fabio Bozza

Received: 5 February 2021
Accepted: 24 February 2021
Published: 15 March 2021

1. Introduction

Automobile pollution accounts for a large percentage of global pollution [1]. This pollution is largely caused by the use of fossil oil derivatives as fuel vehicles.

Spark-ignition and compression-ignition engines are the most widely used as a converter of the internal energy of fuels. They use petrol and diesel, respectively, as fuel. Both types of engines convert the internal energy of the fuels into thermal energy which is used in mechanical form to drive motor vehicles. Generally, heavy-duty vehicles, about 90% of the vehicles in use, are mostly diesel [2]. Self-ignition combustion offers better volumetric efficiency, which means that the diesel engine using this mode of combustion exploits a known energy potential. However, diesel engines present a serious problem for global health [3,4]. The intrinsic chemical composition of this fuel is largely responsible for this phenomenon. The combustion mode fossil fuels should be improved with regard to the level of pollutants emitted. Indeed, the emission of nitrogen oxides would be relatively responsible for pathologies linked to the proliferation of certain types of malignant tumors. In fact, this engine generally operates with poor and non-homogeneous mixtures, which leads to a relatively high level of nitrogen oxides (NO_x) and particulate matter (PM) [1,5,6]. The two major pollutants, particularly nitrogen oxides, are strongly dependent on the temperature of the combustion chamber, which is a function of the cylinder pressure resulting from the high compression ratio imposed by the performance research of this type of engine. The mechanism of formation of these pollutants is explicitly described by the Zeldovich mechanism [1,3,7].

The combustion of diesel engines offers a wide area of investigation [4,8,9]. Experimental work is being carried out. Ducted fuel injection (DFI) is proposed in the literature as a strategy to improve the fuel/gas load mixture of the compression-ignition engine relative to conventional diesel combustion (CDC) [5–7]. The concept of DFI is to inject each fuel spray through a small tube into the combustion chamber to facilitate the creation of a leaner mixture in the self-ignition zone, compared to a fuel jet not surrounded by a duct. The experiments are interesting. Nilsen et al. [5] studied the effects on emissions and engine efficiency using a two-hole injector tip for charging gas mixtures containing 16 and 21% mol oxygen. DFI seems to confirm that it is effective in reducing engine soot emissions. Soot and NOx are reduced with increasing dilution. Christopher [6] conducted an investigation where DFI and CDC were directly compared at each operating point in the study. At the low-load condition, the intake charge dilution was swept to elucidate the soot and NO_X performance of DFI. The authors mentioned that DFI likely has slightly decreased fuel conversion efficiencies relative to CDC. All these experimental studies are interesting and very promising, but they are still in the laboratory domain and require expensive instruments.

For economic reasons, research is increasingly oriented towards modeling engine combustion. The evaluation of wall heat losses, which are responsible for energy losses in thermal engines, is likely to optimize the operation of this type of engine. Modeling is based on control parameters such as rotational speed, compression ratio, engine load, combustion and injection models [7,10–12]. Modeling consists of reproducing physical phenomena using mathematical equations. This method offers a real-time saving and a significant reduction in costs. However, it is important to validate the results obtained from the modeling or simulation systems in order to give the results a formal character.

Zero-dimensional (0D) thermodynamic models of combustion are widely used in the literature. Caligiuri et al. [13] implemented a triple Wiebe model in order to describe and predict the heat release rate and ignition delay of dual-fuel combustion. The Aktar model has been successfully implemented and validated. Prakash correlation relating to the ignition delay was used. A methodological approach based on the prediction of the ignition delay of diesel engines has been implemented. Hariram et al. [14] conducted an investigation, and this experimental and theoretical study addressed the problem related to the effects of beeswax biodiesel mixtures with fossil diesel on combustion parameters. The zero-dimensional thermodynamic model was used as a numerical implementation approach. The variation in cylinder pressure, the net heat release rate and the ignition delay were addressed based on Webe's triple function. The code developed was successfully validated. These models allow evaluating the performance of thermal engines by the use of correlations based on thermodynamic laws.

The advantage of zero-dimensional thermodynamic models is that they are easy to use and do not require large computer memory capacities. The use of thermodynamic postulates makes them accessible and efficient. Thermodynamic 0D models allow the study of both fossil fuel combustion and alternative fuels.

The polluting nature of fossil fuels and the depletable nature and recurrent price fluctuations of fossil energy sources have led researchers for years to focus increasingly on alternative energy sources such as biofuels [4]. However, the production of biofuels has so far remained relatively low in relation to demand, given the many structural constraints, hence the need to maximize energy savings by optimizing the energy converters that use these fuels. Around 60% of energy in thermal engines is lost in several forms. However, engine losses remain a real problem as they account for almost 40% of the energy lost [15–17]. A numerical study of the heat loss through the wall would allow a better understanding of this phenomenon and improve the combustion and, by deduction, the efficiency of engines.

Several published articles have reported on the sensitivity between biodiesel and engine load compared to conventional diesel fuel; however, these do not show the specificity of the impact of the load on the cylinder pressure, taking into account the heat loss through the cylinder wall and the type of fuel used. This reasearch proposes a comparative study of the cylinder pressures generated by the combustion of diesel and neem biodiesel. The study is conducted taking into account the correlations between the heat loss through the cylinder wall. The heat release produced allows representing the phenomenology of the heat loss through the cylinder wall by dissociating the gross heat release rate from the net heat release rate during a combustion phase.

2. Modeling of Zero-Dimensional Thermodynamic Combustion

A modeling of the amount of energy released by combustion and a deduction of the cylinder pressure produced during the closed part of the cycle are proposed.

The chemical formulation of the fuel is the form $C_xH_yO_z$. The complete combustion equation of the fuel is given by Equation (1). The first principle of thermodynamics is applied in Equation (2) [18].

$$C_xH_yO_z + a(O_2 + 3.77N_2) \rightarrow xCO_2 + \frac{y}{2}H_2O + (3.77a)N_2 \tag{1}$$

$$dU = \delta W + \delta Q_W + \sum h_j dm_j \tag{2}$$

where dU represents the variation in the internal energy of the system, δW is the work supplied by the piston to the system, δQ_W is the heat loss through the cylinder wall. The term $\sum h_j dm_j$ represents the energy due to the variation in mass. The heat loss through the cylinder wall is modeled using three characteristic postulates: Woschni, Eichelberg and Hohenberg. By transforming and simplifying Equation (2), the differential Equation (3) is obtained as a function of the crankshaft angle (θ) [19].

$$\begin{cases} \frac{dp}{d\theta} = \frac{\gamma p}{V}\frac{dV}{d\theta} + \frac{(\gamma-1)}{V}\frac{dQ}{d\theta} \\ \frac{dT}{d\theta} = T(\gamma-1)\left[\frac{1}{pV}\frac{dQ}{d\theta} - \frac{1}{V}\frac{dp}{d\theta}\right] \end{cases} \tag{3}$$

where p and T represent the pressure and the cylinder temperature, γ is the specific heat ratio, θ is the crankshaft angle and V is the cylinder volume. The expression of this volume is governed by Equation (3) [19–21].

$$V(\theta) = \frac{\pi D^2 S}{8}(1 - \cos(\theta) + \lambda - \sqrt{\lambda^2 - sin^2(\theta)} + \frac{2}{CR-1}) \tag{4}$$

The variation in the cylinder volume V in relation to the crankshaft angle can be deduced [19].

$$\frac{dV(\theta)}{d\theta} = \frac{\pi D^2 S}{8}(1 - \frac{cos\theta}{\sqrt{\lambda^2 - sin^2\theta}})sin\theta \tag{5}$$

where λ is the ratio of the rod length, D is the cylinder bore, CR is the compression ratio and S is the stroke.

The term $\frac{dQ}{d\theta}$ is the heat release rate of the system expressed as follows [19]:

$$\frac{dQ}{d\theta} = \frac{\gamma}{\gamma-1}p\frac{dV}{d\theta} + \frac{1}{\gamma-1}V\frac{dp}{d\theta} \tag{6}$$

By entering the experimental combustion data into Equation (6), the heat release profile in the engine can be reconstructed as recommended by many authors using the analysis model [11,14,22]. The evolutions of the heat release profile(s) for an engine speed of 1500 rpm correspond to the load ranges: 25%, 50%, 75% and 100%. These loads correspond to 1.1, 2.5, 3.3 and 4.5 kW, respectively.

Four steps make up the modeling of the heat release, which are characteristic of the different terms used: the energy lost at the walls $\frac{dQ_w}{d\theta}$ integrates the heat exchange coefficient of the three models mentioned above; the energy released by combustion takes into account the internal energy of the fuel $\frac{dQ_{comb}}{d\theta}$ and two other terms due to exchanges with the surroundings (inlet and outlet mass flow).

The system is considered closed, and both terms are taken as zero. The system of Equation (7) is proposed.

$$\begin{cases} \frac{dQ_{comb}}{d\theta} = m_{inj} * LCV * \frac{dx_b}{d\theta} \\ \frac{dQ_w}{d\theta} = h_c A(\theta)(T - T_w)\frac{1}{\omega} \end{cases} \tag{7}$$

The terms appearing in Equation (7) are defined by Equation (8) at the heat exchanger surface as a function of the engine geometry. The characteristic fraction of fuel burned as a function of the crankshaft angle variation is given by Equation (9).

$$A(\theta) = \left(\pi * \frac{D^2}{2}\right) + \pi * D * \frac{L}{2}(\lambda + 1 - Cos\theta - \sqrt{\lambda^2 - Sin^2(\theta)} \tag{8}$$

$$x_b = 1 - exp\left[-a\left(\frac{\theta - \theta_0}{\Delta\theta}\right)^{m+1}\right] \tag{9}$$

The fraction of fuel burnt in relation to the crankshaft angle θ is evaluated. m_{inj} is the mass of the injected fuel, T_w is the wall temperature, ω is the engine rotation frequency, LCV is the characteristic lower calorific value of the fuel used and h_c is the convective heat loss coefficient depending on the heat loss model used, expressed in kW/m^{2}K.

The different heat loss coefficients through the walls are proposed [20].

$$h_c = 3.26 * D^{-0.2} * p^{0.8} * T^{-0.55} * W^{0.8} \tag{10}$$

$$h_c = 0.013 * V^{-0.06} * P^{0.8} * T^{-0.4}(V_P + 1.4)^{0.8} \tag{11}$$

$$h_c = 7.799 * 10^{-3} * V_P{}^{\frac{1}{3}} * p^{0.5} * T^{0.5} \tag{12}$$

Equations (10)–(12) represent the coefficients of Woschni, Hohenberg and Eichelberg, respectively. V_P is the average piston speed.

The term W is deduced from Equation (13):

$$W = 2.28 * V_P + C_1 * \frac{V_d * T_a}{p_a * V_a}(p(\theta) - p_m) \tag{13}$$

with C_1 = 0.00324. On the other hand, p_m represents the motored pressure, and subscript "a" is the reference condition

$$V_p = \frac{n * S}{30} \tag{14}$$

where n is the engine speed (rpm).

2.1. Ignition Delay Model

The ignition of the fuel is not spontaneous; it is controlled by the Hardenberg and Haze model whose mathematical formulation is given by relation 14 [23–25]:

$$ID = (0.36 + 0.22\overline{V}_p)exp\left[E_A\left(\frac{1}{R\overline{T}} - \frac{1}{17.190}\right)\left(\frac{21.2}{\overline{p} - 12.4}\right)^{0.63}\right] \tag{15}$$

where E_A is the activation energy, where, R is the universal gas constant.

2.2. Détermining the Heat Release Rate

The net heat release rate (Net HRR) represents the thermal energy useful for engine operation [26–28]. It is transformed into mechanical energy by the piston during combustion inside the combustion chamber. The gross heat release rate (Gross HRR) is, in fact, the thermal energy transformed into mechanical work, and the energy losses in the walls are associated with it. The combustion of the fuel heats the cylinder walls by convection, and a finite amount of calorie is transmitted to the cooling water, constituting a heat loss. A curve representing the difference between the total combustion energy and the useful thermal energy is generated. The characteristic curves of conventional diesel D100 combustion are compared with those of biodiesel B100.

2.3. Experimental Setup

The experimental device is an air-cooled single-cylinder motor of the Lister Peter 0100529-TS1 series type, whose technical characteristics are confined to Table 1. It is a naturally aspirated, four-stroke engine with direct injection. Cooling is by ambient air. This equipment is located at Department of Energetic System and Sustainable Development (DSEE) of *Ecole des Mines de Nantes* (EMN) France. Figure 1 shows the experimental device used during the study.

Table 1. Engine characteristics.

Lister-Petter-01005299-TS1 Série		
Injection Pressure	(Bar)	250
Piston Diameter/Stroke	(mm)	95.3/88.9
Connecting Rod Length	(mm)	165.3
Engine Capacity	(m^3)	630
Compression Ratio	-	18
Injection Timing	(°CA)	15° Before TDC
Engine Power	(kW)	4.5 à 1500 trs/min

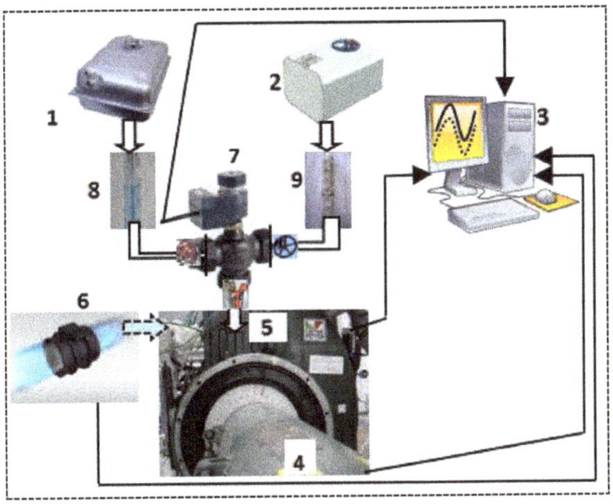

Figure 1. Experimental setup. Legend; 1. Diesel fuel tank; 2. Biodiesel tank; 3. High-frequency acquisition system and computer; 4. Brake dumb dynamometer; 5. Diesel monocylindric motor; 6. Debimeter and air flow sensor; 7. Three-way valve with a purge system; 8. Diesel burette; 9. Biodiesel burette.

The variation in the load was obtained by modifying the effective power of the engine by a dynamometer as well as the quantity of fuel admitted into the cylinders. The engine load was varied at 25%, 50%, 75% and 100%. Effective power of 4.2 kW corresponds to the maximum engine load (100%).

The fuel supply system consists of two different fuel tanks, (1) and (2), that lead to the three-way valve (7). Each way of this valve was equipped with a stop valve. One tank was filled with conventional diesel fuel D100 (1) and the other (2) with the neem oil methyl ester biodiesel B100. The fuel change during the test was conducted in advance by the shutdown of the second fuel supply and the systematic purge by a system incorporated into the three-way valve. It took ten minutes, with the engine running, to begin measurements without the relative accidental risk of mixing the two fuels.

The acquisition of the cylinder pressure was carried out using the Indwin AVL engine rotating at 1500 rpm. The angular position of the crankshaft and the engine speed were obtained via an AVL 364C encoder, attached to the crankshaft. The encoder measures only parameters with frequencies greater than 90 kHz. Table 2 presents the test matrix table, which summarizes the main characteristics of the test carried out, and Table 3 presents the relative errors of the different sensors used.

Table 2. Main characteristics of the test.

Fuels	Engine Speed rpm	Engine Load kW				Injection Timing °CA
Diesel Biodiesel	1500	1.18	2.13	3.38	4.59	15

Table 3. Main characteristics of the test.

Parameters	Errors
Engine torque	±0.1 N.m
Engine speed	±3 rpm
Injection timing	±0.05 °CA
Cylinder pressure	±0.5 of the measured value
LCV	±0.25% of the measured value
Admission air flow	±0.1% of the measured value
Fuel flow	±0.5% of the measured value
Injection pressure	±2 bars
Inlet air temperature	±1.6 °C
Exhaust air temperature	±1.6 °C

2.4. Fuel Characteristics

The experimentation resulting from the work of Ayissi et al. [29] was successfully repeated under the same conditions. The neem oil obtained was characterized in order to calibrate its properties and use it in a heat engine. The characteristic performance values obtained were mostly those found by the authors. These were adopted and are reported in Table 4.

Table 4. Fuel characteristics.

Fuel Properties	(B100)	(D100)
Density (kg/m^3)	883.3	830
Cetane Number	51.3	48
Lower Heating Value LCV (MJ/kg)	39.7	42.5

2.5. Numerical Method

Figure 2 presents a synoptic view of the overall methodological organization of the digital study. The end-of-compression characteristic parameters were considered and retained as well as the characteristics of the model engine. The intrinsic properties of the fuels were taken into account. A numerical code was developed. The characteristic equations of the implemented models were solved by the Runge–Kutta method.

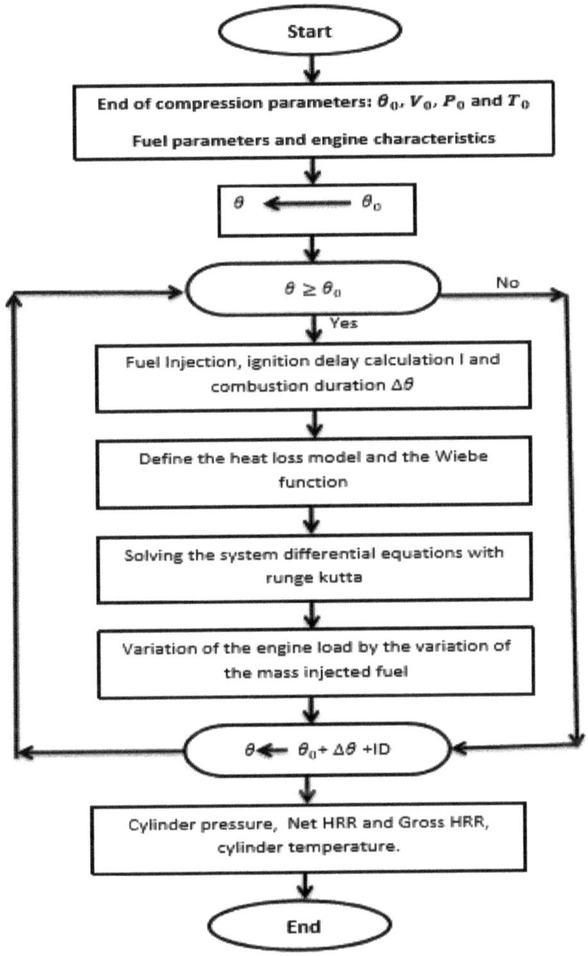

Figure 2. Digital diagram.

3. Results and Discussion

3.1. Experimental Evaluation of Cylinder Pressure

Figure 3 shows the increase in the experimental cylinder pressure based on the crankshaft angle when using B100 and D100 at 100% load. A similarity between the increase in the cylinder pressure curves of biodiesel B100 and D100 is observed. However, there is an estimated pressure increase of 3.3% when using B100 at 100% load. A faster pressure increase in B100 compared to D100 is also observed.

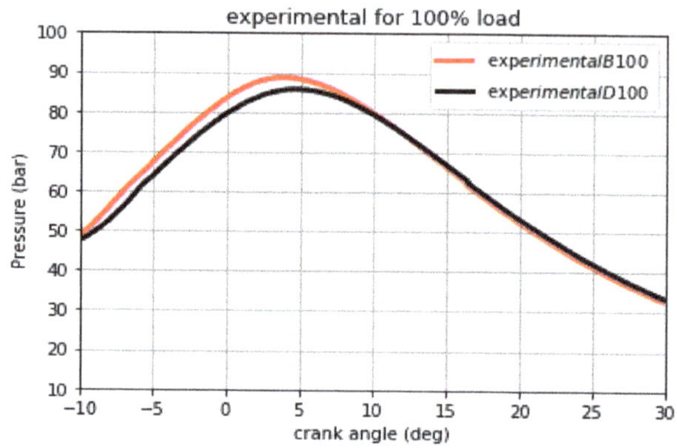

Figure 3. Comparison of experimental pressures of B100 and D100 at 100% load.

High cylinder pressure is characteristic of good fuel vaporization as well as relatively better oxygenation of fuel. This observation has also been made in the literature. Indeed, the studies of Tarabet [18], Mohamed et al. [24] and Evangelos [30] demonstrated this quite clearly. The transesterification process helps reduce the viscosity of biodiesel. The operation of the engine under high loads would sufficiently reduce the effects of the viscosity of biodiesel on the kinetics of combustion. The increase in cylinder pressure during biodiesel combustion is probably due to the relative simplicity of the molecular structure of the hydrocarbons that B100 contains. This molecular structure would be even more advantageous at high load since the high heat inside the cylinder would contribute to the destructuring of the macromolecules of biodiesel, which is relatively less complex after the transesterification process. This predisposition of biodiesel to good combustion could promote the rapid appearance of radicals inside the drops close to the stoichiometry and catalyzed by the high cylinder temperature. This spatial–temporal consideration would suit the premix combustion process with the consequence of the increase in the pressure peak in the cylinders. The intramolecular presence of residual oxygen atoms would be another factor in raising the pressure peak in the cylinders as biodiesel is relatively oxygenated and the kinetics of combustion are easy. This high-pressure peak would be one of the reasons that increases the NOx level in post-combustion gases of an engine fueled with biodiesel generally [5,6,26]. Chiavola et al. [22] and Tarabet [18] mentioned it. The faster pressure increase in B100 biodiesel compared to diesel is characteristic of a short ignition period. Mohamed et al. and Evangelos [20,21,27] made the same observation in similar studies.

3.2. *Cylinder Pressure Evaluation Considering the Heat Loss through the Cylinder Wall Patterns of B100 and D100 at 100% Load*

Figures 4 and 5 show the evolution of the cylinder pressure during the combustion of B100 and D100, respectively. Under the constraint of three heat loss models, namely, Eichelberg, Hohenberg and Woschni, the time evolution of the cylinder pressure is simulated taking into account the heat loss through the cylinder wall and associated with the experimental pressure curve. A relative similarity between the numerical and experimental pressure curves for both B100 and D100 combustions is observed. The peaks characteristic of the temporal variation in the cylinder pressure of the three numerical models are evaluated at 90, 92 and 85 bars, respectively, for the Eichelberg, Hohenberg and Woschni models. The result of the experimental set is 89 bars. The overestimation of cylinder pressures by the implemented models compared to the experimental study is explained by the fact that the numerical models take into account the heat loss through the cylinder wall. The difference observed in the variations in the pressure peaks of the implemented models, although varying from one correlation to another, remains confined to the intervals encountered in the literature.

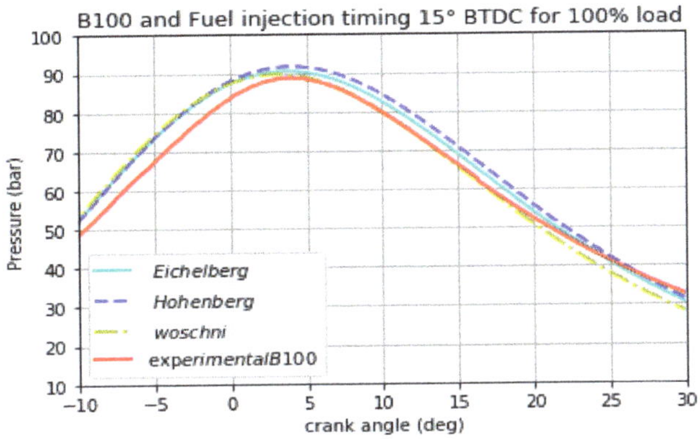

Figure 4. Cylinder pressure of B100 at 100%.

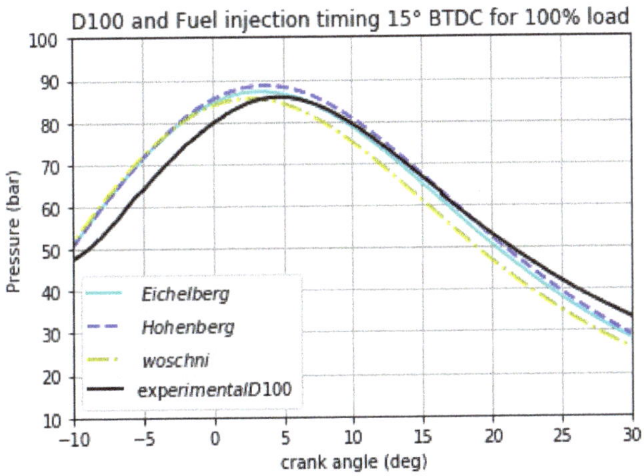

Figure 5. Cylinder pressure of D100 at 100%.

The three wall loss models studied represent fairly well the combustion of the model fuels in the compression-ignition configuration.

In the specific case of B100 combustion illustrated in Figure 4 at 100% load, the characteristic curves of the three models studied keep the same spatial position during the initial 12 °CA of combustion. Between the beginning of the rise in cylinder pressure and its peak, a gap of 7 bars is observed between the cumulative numerical curves and the experimental curve. This gap which benefits the simulated correlations is probably due to the fact that the simulated models take into account the heat loss through the cylinder wall in the physical materialization of the spatiotemporal evolution of the cylinder pressure.

The cylinder pressure of the Woshni model begins a declination around 12 °CA after the start of the rise in the cylinder pressure. The initiated pressure drop falls below the characteristic values of the experimental cylinder pressure. This underestimation of the time evolution of the cylinder pressure by the correlation proposed by Woshni could be assimilated by the fact that it does not take into account the piston speed. The piston displacement would present the cylinder surface at the temperature gradient generated by combustion, which would more or less increase the heat loss through the cylinder wall. Hohenberg's model is the one that retains the highest pressure peak of all the correlations implemented.

In the specific case of Figure 5, it is observed that the curves of the numerical models remain confused during the first 10 °CA before seeing the declination in the characteristic curve of the Woshni model. At this load range, all the models shown decline below the experimental cylinder pressure around 17 °CA. Heat losses at 75% load could be minimized in the post-combustion period. A similar observation is made at 100% load, under the same simulation conditions, although the declination occurs at 25 °CA.

3.3. Cylinder Pressure Evaluation Taking into Account the Heat Loss through the Cylinder Wall Patterns When Engine Is Fueled with B100 at 25%, 50%, 75% and 100% Load

Figures 6–8 show the evolution of the cylinder pressure at 75%, 50% and 25% load, respectively. A decrease in the cylinder pressure proportional to the load is observed at all measuring points. The decrease in the mass of fuel allowed would be at the origin of this observation during the combustion of B100.

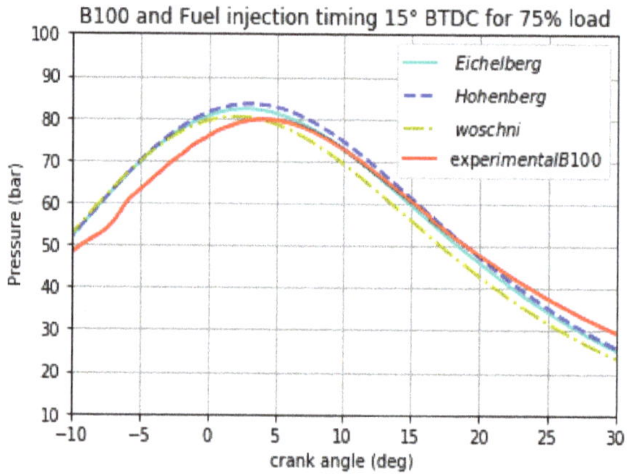

Figure 6. Cylinder pressure of B100 at 75% load.

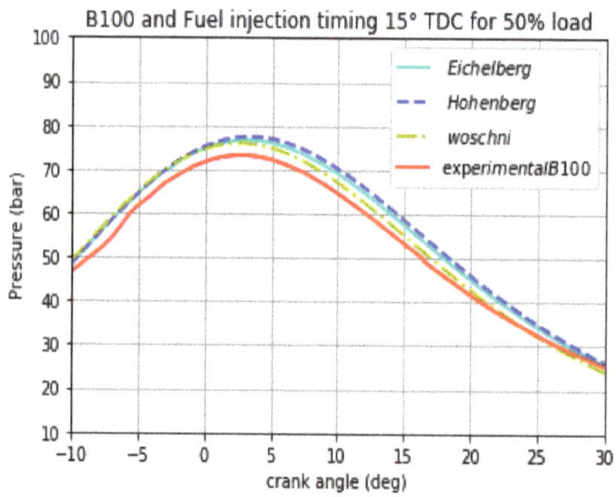

Figure 7. Cylinder pressure of B100 at 50% load.

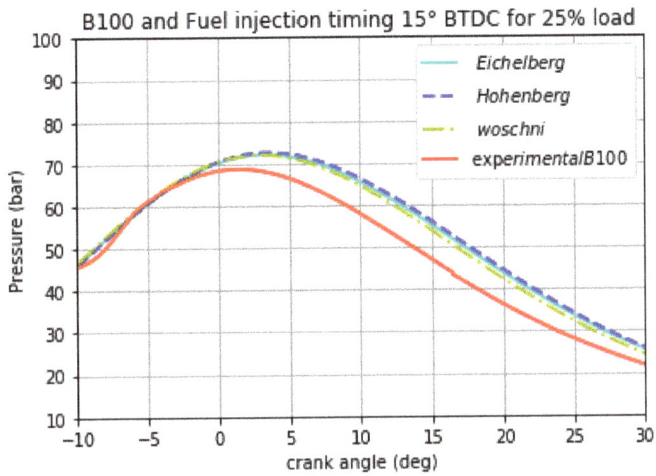

Figure 8. Cylinder pressure of B100 at 25% load.

The experimental pressure decreases by 10% and 6% proportionally to the passage from 100% to 75% load and from 75% to 25% load successively. This drop in experimental pressure is evaluated at 22.5% between 100% and 25% load. This decrease in cylinder pressure could be attributed to the reduction in effective power characterized by the decrease in the quantity of fuel admitted into the cylinders. However, with regard to the evolution of the pressure curves whatever the type of model implemented, the thermal losses would be more important during the ascending phase of the piston.

Considering the specific combustion of B100 at 75% load, the statistical differences between the simulated pressure curves and the experimental curve, between 20 and 30 °CA, are the smallest, all loads combined. This observation visible in Figure 6 would indicate that the heat losses are less important at this phase of engine operation. The rising part of the numerical pressure curves shows the largest deviations from other load points studied. At this load range, the Woschni model underestimates the approximation of the cylinder pressure value.

The 50% load range seems to present the best compromise between the estimated values of wall losses according to the models studied. This load range would be relatively close to the nominal operating rating of the test engine.

Indeed, when compared with other load ranges studied, a better compromise is observed between the evolutions of the combustion models and the experimental curve. The deviations observed during the other load ranges evaluated remain minimized for the Woschni, Eichelberg and Hohenberg correlations.

At 25% load, the three models have curves representing the temporal evolution of the cylinder pressure which are in agreement. This observation can be seen more clearly in Figure 8. This observation could be explained by a small parametric difference between the values of the different correlations implemented.

In general, the Eichelberg and Hohenberg models agree on the time evolution of the cylinder pressure during the tests at 50% and 25% load. This rise in digital cylinder pressure is characteristic of the rise deduced from the heat loss through the cylinder wall during the descending phase of the pressure curves considered. The tendency for heat loss to increase during this phase and at this load range could be due to the fact that the gases have more time to lick the walls because of the reduced speed of the piston in its descending phase.

Observation at high loads shows that the Eichelberg and Hohenberg models show a slight difference in the approximation of the cylinder pressure evolution. However, this characteristic deviation remains less than 0.6 bar regardless of the time position of the piston. The Woschni model seems to underestimate the time value of the heat loss through the cylinder wall regardless of the engine load point compared to the other two correlations implemented.

3.4. Evaluation of Heat Release

The heat release at different engine loads at 100%, 75% and 50% is evaluated. The heat release curves of the different fuel models D100 and B100 have characteristic curves of combustion by auto-ignition.

Whatever the load range or the fuel tested, the evolution profiles of net heat release rate (Net HRR) and gross heat release rate (Gross HRR) during the premix pressure rise phase remain the same. A relatively significant difference is observed between Net HRR and Gross HRR when compared with D100 diesel. These differences are evaluated at 10, 10, and 20 J/°CA for 50%, 75% and 100% load, respectively. These differences would mean that heat losses would be greater during combustion of B100 biodiesel. Heat losses in the premix phase would be minimized based on information from the study.

3.4.1. Heat Release Curve for 100% Load

Figures 9 and 10 show the different heat release profiles of D100 diesel and B100 biodiesel, respectively, for a 100% load range, equivalent to 4.5 kW. The Net HRR converted by the engine to mechanical energy when B100 is used as a test fuel is approximately 30 J/°CA. Between 0 and 10 °CA, this biofuel provides an estimated energy of 17.32 J/°CA compared to 21.31 J/°CA for D100. No fundamental difference between the different Gross HRR values of the two fuels over this load range is observed. B100 biodiesel would have better performance at high loads. This behavior is probably due to the oxygenation of the molecule and the structural simplicity of its chemical composition [23–26,30]. At high temperatures, the relatively higher viscosity of biodiesel is no longer fundamentally a handicap for the combustion of this fuel. This would justify its relatively good combustion.

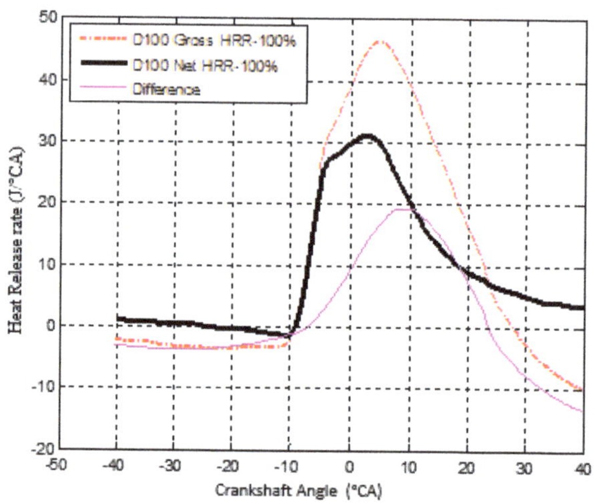

Figure 9. Profile of net heat release rate (Net HRR) evolution and gross heat release rate (Gross HRR) for engine speed n = 1500 rpm at 100% load according to D100 combustion.

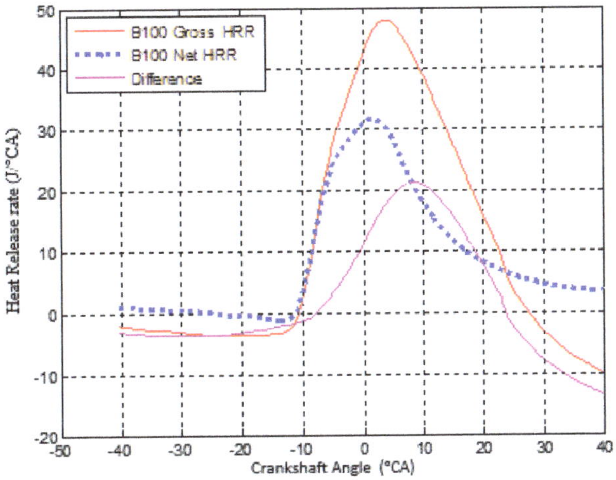

Figure 10. Profile of net heat release rate (Net HRR) evolution and gross heat release rate (Gross HRR) for engine speed n = 1500 rpm at 100% load according to B100 combustion.

3.4.2. Heat Release Curves Net HRR and Gross HRR for 75% Load

Figures 11 and 12 show the different profiles the net heat release rate (Net HRR) and the gross heat release rate (Gross HRR) for an engine speed of 1500 rpm at 75% load. Between 0 and 10 °CA, the combustions of B100 and D100 produce 13.20 and 14.20 J/°CA, respectively. The heat loss through the cylinder wall as a function of the crankshaft position is evaluated at about 10 J/°CA during B100 combustion, representing 20% higher than during the combustion of D100 at this range of load. A similar observation was made by Pankaj et al. [31] and Muhamed et al. [28]. These authors showed that at a range of 75%, the ignition delay of B100 and D100 increases. This increase of about 2 °CA in the ignition delay is significant and has also been observed by Muhamed et al. [27] at this range of load.

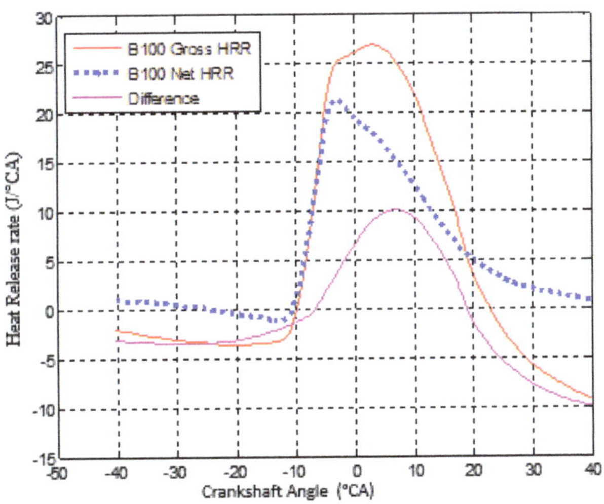

Figure 11. Profile of net heat release rate (Net HRR) evolution and gross heat release rate (Gross HRR) for engine speed n = 1500 rpm at 75% load according to the B100 combustion analysis model.

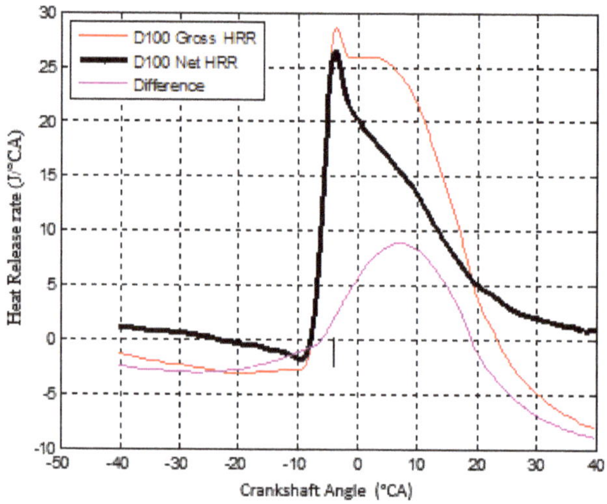

Figure 12. Profile of net heat release rate (Net HRR) evolution and gross heat release rate (Gross HRR) for engine speed n = 1500 rpm at 75% load according to D100 combustion.

3.4.3. The Net HRR and Gross HRR for 50% Load

Figures 13 and 14 show the different profiles of the evolution of the net heat release rate (Net HRR) and the gross heat release rate (Gross HRR) for an engine speed of 1500 rpm at 50% load. These two figures indicate that the diffusion combustion phase of B100 and D100 starts with a Net HRR of 20 and 21 J/°CA, respectively. The heat loss through the cylinder wall (curve representing the difference) is estimated at 10 for B100 and 8 J/°CA for D100. At 10 °CA, the Net HRR value of B100 is 12 and 13.5 J/°CA for D100. The heat loss through the cylinder wall values are as follows: B100 (10 J/°CA), D100 (8 J/°CA).

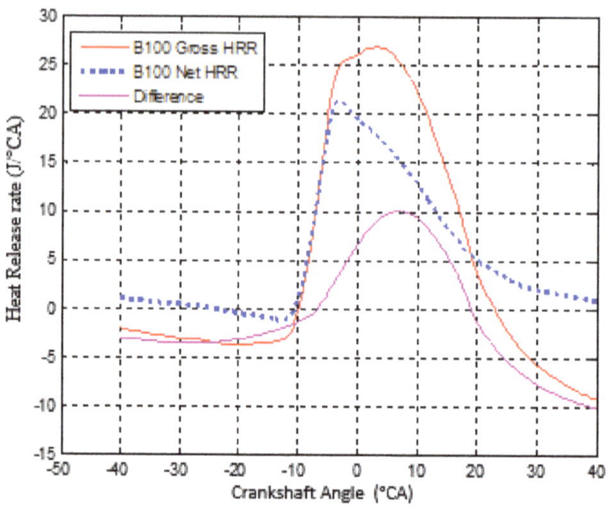

Figure 13. Profile of net heat release rate (Net HRR) evolution and gross heat release rate (Gross HRR) for engine speed n = 1500 rpm at 50% load according to B100 combustion.

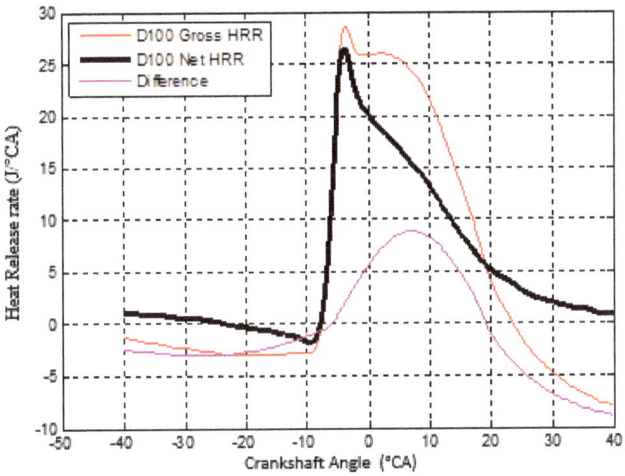

Figure 14. Profile of net heat release rate (Net HRR) evolution and gross heat release rate (Gross HRR) for engine speed n = 1500 rpm at 50% load according to D100 combustion.

4. Conclusions

An experimental investigation coupled with a thermodynamic 0D simulation of a neem-based biodiesel, a conventional diesel and their respective different heat releases was carried out. The different correlations of Woschni, Eichelberg and Hohenberg served as numerical models in the systemic simulation of combustion phenomenology. The Woshni correlation best represents the peak of the cylinder pressure comparison made with other implemented models. This correlation, however, seems to relatively underestimate the value of the cylinder pressure after bottom dead center. The Eichelberg and Hohenberg models represent the spatial–temporal evolution of the cylinder pressure in the same way. Implementation of other heat losses through cylinder wall models such as Zainal or Sitkey could be exploited for greater openness to the field of investigation. The investigation of the cylinder pressure confirmed that the load rate has an influence on the value of the peak of the cylinder pressure. The study presented provides a better understanding of the phenomenology of parietal thermal losses from the combustion of alternative fuels and its impact on the performance of combustion ignition engines. The cylinder pressure in the case of the combustion of biodiesel B100 is estimated at 89 bars against 86 bars for diesel D100; a gap of 3.3% to the advantage of B100 is thus observed when using this biofuel at 100% load. A faster pressure increase in B100 compared to D100 is also observed. A study of a parametric variation in the ignition delay associated with this study could better explain its influence on the comparative evolution of the peak cylinder pressure depending on the nature of the fuel.

Author Contributions: Conceptualization, D.R.K.L. and Z.M.A.; methodology, M.O.; software, D.R.K.L. and Z.M.A. and M.H.B.; validation, E.F.H.P., M.O. and Z.M.A.; formal analysis, Z.M.A.; investigation, Z.M.A. and D.R.K.L.; resources, M.O.; data curation, D.R.K.L.; writing—original draft preparation, Z.M.A. and D.R.K.L.; writing—review and editing, Z.M.A. and D.R.K.L.; visualization, M.H.B.; supervision, E.F.H.P.; project administration, M.O.; funding acquisition, Z.M.A. and D.R.K.L. All authors have read and agreed to the published version of the manuscript.

Funding: This research received no external funding.

Institutional Review Board Statement: Ethical review and approval were waived for this study, due to the fact that the studies not involve humans or animals.

Informed Consent Statement: Patient consent was waived due to the reason that the studies not involving humans.

Acknowledgments: We thank the Department of Energetic System and Sustainable Development of *Ecole des Mines de Nantes* EMN for giving us the experimental device; the G.C.C. team for always encouraging us in combustion and simulation aspects; and the team members of Energy, Electric and Electronic Systems, University of Yaoundé 1, for welcoming and assisting us.

Conflicts of Interest: The authors declare that there is no conflict of interests regarding the publication of this paper.

Abbreviations

θ	Crank angle (°CA)
$\Delta\theta$	Total combustion duration (°CA)
θ_0	Start of combustion (°CA)
x	Burnt mass fraction
W	Work done (KJ)
U	Internal energy
CR	Compression ratio
c_v	Specific heat at constant volume
h_c	Heat transfer coefficient (W. m^{-2}.K^{-1})
m	Masse (Kg)
R	Gas constant
R	Gas constant
Q	Heat transfer (KJ)
V	Volume (m^3)
p	Pressure (Bar)
T	Temperature (K)
$A(\theta)$	Area exposed to heat transfer (m^2)
D	Cylinder bore (m)
S	Stroke (m)
n	Engine speed
ω	Angular velocity (rad/s)
V_p	Mean piston speed (rad/s)
TDC	Top dead center (deg)
BTDC	Before TDC (deg)
BDC	Bottom dead center (deg)
HRR	Heat release rate
°CA	Degree crank angle

References

1. Park, J.; Choi, I.; Oh, J.; Lee, C. Nitrogen Oxides and Particulate Matter from Marine Diesel Oil (MDO), Emulsified MDO, and Dimethyl Ether Fuels in Auxiliary Marine Engines. *J. Mar. Sci. Eng.* **2020**, *8*, 322. [CrossRef]
2. Moussou, J. Caractérisation Expérimentale du Flux Thermique Transitoire Pariétal Pour Différents Modes de Combustion. Ph.D. Thesis, Ecole Nationale Supérieure de Mécanique et d'Aérotechique-Poitiers, Chasseneuil-du-Poitou, France, 2019.
3. Loschiavo, L. Diesel Particulate Matter & Occupational Health Issue. Prepared by AIOH Exposure Standards Committee Update. 2013. Available online: https://gastech.com/files/dpm/Diesel-Particulate-Matter-and-Occupational-Health-Issues.pdf (accessed on 23 January 2020).
4. U.S. Department of Energy. *Clean Cities Alternative Fuel Price Report*; U.S. Department of Energy: Washington, DC, USA, 2018; p. 4. Available online: https://afdc.energy.gov/publications/search/category (accessed on 23 January 2020).
5. Nilsen, C.W.; Biles, D.E.; Mueller, C.J. Using Ducted Fuel Injection to Attenuate Soot Formation in a Mixing-Controlled Compression-Ignition Engine. *SAE Int. J. Engines* **2019**, *12*, 309–322. [CrossRef]
6. Nilsen, C.W.; Biles, D.E.; Yraguen, B.F.; Mueller, C.J. Ducted Fuel Injection vs. Conventional Diesel Combustion: Extending the Load Range in an Optical Engine with a Four-Orifice Fuel Injector. *SAE Int. J. Engines* **2020**, *14*. [CrossRef]
7. Gehmlich, R.K.; Mueller, C.J.; Ruth, D.J.; Nilsen, C.W.; Skeen, S.A.; Manin, J. Using Ducted Fuel Injection to Attenuate or Prevent Soot Formation in Mixing-Controlled Combustion Strategies for Engine Applications. *Appl. Energy* **2018**, *226*, 1169–1186. [CrossRef]

8. Dhole, A.; Yarasu, R.; Lata, D. Investigations on the combustion duration and ignition delay period of a dual fuel diesel engine with hydrogen and producer gas as secondary fuels. *Appl. Therm. Eng.* **2016**, *107*, 524–532. [CrossRef]
9. Abhishek, A. Modélisation Mathématique de la Formation des NOX, et de la Volatilisation des Métaux Lourds Lors de l'Incinération sur Grille Ménagères. Ph.D. Thesis, Institut National Polytechnique de Lorraine, Lorraine, France, 2008.
10. Di Blasio, G.; Bonura, G.; Frusteri, F.; Beatrice, C.; Cannilla, C.; Viscardi, M. Experimental Characterization of Diesel Combustion Using Glycerol Derived Ethers Mixtures. *SAE Int. J. Fuels Lubr.* **2013**, *6*, 940–950. [CrossRef]
11. Bolla, M.; Chishty, M.A.; Hawkes, E.R.; Chan, Q.N.; Kook, S. Influence of turbulent fluctuations on radiation heat transfer, NO and soot formation under ECN Spray A conditions. *Proc. Combust. Inst.* **2017**, *36*, 3551–3558. [CrossRef]
12. Park, S.H.; Yoon, S.H. Effect of dual-fuel combustion strategies on combustion and emission characteristics in reactivity controlled compression ignition (RCCI) engine. *Fuel* **2016**, *181*, 310–318. [CrossRef]
13. Caligiur, C.; Renzi, M.; Batratieri, M. Development of a 0D Thermodynamics Combustion Simulation Tool for a Dual Fuel Diesel-Producer Gas Engine. In Proceedings of the 8th European Combustion Meeting, Dubrovnik, Croatia, 18–21 April 2017. Available online: https://www.researchgate.net/publication/320244655 (accessed on 23 January 2020).
14. Korsunovs, A.; Campean, F.; Pant, G.; Garcia-Afonso, O.; Tunc, E. Evaluation of zero-dimensional stochastic reactor modelling for a Diesel engine application. *Int. J. Engine Res.* **2020**, *21*, 592–609. [CrossRef]
15. Binder, C.; Nada, F.A.; Richter, M.; Cronhjort, A.; Norling, D. Heat Loss Analysis of a Steel Piston and a YSZ Coated Piston in a Heavy-Duty Diesel Engine Using Phosphor Thermometry Measurements. *SAE Int. J. Engines* **2017**, *10*, 1954–1968. [CrossRef]
16. Dinescu, D. Modélisation des Moteurs Thermiques pour l'Evaluation des Stratégies de Contrôle Moteur. Ph.D. Thesis, Ecole Polytechnique de l'Université de Nantes, Nantes, France, 2010.
17. Theotokatos, G.; Stoumpos, S.; Bolbot, V.; Boulougouris, E. Simulation-based investigation of a marine dual-fuel engine. *J. Mar. Eng. Technol.* **2020**, *19*, 5–16. [CrossRef]
18. Tarabet, L. Etude de la Combustion d'un Biocarburant Innovant dans les Moteurs à Combustion Interne de Véhicules. Ph.D. Thesis, Ecole des Mines de Nantes, Nantes, France, 2012.
19. Hariram, V.; Bharathwaaj, R. Application of zero-dimensional thermodynamic model for predicting combustion parameters of CI engine fuelled with biodiesel-diesel blends. *Alex. Eng. J.* **2016**, *55*, 3345–3354. [CrossRef]
20. Pise, A.N.; Deshmukh, M.M. Second Law Analysis of Compression Ignition Engine fuelled with Ethanol Blending by Using Heat Transfer Models. *Int. Res. J. Eng. Technol.* **2017**, *5*, 1279–1289.
21. Rostami, S.; Ghobadian, B.; Kiani, M.K.D. Effect of the Injection Timing on the Performance of a Diesel Engine Using Diesel-Biodiesel Blends. *Int. J. Automot. Mech. Eng.* **2014**, *10*, 1945–1958. [CrossRef]
22. Chiavola, O.; Recco, E. Emission Performance of a Diesel Engine Fuelled with Petrol Diesel, Green Diesel, and Waste Cooking Oil Blends. *J. Combust.* **2018**. [CrossRef]
23. Abbe, C.V.N.; Nzengwa, R.; Danwe, R.; Ayissi, Z.M.; Obonou, M. A study on the 0D phenomenological model for diesel engine simulation: Application to combustion of Neem methyl esther biodiesel. *Energy Convers. Manag.* **2015**, *89*, 568–576. [CrossRef]
24. Shehata, M.S. Emissions, performance and cylinder pressure of diesel engine fuelled by biodiesel fuel. *Fuel* **2013**, *112*, 513–522. [CrossRef]
25. Mohamed, M.; Chee, K.; Ali, F.; Mohamed, S.; Osayed, A.; Abel, F. Diesel Engine Performance, Emissions and Combustion Characteristics of Biodiesel and Its Blends Derived from Catalytic Pyrolysis of Waste Cooking Oil. *Energies* **2020**, *13*, 5708. [CrossRef]
26. Zheng, J.; Caton, J.A. Second law analysis of a low temperature combustion diesel engine: Effect of injection timing and exhaust gas recirculation. *Energy* **2012**, *38*, 78–84. [CrossRef]
27. Makode, C.R.; Deshmukh, M.M. Second Law Analysis of Diesel Engine by Comparing Different Heat Transfer Models. *Int. J. Eng. Res. Technol.* **2016**, *4*, 2–4.
28. Qasim, M.; Ansari, T.M.; Hussain, M. Combustion, Performance, and Emission Evaluation of a Diesel Engine with Biodiesel Like Fuel Blends Derived From a Mixture of Pakistani Waste Canola and Waste Transformer Oils. *Energies* **2017**, *10*, 1023. [CrossRef]
29. Ayissi, Z.M.; Ayina, L.M.O.; Salem, C.; Jean, G.; Obonou, M. Experimental studies on the combustion characteristics and performance of a direct injection engine fueled with Neem Methyl Ester Biodiesel. *Energy Convers. Manag.* **2010**, *51*, 2985–2992.
30. Giakoumis, E.G. Diesel and Spark Ignition Engines Emissions and After-Treatment Control: Research and Advancements. *Energies* **2017**, *10*, 1882. [CrossRef]
31. Pankaj, S.; Nitin, M.; Subhash, L. Investigation of Combustion Characteristics of a Cottonseed Biodiesel Fuelled Diesel Engine. *Procedia Technol.* **2016**, *25*, 1049–1055.

energies

Article

A New Method to Determine the Impact of Individual Field Quantities on Cycle-to-Cycle Variations in a Spark-Ignited Gas Engine

Clemens Gößnitzer [1,*] and Shawn Givler [2]

1 LEC GmbH, Inffeldgasse 19, 8010 Graz, Austria
2 INNIO Waukesha Gas Engines Inc., 1101 West St. Paul Avenue, Waukesha, WI 53188, USA; shawn.givler@innio.com
* Correspondence: clemens.goessnitzer@lec.tugraz.at

Abstract: Cycle-to-cycle variations (CCV) in spark-ignited (SI) engines impose performance limitations and in the extreme limit can lead to very strong, potentially damaging cycles. Thus, CCV force sub-optimal engine operating conditions. A deeper understanding of CCV is key to enabling control strategies, improving engine design and reducing the negative impact of CCV on engine operation. This paper presents a new simulation strategy which allows investigation of the impact of individual physical quantities (e.g., flow field or turbulence quantities) on CCV separately. As a first step, multi-cycle unsteady Reynolds-averaged Navier–Stokes (uRANS) computational fluid dynamics (CFD) simulations of a spark-ignited natural gas engine are performed. For each cycle, simulation results just prior to each spark timing are taken. Next, simulation results from different cycles are combined: one quantity, e.g., the flow field, is extracted from a snapshot of one given cycle, and all other quantities are taken from a snapshot from a different cycle. Such a combination yields a new snapshot. With the combined snapshot, the simulation is continued until the end of combustion. The results obtained with combined snapshots show that the velocity field seems to have the highest impact on CCV. Turbulence intensity, quantified by the turbulent kinetic energy and turbulent kinetic energy dissipation rate, has a similar value for all snapshots. Thus, their impact on CCV is small compared to the flow field. This novel methodology is very flexible and allows investigation of the sources of CCV which have been difficult to investigate in the past.

Keywords: internal combustion engine; combustion; CFD; RANS simulation; cycle-to-cycle variations

Citation: Gößnitzer, C.; Givler, S. A New Method to Determine the Impact of Individual Field Quantities on Cycle-to-Cycle Variations in a Spark-Ignited Gas Engine. *Energies* **2021**, *14*, 4136. https://doi.org/10.3390/en14144136

Academic Editor: Vincenzo De Bellis

Received: 7 May 2021
Accepted: 5 July 2021
Published: 8 July 2021

1. Introduction

With increasing legislative pressures to reduce reciprocating engine emissions and both governmental and consumer pressures for higher fuel efficiencies, reciprocating engine manufacturers have long been pursuing engine technologies that simultaneously push the envelope of engine performance, efficiency and emissions reductions. For spark-ignited (SI) engines, such pressures have resulted in the development of a wide array air charging and fueling engine technologies that are simultaneously optimized through complex engine control strategies to deliver on one or more aspects of the ultimate goal of highly efficient, high power density, low emissions engines. One of the challenges faced with tailoring the multi-dimensional engine operating envelope is the advanced combustion regimes that are being developed as a part of the overall engine operating recipes tailored to meet these challenges. Use of lean burn combustion and exhaust gas recirculation are two common approaches to improving fuel efficiency while reducing emissions; however, such strategies also introduce significant combustion sensitivities to the in-cylinder physics resulting in combustion regimes which are more susceptible to cycle-to-cycle variation (CCV) as the result of variations the ignition kernel development, kernel growth rates and/or the final fully developed turbulent flame as summarized by the reviews by Young [1], Ozdor et al.[2] and Maurya [3] and extensively noted in experimental studies [3–8]. After understanding the causes of CCV, the ultimate goal is the development of control strategies to reduce the magnitude of CCV [9–11].

Prior research has well documented the many factors that may promote CCV and these may be categorized as physico-chemical sources—flow field, trapped fuel mass, turbulence, dilution, stoichiometry and fuel reactivity and/or compsition; geometric sources—compression ratio, intake port geometry, spark plug location, spark plug electrode(s) orientation and spark gap; and operational sources—engine speed, load and ignition system performance [1–4,12,13]. Many of the experimental works quantify CCV by computing the coefficient of variation (CoV) in one or more easily measured metrics including peak cylinder pressure, location of peak firing-pressure, indicated mean effective pressure (IMEP) and/or maximum rate of pressure rise or fuel burning rates with peak cylinder pressure and IMEP being the most common. While these metrics are often easily measured, they tend to lump the many possible driving sources of CCV into a single net contributing factor making it challenging to delineate individual contributions of CCV amongst the possible sources. As a result, recent experimental work has focused on precise measurement of the development of the ignition kernel as the point of 2% of burn fuel mass (MFB2) and kernel growth rate as 0–10% (MFB10) as these two phases of the combustion event tend to be more sensitive to the CCV sources than the fully turbulent flame [5,14,15]. Unfortunately, these metrics are also more prone to measurement error due to the increased level of accuracy required in the cylinder pressure measurement and are complicated by the determination of the cumulative burn mass.

To unravel the complexities associated with the aforementioned strategies for identifying individual sources of CCV in SI engines, researchers have applied a variety of statistical and non-statistical analysis methods of the decomposing measured in-cylinder flow fields into its mean (large-scale) and turbulent (small-scale) components as summarized by Maurya [3] and Sadeghi et al. [16]. The results of such methods are often difficult to interpret from a physical perspective and, as indicated by Sadeghi [16] and Roudnitzky et al. [17], most of the methods invoke ad hoc assumptions that can impact the results of the decomposition, raising questions about the usefulness of such methods.

The early works of Venmorel et al. [18] and Liu et al. [19] and recently the works of Li et al. [20], Richard et. al. [21], Netzer et al. [22] and Truffin et al. [23] have utilized CFD simulation tools to understand the sources of CCV under a variety of engine operating conditions using LES turbulence modeling techniques under the pretense that they may be required in order to resolve the turbulence scales necessary to capture the impacts of turbulent flow field on both early flame kernel development and late stage turbulent flame progression. In their review, Lauer and Frühhaber [24] give an overview of current best practices in science and industry for turbulence modeling in combustion simulations. On the other side, the work of Scarcelli et al. [25] has shown that unsteady RANS (uRANS) turbulence models, with sufficient and practical grid resolutions, can capture many of the driving metrics of CCV in SI engine combustion. Additionally, modifications to the turbulence model used in this paper were proposed in the past [26]. To this end, this work proposes a novel approach for the investigation of the individual contributions of CCV using a detailed uRANS CFD simulation methodology.

2. Materials and Methods

2.1. Engine Setup

The engine used for the simulation is an INNIO Jenbacher natural gas four valve engine with premixed gas–air mixture and roughly 2.4 L of displaced volume. Bore and stroke are 135 mm and 170 mm, respectively. The engine is operated at a speed of 1500 rpm.

2.2. CFD Simulation Setup

The three-dimensional CFD simulations are performed with the CONVERGE CFD solver (version 3.0.13; Convergent Science; Madison, WI, USA). The mesh size in the intake and exhaust ports is 2 mm and 1 mm in the cylinder, respectively. Near-wall refinements and adaptive mesh refinement (AMR) for the flame front is employed, leading to 0.25 mm cells at the flame front location. Turbulence is modeled by the two-equation RANS k-ϵ RNG turbulence model. Ignition is modeled by an energy deposition at the spark plug gap of 30 mj each for 0.6 CAD for breakdown phase and 7.6 CAD for arc-glow phase. Combustion is modeled using the detailed chemistry approach and the GRI Mech 3.0 reaction mechanism [27] for natural gas combustion. An adaptive time step marching scheme is used to maintain a maximum Courant–Friedrichs–Lewy (CFL) limit within the domain to 2 during the gas exchange, and 4 during compression, except during combustion, a constant time step of 10^{-6} s is used resulting in maximum advection and diffusion CFL numbers of less than 1.0. Total pressure and temperature are prescribed at the inlet to the intake port, and static pressure is prescribed at the outlet of the exhaust port. Additionally, turbulence intensity of 2% and a turbulence length scale of 3 mm were specified at the inlet to calculate the boundary conditions for k and ε.

2.3. Multi-Cycle Simulations

A total of eleven consecutive engine cycles are simulated, using periodic (one engine cycle) boundary conditions with first cycle discarded due to initialization effects. The location and duration of the ignition are not changed. At 5 CAD and 0.5 CAD before spark timing, snapshots of the simulation are saved allowing for simulations to be restarted and/or implementation of the recombination strategy as described in the next section. The uRANS approach is in line with past work to predict CCV of internal combustion engines [28–30].

2.4. Combination of Simulation Snapshots

The restart files are a snapshot of the simulation at a certain time and allow one to restart the simulation from this time with minimal or no change in results as shown in Figures A1 and A2. The process of combining fields from different restart files is not supported by the CFD code and therefore a small Python script is used to combine fields from different restart files into a single restart file, a combined snapshot. Figure 1 visualizes the general idea of the recombinations. The restart files are HDF5 files and can be read and edited using standard tools from the `h5py` [31] Python package.

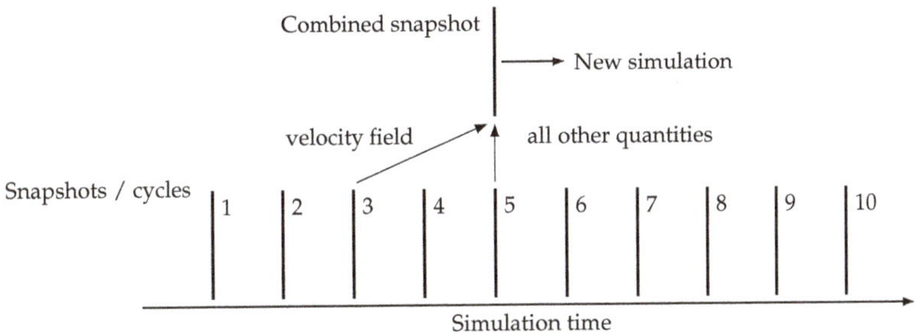

Figure 1. One possible combination of restart files: Here, the flow field is extracted from the third cycle and combined with all other quantities of the fifth cycle. The simulation then starts at the time when the fifth snapshot was taken.

It should be noted that while restart files (i.e., snapshots) are written at the same crank angle degree before spark timing of each engine cycle and the geometry and position of the piston are exactly the same, the mesh is different as a result of AMR being used to resolve the velocity field during gas scavenging. Thus, the combination algorithm accounts for local variances in mesh resolution combining the fields. For example, starting from a base snapshot where a majority of the instantaneous field quantities will be used without modification, to replace the instantaneous velocity fields with the values from another snapshot, here referred to as the velocity snapshot, the following procedure is applied:

1. Extract the mesh data from both the base and velocity snapshot.
2. Extract the velocity field from the velocity snapshot.
3. Interpolate the velocity field, which is defined on the mesh of the velocity snapshot, onto the mesh of the base snapshot. This is done by looping over all the cells of the base snapshot and finding the value of the velocity field of the cell of the velocity snapshot which is nearest to the cell of the base snapshot.
4. Write the newly interpolated field, which is now defined on the mesh of the base snapshot mesh, into the corresponding data array. Thus, the original, base velocity field is replaced with the exchange dataset.

3. Results

Due to confidentiality, all data reported here is normalized: The in-cylinder peak pressures are normalized to the minimum and maximum peak pressure, i.e., a peak pressure of 100% and 0% correspond to the highest and lowest peak pressure, respectively, and the reported absolute crank angle degrees are normalized with respect to the ignition timing, i.e., all reported absolute crank angle degrees in this paper are relative to the ignition.

The multi-cycle uRANS simulation yields ten different in-cylinder pressure traces as shown in Figure 2. The in-cylinder peak pressure has a coefficient of variation of 1.15%. The main differences between the cycles are in the peak-pressure and $MFB_{50\%}$. The differences from cycle to cycle in ignition delay and IMEP are significantly lower and are not considered in the following sections.

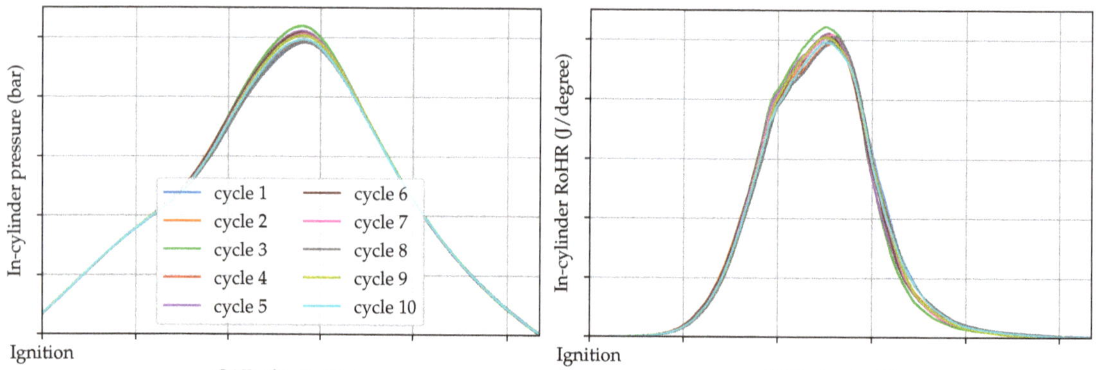

(**a**) Pressure. (**b**) Rate of heat release.

Figure 2. In-cylinder results for all ten cycles.

To ensure that restarting the simulation does not change the combustion result, a dry run is performed where no datasets were exchanged and the same combustion event is re-run using the restart file. These tests show that no significant difference between results from original and restart simulations existed as shown in Figures A1 and A2 in Appendix A.

In total, simulations for 16 different combinations are performed and are summarized in Table 1. For the purpose of this work not all cycles are combined with all other cycles. Instead, the two extreme cycles with the minimum and maximum peak pressure, cycles 3 and 8, respectively, are selected for restart file combination with the cylinder pressure and rate of heat release shown in Figure 3. Additionally, this paper only considers the velocity and turbulence quantities as exchange datasets.

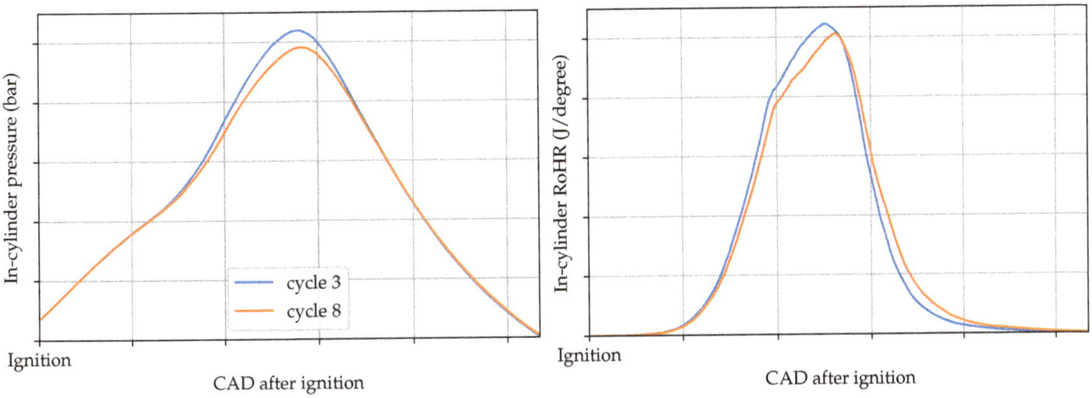

(a) Pressure. (b) Rate of heat release.

Figure 3. In-cylinder results for the two cycles with the lowest and highest pressure.

Table 2 summarizes results for four combustion metrics for 16 simulations using the combinations listed in Table 1. Respectively, these metrics are peak cylinder pressure ($p_{\max}$), crank angle degree after ignition of peak cylinder pressure ($CAD_{p_{\max}}$), crank angle degree after ignition for 50% fuel mass fraction burned ($MFB_{50\%}$) and crank angle duration from 10–90% mass fraction burned ($MFB_{10–90\%}$).

Table 1. Overview of the different restart combinations.

Comb.	Base	Exchange	Snapshot Time	Exchanged Datasets
1	8	3	5.0 CAD b. ign.	velocity
2	3	8	5.0 CAD b. ign.	velocity
3	8	3	5.0 CAD b. ign.	turbulence (k and ϵ)
4	3	8	5.0 CAD b. ign.	turbulence (k and ϵ)
5	8	3	5.0 CAD b. ign.	velocity close to spark plug (1 cm sphere)
6	8	3	5.0 CAD b. ign.	velocity close to spark plug (2 cm sphere)
7	8	3	5.0 CAD b. ign.	velocity close to spark plug (5 cm sphere)
8	8	3	0.5 CAD b. ign.	velocity close to spark plug (1 cm sphere)
9	8	3	0.5 CAD b. ign.	velocity close to spark plug (2 cm sphere)
10	8	3	0.5 CAD b. ign.	velocity close to spark plug (5 cm sphere)
11	3	8	5.0 CAD b. ign.	velocity close to spark plug (1 cm sphere)
12	3	8	5.0 CAD b. ign.	velocity close to spark plug (2 cm sphere)
13	3	8	5.0 CAD b. ign.	velocity close to spark plug (5 cm sphere)
14	3	8	0.5 CAD b. ign.	velocity close to spark plug (1 cm sphere)
15	3	8	0.5 CAD b. ign.	velocity close to spark plug (2 cm sphere)
16	3	8	0.5 CAD b. ign.	velocity close to spark plug (5 cm sphere)

Excluding maximum cylinder pressure, the table shows both absolute values and the percentile in the span from minimum to maximum of the observed range. As previously noted, cycle 3 has the highest peak cylinder pressure and is thus scaled to 100% while cycle 8 has the lowest and is scaled to 0%. Due to the inverse relationship between the combustion metrics and peak cylinder pressure, cycle 3 also has the minimum in the span for the remaining metrics and therefore each of these are scaled to 0%. Similarly, since cycle 8 has the lowest maximum cylinder pressure it also has the maximum in the span for the other metrics and therefore each of these are scaled to 100%.

Figure 4 presents the results from Table 2 showing the following: (**a**) 10–90% mass fraction burned versus 50% mass fraction burn, (**b**) peak cylinder pressure versus location of peak pressure, (**c**) 50% mass fraction burned versus peak cylinder pressure and (**d**) 10–90% mass fraction burned versus peak cylinder pressure. In each figure the results for cycle 3 and 8 are plotted as colored square symbols (blue—cycle 3, orange—cycle 8) and denoted as base. Since cycles 3 and 8 represent either the maximum or minimum observed combustion metric, they fall at the corners within each figure with cycle 3 always at lower left or right and cycle 8 at upper left or right. Results for the 16 simulated combinations are plotted and generally follow a linear relationship between the two base cycle results.

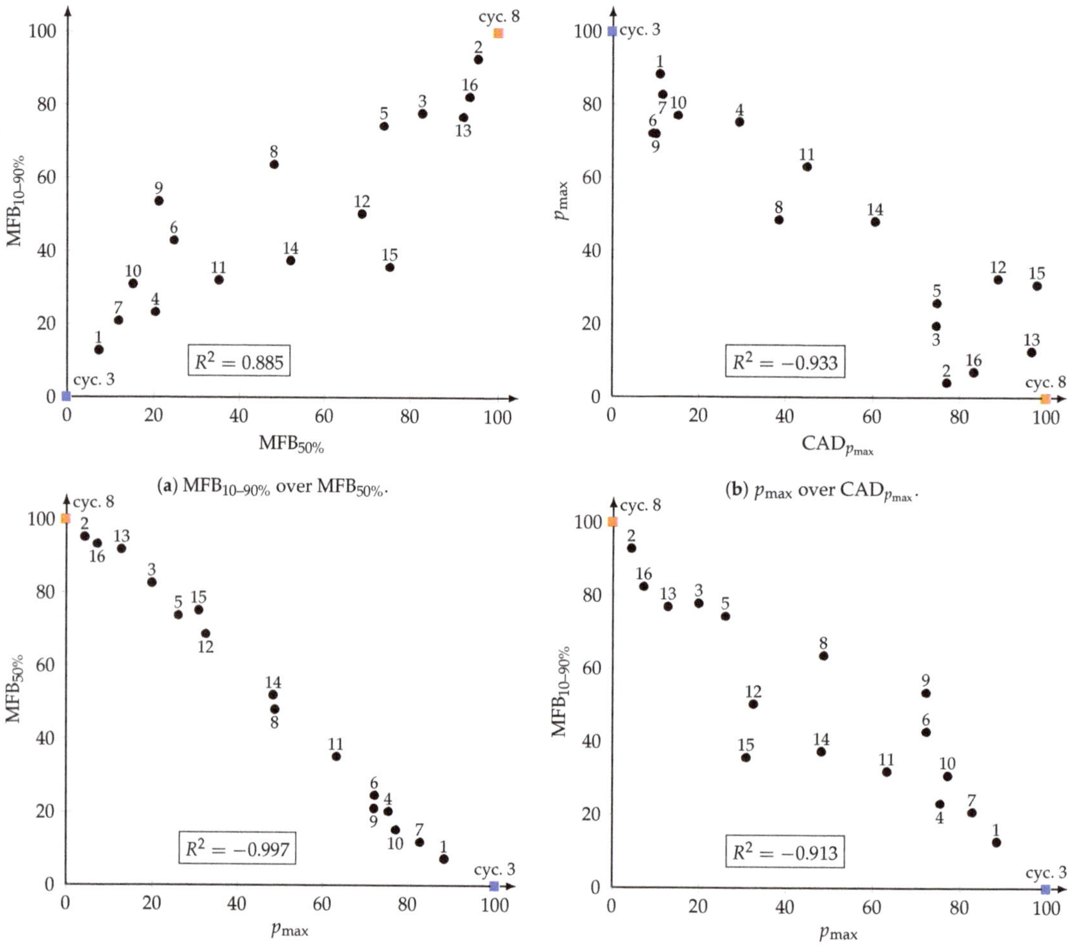

(**a**) $MFB_{10–90\%}$ over $MFB_{50\%}$. (**b**) p_{max} over $CAD_{p_{max}}$.

(**c**) $MFB_{50\%}$ over p_{max}. (**d**) $MFB_{10–90\%}$ over p_{max}.

Figure 4. Visualization of combustion charactersitics. All values are relative to the two extreme cycles. The numbers above or below the markers correspond to the combination number.

Table 2. Normalized maximum pressure, CAD after ignition of its occurrence, CAD after ignition of the mass fraction burnt 50% and combustion duration ($MFB_{10–90\%}$).

Case	p_{max}	$CAD_{p_{max}}$		$MFB_{50\%}$		$MFB_{10–90\%}$	
Cycle 3	100.0%	36.51	0.0%	31.30	0.0%	17.96	0.0%
Cycle 8	0.0%	37.03	100.0%	32.20	100.0%	19.21	100.0%
Comb. 1	88.3%	36.57	10.9%	31.36	7.3%	18.11	12.8%
Comb. 2	4.2%	36.91	77.0%	32.15	95.2%	19.12	92.8%
Comb. 3	19.8%	36.90	74.6%	32.04	82.6%	18.93	77.8%
Comb. 4	75.3%	36.66	29.3%	31.48	20.3%	18.24	23.2%
Comb. 5	25.9%	36.90	74.7%	31.96	73.7%	18.88	74.3%
Comb. 6	72.1%	36.56	9.3%	31.52	24.7%	18.49	42.9%
Comb. 7	82.7%	36.57	11.5%	31.40	11.8%	18.22	20.8%
Comb. 8	48.7%	36.71	38.5%	31.73	48.0%	18.75	63.6%
Comb. 9	72.0%	36.56	10.0%	31.49	21.1%	18.62	53.5%
Comb. 10	77.1%	36.59	15.1%	31.43	15.2%	18.34	30.9%
Comb. 11	63.2%	36.74	44.9%	31.61	35.2%	18.35	31.9%
Comb. 12	32.5%	36.97	88.8%	31.92	68.7%	18.58	50.3%
Comb. 13	12.7%	37.01	96.6%	32.12	91.9%	18.92	76.9%
Comb. 14	48.3%	36.82	60.5%	31.76	52.0%	18.42	37.4%
Comb. 15	30.9%	37.02	97.8%	31.97	75.2%	18.40	35.7%
Comb. 16	7.1%	36.94	83.2%	32.14	93.3%	18.98	82.3%

3.1. Impact of the Velocity Field on the Combustion

The first two combinations, combination 1 and combination 2, extract the velocity of one of the selected cycles and combine it with all other quantities, excluding the velocity field, from the other cycle. Figure 5 shows the in-cylinder pressure traces and rate of heat releases of the two original cycles and the two combinations. The figure shows in-cylinder pressure is very close to the simulation from which the velocity field is used. The peak cylinder pressure for combination 1 (base of cycle 8) is increased from 0% to 88.3% and decreased for combination 2 (base of cycle 3) from 100% to 4.2%.

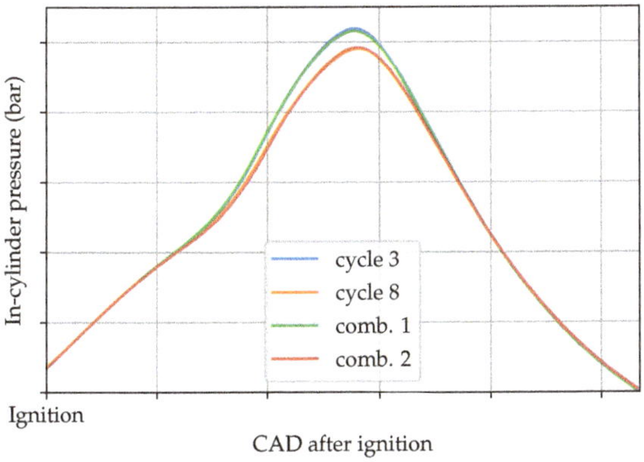

(**a**) Pressure.

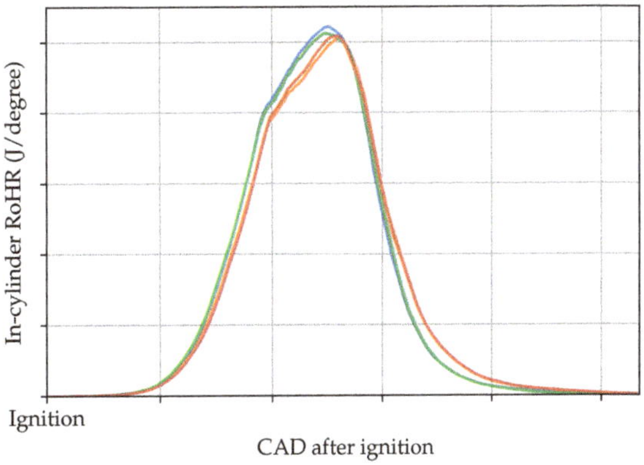

(**b**) Rate of heat release.

Figure 5. In-cylinder results for combinations 1 and 2.

Of particular importance are the results for combinations 1 and 2 which exchange only the velocity field of the base cycles while retaining the remaining simulation results. Simply by exchanging the velocity field in the simulation snapshot of cycle 3 with the velocity field of cycle 8 moves the crank angle combustion metrics ($CAD_{p_{max}}$, $MFB_{50\%}$) of cycle 3 from 0% (earliest) to the 75 to 90th (late) percentile and close to cycle 8's metrics.

This suggests that the velocity field of cycle 8 plays a significant role in a slowed ignition kernel development and/or turbulent flame propagation.

Similarly, exchanging the velocity field of cycle 3 in the simulation snapshot of cycle 8 moves the crank angle combustion metrics of cycle 8 from 100% (latest) downward to the 5 to 10th (early) percentile suggesting that the velocity field of cycle 3 plays a significant role in an accelerated ignition kernel development and/or turbulent flame propagation. Naturally, this leads to the question of what it is specifically about these two flow fields that appear to drive either a faster or slower combustion regime. This question is out of scope for the present work.

3.2. *Impact of the Turbulence on the Combustion*

Next, the impact of turbulence, the turbulent kinetic energy k and the rate of dissipation of turbulent energy ε are investigated. Figure 6 shows the combinations 3 and 4 which exchange only the turbulence field between cycles 3 and 8. The figure shows that the turbulence fields at ignition timing do not play a major role in the combustion event.

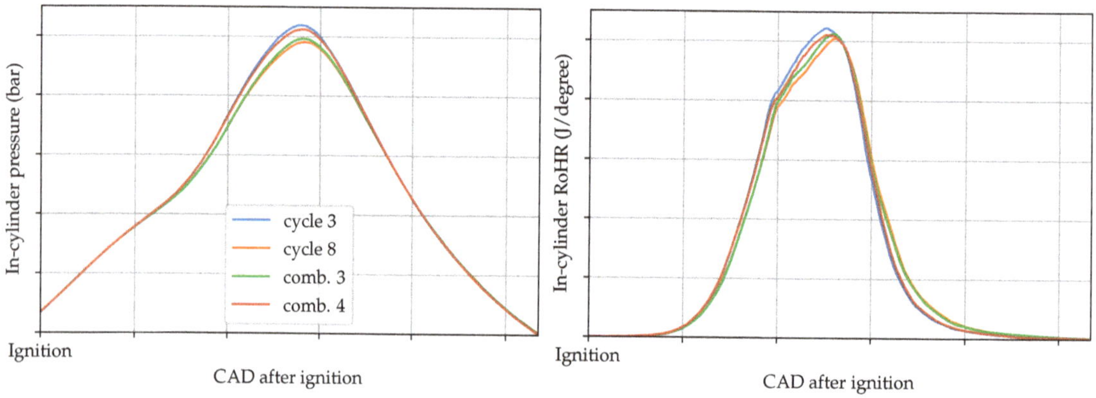

(**a**) Pressure. (**b**) Rate of heat release.

Figure 6. In-cylinder results for combinations 3 and 4.

However, it is arguable that exchanging the turbulence quantities alone is a proper assessment of the impact of the turbulence on combustion in the context of a two equation uRANS turbulence model, since velocity and turbulence fields have a complex coupling. Figure 7 shows that an immediate relaxation of the turbulence to their prior values does not occur even when the velocity field remains unchanged. Figure 7a shows that the turbulence intensity of combination 1 is much closer to cycle 8 during the ignition phase. This seems plausible since k and ε are sourced from cycle 8, and only the flow field comes from cycle 3. Similar results and conclusions can be drawn for the other three combinations that are shown in Figure 7b for combinations 2, 3 and 4.

3.3. *Impact of the Flow Field near the Spark Plug on the Combustion*

Since the first four combinations show that the flow field seems to have the most impact on the combustion, a new combination strategy is considered. For the next combinations, only the flow field in the vicinity of the ignition location (spherical volume with different radii: 1 cm, 2 cm and 5 cm) is extracted and combined with the flow field in the rest of the domain and all other quantities of the other cycle. To assess the impact of changes in the flow field from the initial snapshot to the spark event the time before ignition when the snapshots are taken is varied from 5 CAD to 0.5 CAD.

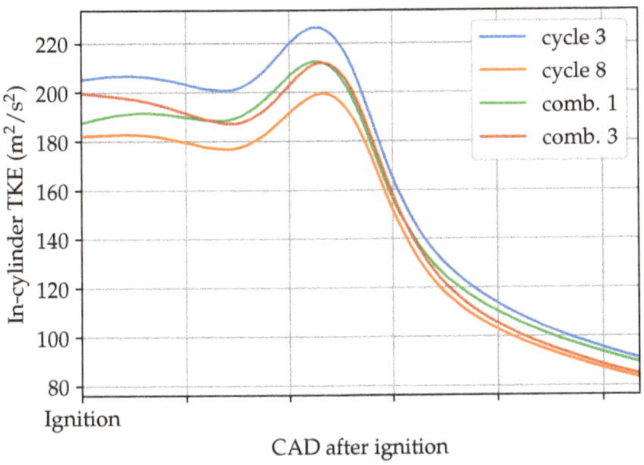

(a) Combinations 1 and 3.

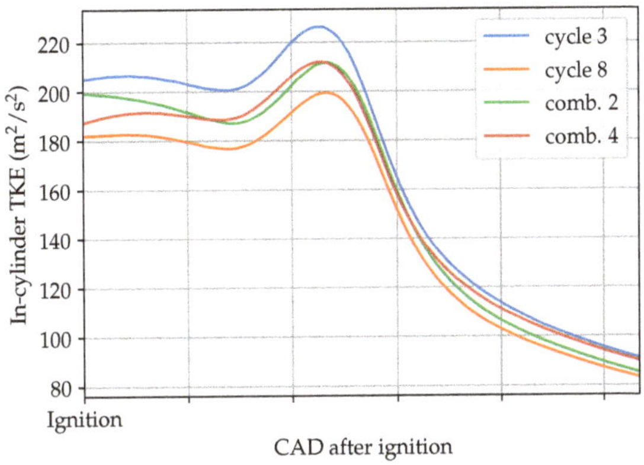

(b) Combinations 2 and 4.

Figure 7. Average in-cylinder turbulent kinetic energy.

Combinations 5, 6 and 7 use the flow field in a spherical volume around the ignition location of cycle 3 and all other quantities and the flow field outside of said volume of cycle 8. The in-cylinder pressure and rate of heat release are shown in Figure 8. The time before ignition when the snapshot is taken is 5 CAD. The figure shows that the larger the spherical volume the closer the cylinder pressure is to cycle where the velocity field is sourced with the peak cylinder pressures for combinations 5, 6 and 7 of 25.9%, 72.1% and 82.7% for radii 1 cm, 2 cm and 5 cm, respectively with 100% corresponding to the cycle from which the near spark plug flow field is extracted.

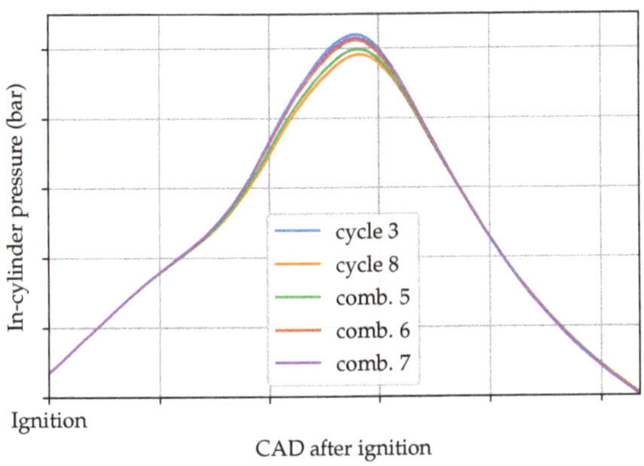

(a) Pressure.

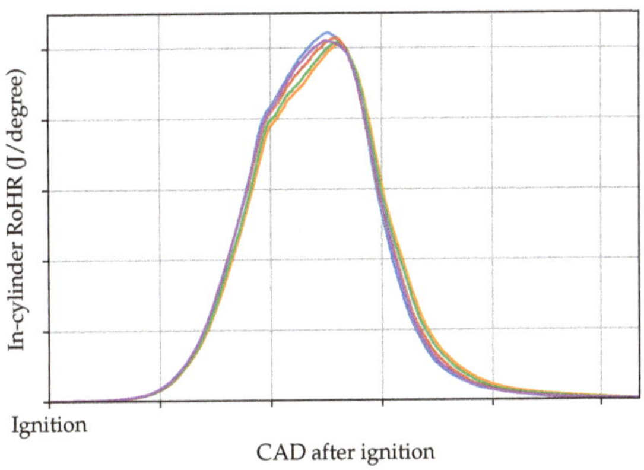

(b) Rate of heat release.

Figure 8. In-cylinder results for combinations 5, 6 and 7.

Figure 9 shows the in-cylinder pressure and rate of heat release of combinations 8, 9 and 10, which are similar to combinations 6, 7 and 8, respectively; however, the time of the snapshot is shifted closer to spark timing from 5 CAD to 0.5 CAD before ignition. The impact of this shift is most pronounced on the results using a 1 cm sphere extraction volume. In this case the in-cylinder peak pressure increased from 25.9% to 48.7%. The other two combinations only show a minor change in peak pressure, where combination 9 has almost the same peak pressure as combination 6 (72.0% vs. 72.1%) and combination 10 even shows a significantly lower peak pressure than combination 7 (82.7% vs. 77.1%). This implies that the flow field between the 2 cm and 5 cm spheres only plays a minor role for turbulent flame propagation and that most contributions to CCV are due to the flow field inside the 1 cm sphere.

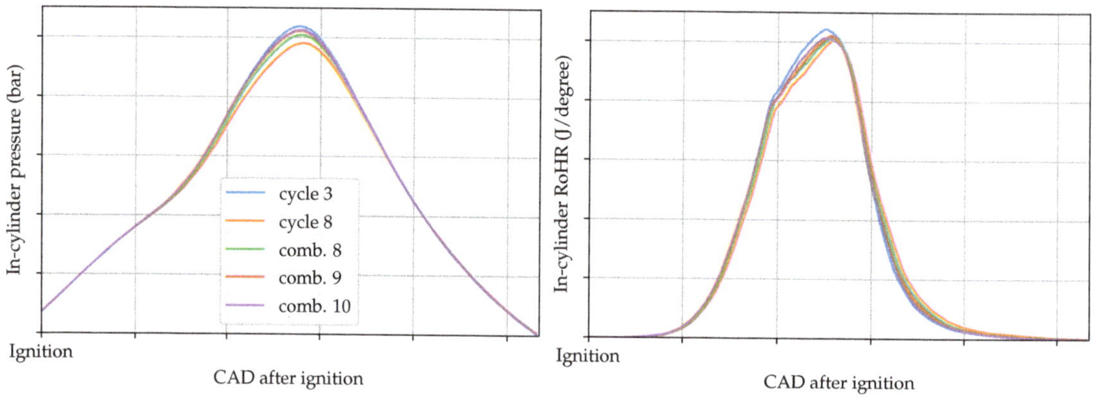

(a) Pressure. (b) Rate of heat release.

Figure 9. In-cylinder results for combinations 8, 9 and 10.

Combinations 11, 12 and 13, shown in Figure 10, are similar to combinations 5, 6 and 7; only the original cycles are exchanged, i.e., a peak pressure of 0% corresponds to the cycle where the near-ignition flow field comes from. Again, a larger sphere of exchange region means a lower peak pressure, going from 63.2% for the 1 cm sphere to 32.5% for the 2 cm sphere and as low as 12.7% for the largest sphere of 5 cm.

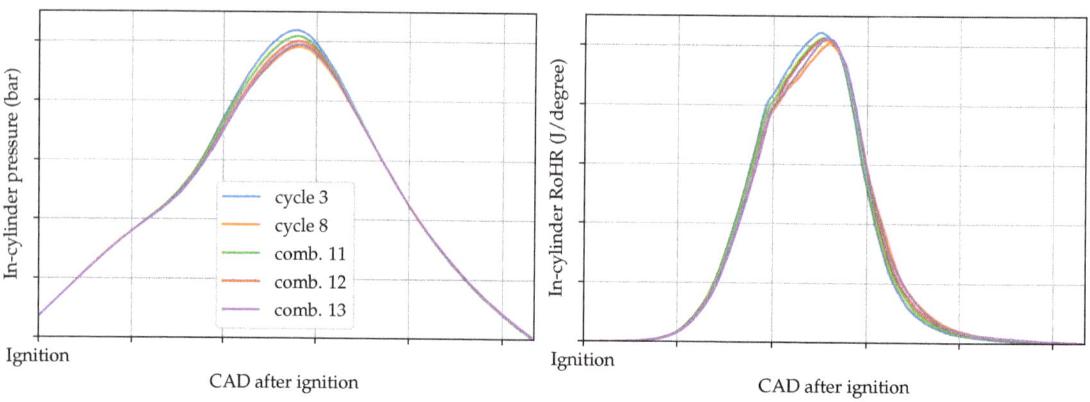

(a) Pressure. (b) Rate of heat release.

Figure 10. In-cylinder results for combinations 11, 12 and 13.

Combinations 14, 15 and 16 are equal to combinations 11, 12 and 13 with the exception of the time when the snapshot is taken being shifted to 0.5 CAD before ignition. This leads to a lower peak pressure, shown in Figure 11, which means that these combinations are closer to the 0% peak pressure cycles than the combinations with an earlier snapshot time.

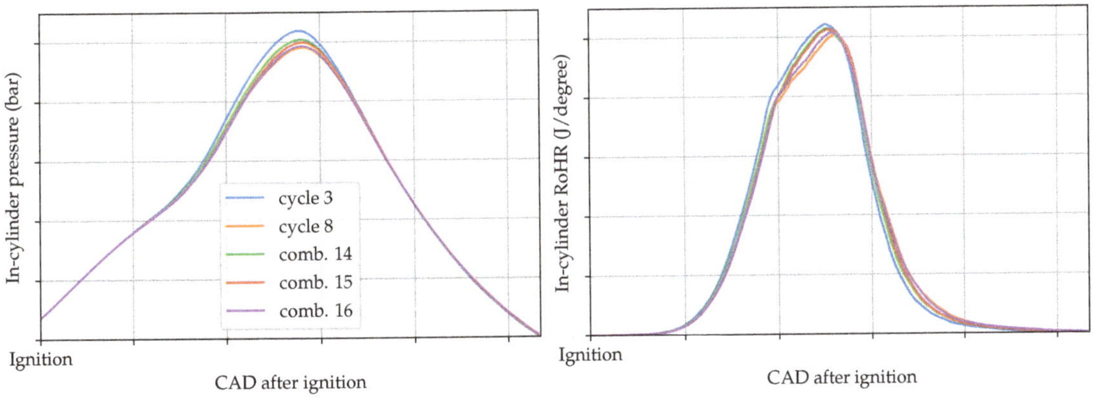

(a) Pressure. (b) Rate of heat release.

Figure 11. In-cylinder results for combinations 14, 15 and 16.

4. Summary and Conclusions

A novel method for quantifying the individual flow field parameters that impact CCV in the multi-cycle CFD simulation of spark-ignited engines has been presented. The method has been applied to a spark-ignited lean burn natural gas engine and reveals that for this engine and the operating point examined the velocity field has the largest impact on the combustion behavior. Exchanging velocity fields between the slowest/fastest combusting cycles can advance/retard crank angle combustion metrics such as location of peak pressure, 50% mass fraction burned and 10–90% mass fraction burned by 75–90% of the observed span. Applying the methodology to subsets of the physical domain in proximity to the spark plug shows that the velocity field in this region also has a significant impact on the ignition kernel and combustion event.

The methodology indicates that the turbulence field (k and ε) for these uRANS simulations plays a significantly minor role as compared to the velocity field. However, it is acknowledged that proper evaluation of the impacts of turbulence may be significantly limited due to the rapid relaxation of the turbulence field when being introduced into a new velocity field, as these systems are not decoupled. As shown in Section 3.2, this relaxation does not take place at least for the average in-cylinder turbulent kinetic energy. Instead, the resulting k field is in between the two extreme cycles.

The cycle-to-cycle variations are rather low (CoV of p_{max} is 1.15%) in this study and this may be attributed to using the RANS turbulence modeling approach, the same boundary conditions for each engine cycle, same ignition location and/or the same ignition energy for each engine cycle. Regardless, the methodology presented is capable of identifying and quantifying the significance of various drivers for slow and fast combustion events, suggesting it can provide greater insight in instances of much larger CCV.

Author Contributions: Conceptualization, both authors; methodology, C.G.; software, both authors; validation, both authors; investigation, C.G.; resources, C.G.; data curation, C.G.; writing—original draft preparation, both authors; writing—review and editing, S.G.; visualization, C.G. Both authors have read and agreed to the published version of the manuscript.

Funding: Open Access Funding by the Graz University of Technology.

Acknowledgments: C.G. acknowledges the financial support of the "COMET—Competence Centres for Excellent Technologies" Programme of the Austrian Federal Ministry for Climate Action, Environment, Energy, Mobility, Innovation and Technology (BMK) and the Federal Ministry for Digital and Economic Affairs (BMDW) and the Provinces of Styria, Tyrol and Vienna for the COMET Centre (K1) LEC EvoLET. The COMET Programme is managed by the Austrian Research Promotion Agency (FFG).

Conflicts of Interest: The authors declare no conflict of interest. The funders had no role in the design of the study; in the collection, analyses, or interpretation of data; in the writing of the manuscript, or in the decision to publish the results.

Abbreviations

The following abbreviations are used in this manuscript:

CoV	coefficient of variation
CCV	cycle-to-cycle variations
CFL	Courant–Friedrichs–Lewy
SI	spark-ignited
RANS	Renolds-averaged Navier–Stokes equations
CFD	computational fluid dynamics
AMR	adaptive mesh refinement
CAD	crank angle degree
MFB	mass fraction burned

Appendix A

Results from an ordinary restart, i.e., no datasets are exchanged, showing that restarting the simulation has no impact on the combustion behavior.

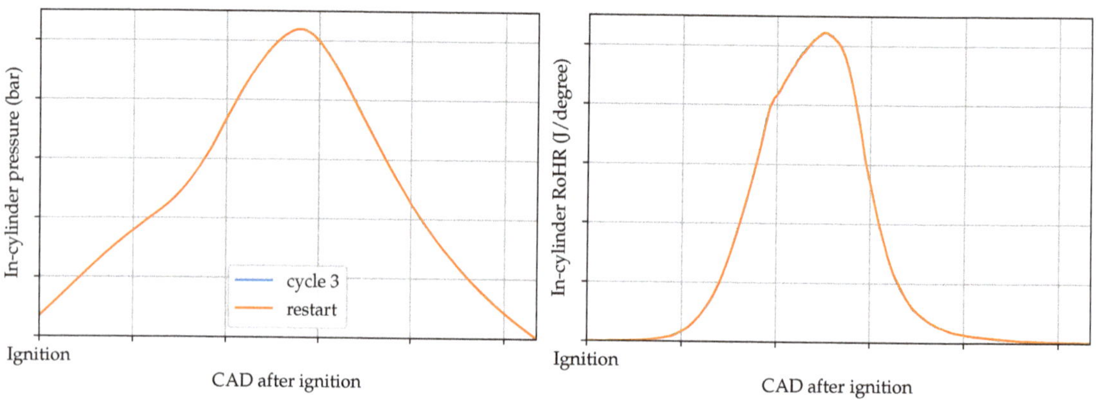

(**a**) Pressure. (**b**) Rate of heat release.

Figure A1. In-cylinder results for cycle 3 and restart of cycle 3.

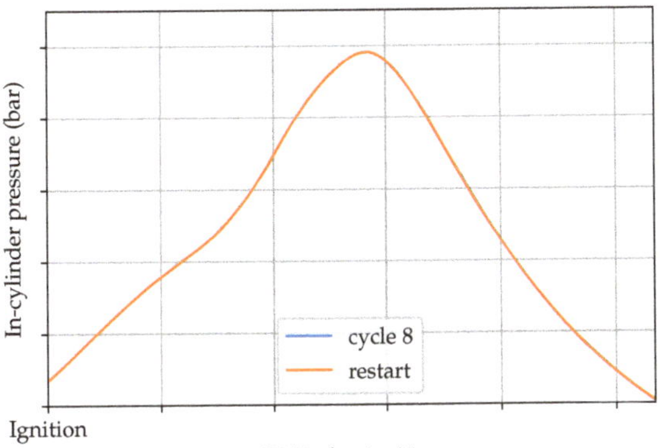

(**a**) Pressure.

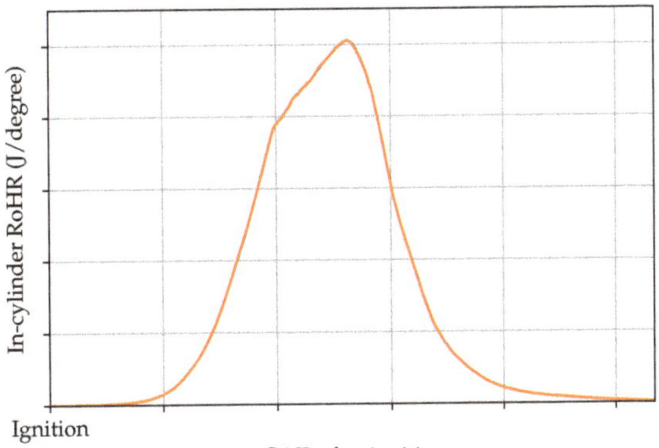

(**b**) Rate of heat release.

Figure A2. In-cylinder results for cycle 8 and restart of cycle 8.

References

1. Young, M.B. Cyclic Dispersion in the Homogeneous-Charge Spark-Ignition Engine—A Literature Survey. *SAE Trans.* **1981**, *90*, 49–73. [CrossRef]
2. Ozdor, N.; Dulger, M.; Sher, E. Cyclic Variability in Spark Ignition Engines A Literature Survey. *SAE Trans.* **1994**, *103*, 1514–1552. [CrossRef]
3. Maurya, R.K. Combustion Stability Analysis. In *Reciprocating Engine Combustion Diagnostics: In-Cylinder Pressure Measurement and Analysis*; Springer: Cham, Switzerland, 2019; pp. 361–459. [CrossRef]
4. Matekunas, F.A. Modes and Measures of Cyclic Combustion Variability. *SAE Trans.* **1983**, *92*, 1139–1156. [CrossRef]
5. Johansson, B. Cycle to Cycle Variations in S.I. Engines—The Effects of Fluid Flow and Gas Composition in the Vicinity of the Spark Plug on Early Combustion. *SAE Trans.* **1996**, *105*, 2286–2296. [CrossRef]
6. Han, S.B. Cycle-to-cycle variations under cylinder-pressure-based combustion analysis in spark ignition engines. *KSME Int. J.* **2000**, *14*, 1151–1158. [CrossRef]
7. Reuss, D.L. Cyclic Variability of Large-Scale Turbulent Structures in Directed and Undirected IC Engine Flows. *SAE Trans.* **2000**, *109*, 125–145. [CrossRef]
8. Abraham, P.S.; Yang, X.; Gupta, S.; Kuo, T.W.; Reuss, D.L.; Sick, V. Flow-pattern switching in a motored spark ignition engine. *Int. J. Engine Res.* **2015**, *16*, 323–339. [CrossRef]
9. Mahmoudzadeh Andwari, A.; Pesyridis, A.; Esfahanian, V.; Said, M.F.M. Combustion and Emission Enhancement of a Spark Ignition Two-Stroke Cycle Engine Utilizing Internal and External Exhaust Gas Recirculation Approach at Low-Load Operation. *Energies* **2019**, *12*, 609. [CrossRef]
10. Marchitto, L.; Tornatore, C.; Teodosio, L. Individual Cylinder Combustion Optimization to Improve Performance and Fuel Consumption of a Small Turbocharged SI Engine. *Energies* **2020**, *13*, 5548. [CrossRef]
11. Omanovic, A.; Zsiga, N.; Soltic, P.; Onder, C. Increased Internal Combustion Engine Efficiency with Optimized Valve Timings in Extended Stroke Operation. *Energies* **2021**, *14*, 2750. [CrossRef]
12. El-Adawy, M.; Heikal, M.R.; Aziz, A.; Rashid, A.; Adam, I.K.; Ismael, M.A.; Babiker, M.E.; Baharom, M.B.; Firmansyah; Abidin, E.Z.Z. On the Application of Proper Orthogonal Decomposition (POD) for In-Cylinder Flow Analysis. *Energies* **2018**, *11*, 2261. [CrossRef]
13. Martinez-Boggio, S.; Merola, S.; Teixeira Lacava, P.; Irimescu, A.; Curto-Risso, P. Effect of Fuel and Air Dilution on Syngas Combustion in an Optical SI Engine. *Energies* **2019**, *12*, 1566. [CrossRef]
14. Schiffmann, P. Root Causes of Cyclic Variability of Early Flame Kernels in Spark Ignited Engines. Ph.D. Thesis, University of Michigan, Ann Arbor, MI, USA, 2016.
15. Schiffmann, P.; Reuss, D.L.; Sick, V. Empirical investigation of spark-ignited flame-initiation cycle-to-cycle variability in a homogeneous charge reciprocating engine. *Int. J. Engine Res.* **2018**, *19*, 491–508. [CrossRef]
16. Sadeghi, M.; Truffin, K.; Peterson, B.; Böhm, B.; Jay, S. Development and Application of Bivariate 2D-EMD for the Analysis of Instantaneous Flow Structures and Cycle-to-Cycle Variations of In-cylinder Flow. *Flow Turbul. Combust.* **2021**, *106*, 231–259. [CrossRef]
17. Roudnitzky, S.; Druault, P.; Guibert, P. Proper orthogonal decomposition of in-cylinder engine flow into mean component, coherent structures and random Gaussian fluctuations. *J. Turbul.* **2006**, *7*, N70. [CrossRef]
18. Vermorel, O.; Richard, S.; Colin, O.; Angelberger, C.; Benkenida, A.; Veynante, D. Towards the understanding of cyclic variability in a spark ignited engine using multi-cycle LES. *Combust. Flame* **2009**, *156*, 1525–1541. [CrossRef]

19. Liu, K.; Haworth, D.C.; Yang, X.; Gopalakrishnan, V. Large-eddy simulation of motored flow in a two-valve piston engine: POD analysis and cycle-to-cycle variations. *Flow Turbul. Combust.* **2013**, *91*, 373–403. [CrossRef]
20. Li, W.; Li, Y.; Wang, T.; Jia, M.; Che, Z.; Liu, D. Investigation of the effect of the in-cylinder tumble motion on cycle-to-cycle variations in a direct injection spark ignition (DISI) engine using large eddy simulation (LES). *Flow Turbul. Combust.* **2017**, *98*, 601–631. [CrossRef]
21. Richard, S.; Dulbecco, A.; Angelberger, C.; Truffin, K. Invited Review: Development of a one-dimensional computational fluid dynamics modeling approach to predict cycle-to-cycle variability in spark-ignition engines based on physical understanding acquired from large-eddy simulation. *Int. J. Engine Res.* **2015**, *16*, 379–402. [CrossRef]
22. Netzer, C.; Pasternak, M.; Seidel, L.; Ravet, F.; Mauss, F. Computationally efficient prediction of cycle-to-cycle variations in spark-ignition engines. *Int. J. Engine Res.* **2020**, *21*, 649–663. [CrossRef]
23. Truffin, K.; Angelberger, C.; Richard, S.; Pera, C. Using large-eddy simulation and multivariate analysis to understand the sources of combustion cyclic variability in a spark-ignition engine. *Combust. Flame* **2015**, *162*, 4371–4390. [CrossRef]
24. Lauer, T.; Frühhaber, J. Towards a Predictive Simulation of Turbulent Combustion?—An Assessment for Large Internal Combustion Engines. *Energies* **2021**, *14*, 43. [CrossRef]
25. Scarcelli, R.; Richards, K.; Pomraning, E.; Senecal, P.K.; Wallner, T.; Sevik, J. Cycle-to-Cycle Variations in Multi-Cycle Engine RANS Simulations. In Proceedings of the SAE 2016 World Congress and Exhibition. SAE International, Grand Rapids, MI, USA, 7–10 September 2016. [CrossRef]
26. Krastev, V.K.; Silvestri, L.; Falcucci, G. A Modified Version of the RNG k–ε Turbulence Model for the Scale-Resolving Simulation of Internal Combustion Engines. *Energies* **2017**, *10*, 2116. [CrossRef]
27. Smith, G.P.; Golden, D.M.; Frenklach, M.; Moriarty, N.W.; Eiteneer, B.; Goldenberg, M.; Bowman, C.T.; Hanson, R.K.; Song, S.; William C.; et al. GRI Mech 3.0. 1999. Available online: http://combustion.berkeley.edu/gri-mech/version30/text30.html (accessed on 26 April 2021)
28. Scarcelli, R.; Sevik, J.; Wallner, T.; Richards, K.; Pomraning, E.; Senecal, P.K. Capturing Cyclic Variability in Exhaust Gas Recirculation Dilute Spark-Ignition Combustion Using Multicycle RANS. *J. Eng. Gas Turbines Power* **2016**, *138*, 112803. [CrossRef]
29. Kosmadakis, G.M.; Rakopoulos, C.D. A Fast CFD-Based Methodology for Determining the Cyclic Variability and Its Effects on Performance and Emissions of Spark-Ignition Engines. *Energies* **2019**, *12*, 4131. [CrossRef]
30. Kosmadakis, G.; Rakopoulos, D.; Arroyo, J.; Moreno, F.; Muñoz, M.; Rakopoulos, C. CFD-based method with an improved ignition model for estimating cyclic variability in a spark-ignition engine fueled with methane. *Energy Convers. Manag.* **2018**, *174*, 769–778. [CrossRef]
31. h5py Developers. HDF5 for Python. 2021. Available online: https://www.h5py.org/ (accessed on 28 April 2021)

energies

MDPI

Article

Development and Experimental Validation of an Adaptive, Piston-Damage-Based Combustion Control System for SI Engines: Part 2—Implementation of Adaptive Strategies

Alessandro Brusa [1,*], Nicolò Cavina [1], Nahuel Rojo [1], Jacopo Mecagni [1], Enrico Corti [1], Davide Moro [1], Matteo Cucchi [2] and Nicola Silvestri [2]

1 Department of Industrial Engineering, School of Engineering and Architecture, University of Bologna, 40126 Bologna, Italy; nicolo.cavina@unibo.it (N.C.); nahuel.rojo2@unibo.it (N.R.); jacopo.mecagni2@unibo.it (J.M.); enrico.corti2@unibo.it (E.C.); davide.moro@unibo.it (D.M.)
2 Ferrari S.p.A., 41053 Maranello MO, Italy; Matteo.Cucchi@ferrari.com (M.C.); Nicola.Silvestri@ferrari.com (N.S.)
* Correspondence: alessandro.brusa6@unibo.it

Abstract: This work focuses on the implementation of innovative adaptive strategies and a closed-loop chain in a piston-damage-based combustion controller. In the previous paper (Part 1), implemented models and the open loop algorithm are described and validated by reproducing some vehicle maneuvers at the engine test cell. Such controller is further improved by implementing self-learning algorithms based on the analytical formulations of knock and the combustion model, to update the fuel Research Octane Number (RON) and the relationship between the combustion phase and the spark timing in real-time. These strategies are based on the availability of an on-board indicating system for the estimation of both the knock intensity and the combustion phase index. The equations used to develop the adaptive strategies are described in detail. A closed-loop chain is then added, and the complete controller is finally implemented in a Rapid Control Prototyping (RCP) device. The controller is validated with specific tests defined to verify the robustness and the accuracy of the adaptive strategies. Results of the online validation process are presented in the last part of the paper and the accuracy of the complete controller is finally demonstrated. Indeed, error between the cyclic and the target combustion phase index is within the range ±0.5 Crank Angle degrees (°CA), while the error between the measured and the calculated maximum in-cylinder pressure is included in the range ±5 bar, even when fuel RON or spark advance map is changing.

Keywords: knock; combustion; efficiency improvement; CO_2 emissions; control; adaptive strategy

Citation: Brusa, A.; Cavina, N.; Rojo, N.; Mecagni, J.; Corti, E.; Moro, D.; Cucchi, M.; Silvestri, N. Development and Experimental Validation of an Adaptive, Piston-Damage-Based Combustion Control System for SI Engines: Part 2—Implementation of Adaptive Strategies. *Energies* **2021**, *14*, 5342. https://doi.org/10.3390/en14175342

Academic Editor: Suhan Park

Received: 21 June 2021
Accepted: 23 August 2021
Published: 27 August 2021

1. Introduction

In this paper, the combustion controller described in the Part 1 is further developed and validated. Such controller can be considered as a significant step in combustion management, because of the definition of a target value of Maximum Amplitude of Pressure Oscillations (MAPO) deriving from the estimation of the piston damage admissible in a certain amount of time and the model-based open-loop chain, that allows to accurately manage the combustion phase cycle-by-cycle even under transient conditions. In this way, the knock intensity threshold is defined as a tangible piston damage index and, outside the range of the knock-limited engine operating conditions, the control algorithm keeps the combustion phase close to the maximum efficiency value, integrating knock and combustion control in a single algorithm.

With the knock control strategy described in this work, the authors are proposing a solution for the following issues that characterize standard knock control strategies:

1. The definition of a threshold for the statistical knock indexes that are defined independently from the consequences induced by knocking events

2. The implementation of closed-loop controllers that are unsuitable to control an extremely variable engine parameter, such as a high knock intensity percentile, especially when the engine is operated under fast transient conditions.

In the Part 1 of the whole activity a robust and accurate open-loop controller is developed and validated, and the main implemented models are described, including the control-oriented piston damage model [1] and the analytical knock model [2]. Other important models introduced in the open-loop controller are the analytical maximum in-cylinder pressure (Pmax) and 50% of mass of fuel burned (MFB50) ones, described by the authors in [3]. Knock and Pmax models are further validated in this work through the comparison of the model coefficients calibrated for two different Spark Ignition (SI) engines. Moreover, the reversed analytical knock model and the MFB50 ones are further developed implementing two adaptive strategies to automatically learn the fuel RON of vehicle tank and the parabolic relationship between MFB50 and Spark Advance (SA) for a certain engine point. Indeed, such quantities were calibrated and kept fixed with the previous open-loop controller. The self-learning strategies developed by the authors guarantee an always up-to-date fuel RON value and MFB50-SA functions. Nevertheless, such algorithms need the cycle-by-cycle value of knock intensity (instantaneous MAPO) and combustion phase (instantaneous MFB50), such as the closed-loop chain, that reacts only when the damage speed overcomes a safety threshold. Indeed, as discussed in the Part 1, a pure open-loop controller cannot include the effects of unpredictable events, such as the engine or exhaust system aging or extreme environmental conditions, on the calculation of cyclic MAPO and MFB50. For such reason, the adaptive strategies and a protective Proportional Integral (PI) controller are introduced in this paper. Hence, the proposed control strategy needs an on-board system for the knock intensity and the combustion phase estimation.

Nowadays, the need of an on-board indicating system or a solution for the calculation of both the knock intensity and the combustion phase indexes has not been considered as a restrictive limitation. Indeed, recent anti-pollution policies are forcing the engine manufacturers to develop new solutions to control innovative combustion processes, such as the latest applications of Low Temperature Combustion and those that involve novel types of fuel [4–6]. Such systems are often affected by poor combustion stability and the risk of both knock or misfire imposes the in-cylinder pressure sensing. Several solutions are available in literature. For instance, the accelerometer is considered as a viable solution not only for knock intensity estimation, but also for the MFB50 calculation [7,8]. The ionization current is a valid and robust technology for the in-cylinder pressure signal estimation [9]. A third method is that based on the acoustic emissions that can be measured with microphones [10–12]. Finally, some low-cost force transducers have been developed for the in-cylinder pressure sensing. Using piezoelectric materials, the forces acting on the engine head may be partially related to the pressure of the gases inside the cylinder. Thus, instrumenting the head stud bolt [13] or using a piezoelectric spark plug washer and acquiring at high frequency the electric output signal, the in-cylinder pressure variation can be estimated. Alma Automotive recently developed a piezoelectric spark plug washer that achieved very promising results [14].

These considerations about the in-cylinder pressure sensing techniques for the on-board applications legitimize the development of accurate and reliable combustion control strategies based on the availability of some combustion indexes, that until a few years ago were available only at the test bench. Moreover, some cheap and reliable piezoelectric sensors for the on-board installation are coming to the market, even if main features of these technologies will be published once they are patented.

With the knock control strategy proposed in this work, the authors are thus proposing a solution for both issues listed above:

1. The target value of the MAPO percentile derives from the piston damage admissible in a certain amount of time (i.e., a target piston damage speed) and this means it is defined according with the concrete relationship between knock intensity and

the induced damage. The Pmax and MFB50 model convert that value into a target combustion angular phase
2. The open loop chain allows to achieve the target MFB50 calculating the SA needed just before each cycle. The controller is thought to be cycle-resolved, and this makes such algorithm suitable to accurately manage the combustion phase cycle-by-cycle even under transient conditions.

As mentioned above, outside the engine knock-limited operating field, the MFB50 model continues to calculate the SA that keeps the combustion phase close to the target value. This condition can be easily identified by calibrating a saturation value equal to the Maximum Brake Torque (MBT) combustion phase. When the target MFB50 (the value that guarantees the target piston damage speed) is lower than such calibrated value, the MBT combustion phase becomes the current target. Moreover, the adaptive strategies update the coefficients of knock and MFB50 models, using measured combustion indexes. The novelty of this work is thus the development and validation of adaptive strategies for the real-time updating of the fuel RON and the analytical relationship between MFB50 and ΔSA.

In the first part of this paper, the experimental setup and the database used for the controller calibration and validation are described. The complete combustion controller is described in detail, showing the models' mathematical reformulations needed for their coupling with the self-learning strategies. The controller is then implemented in a Rapid Control Prototyping (RCP) device, and it is validated online at the engine test bench. Tests are specifically designed to demonstrate the robustness and the accuracy of the novel adaptive strategies.

2. Experimental Setup

The model calibration is carried out both on an 8-cylinder and a 4-cylinder Gasoline Direct Injection (GDI), Turbo-Charged (TC) production engine, equipped with a piezoelectric pressure transducer for each cylinder. While the former is a high-performance engine, the latter is equipping a standard passenger car. The comparison of the calibrated coefficients for both engines confirms the robustness of the described modeling approach for the definition of accurate and reliable control-oriented combustion models. The controller validation is instead performed on the 8-cylinder engine only, for which the controller was developed. Such engine is the same considered in the Part 1 of this work, and the complete combustion controller is finally implemented on an RCP machine connected to the Engine Control Unit (ECU) of the 8-cylinder engine. On the other hand, the 4-cylinder engine is used to further validate the knock and Pmax models, but it is not used as an application for the damage model and the complete combustion controller. A description of the variables and the time needed to recalibrate such models for a different engine is provided in the Part 1 of this activity, but all the models included in the control strategy can be ported even on the 4-cylinder engine without modifying their formulation. The main characteristics of the 8-cylinder and the 4-cylinder engine are reported in Tables 1 and 2, respectively, while the layout of the development network used to test the complete controller via RCP is shown in Figure 1. The combustion chamber layout of both engines has a center-mounted spark plug and a side-mounted in-chamber fuel injector. The indicating system is composed of CHAMPION charge amplifiers and OBI-M2 indicating system, both provided by Alma Automotive S.r.l. (Bologna, Italy), that are used for the signal conditioning and the data acquisition, respectively. Such system records the in-cylinder pressure signals at 200 kHz. The calculation of the MFB50, Indicated Mean Effective Pressure (IMEP) and Pmax combustion indexes uses the low-pass filtered pressure trace, while the same signal is windowed and band-pass filtered (the specific cut-off frequencies and the angular window cannot be disclosed for confidentiality reasons) to calculate MAPO. As in the previous paper (Part 1), the nomenclature "MAPO" indicates cyclic value of such index, while, if it refers to a percentile value, it is always specified or it appears as "MAPO98" (for the 98th percentile).

Table 1. The 8-cylinder engine characteristics.

Displaced Volume	3855 cc (8 Cylinder)
Stroke	82 mm
Bore	86.5 mm
Connecting rod	143 mm
Compression ratio	9.45:1
Number of valves per cylinder	4

Table 2. The 4-cylinder engine characteristics.

Displaced Volume	1389.9 cc (4 Cylinder)
Stroke	75.6 mm
Bore	76.5 mm
Connecting rod	144 mm
Compression ratio	10:1
Number of valves per cylinder	4

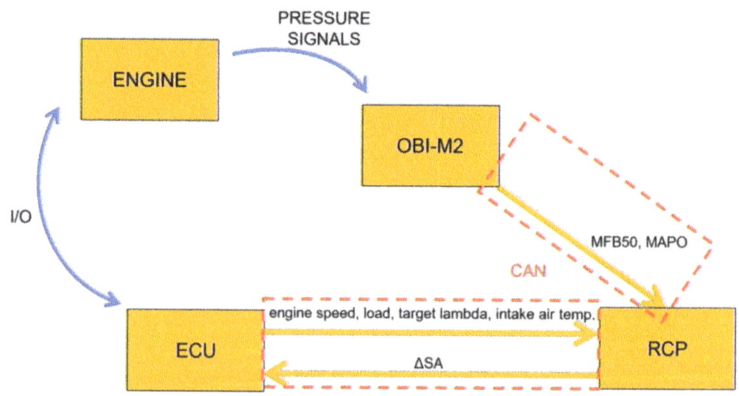

Figure 1. Functional layout of the communication loop for the 8-cylinder engine. For each communication line the main transmitted variables are indicated.

As mentioned above, the combustion controller is tested on a Real Time (RT) machine for RCP application that communicates with the ECU and the indicating system via the Controller Area Network (CAN) communication protocol. For each cylinder, the controller calculates SA corrections with respect to the mapped value for any engine operating point, and it receives the main parameters (such as the engine speed, load, target lambda, etc.) and the combustion indexes from the ECU and OBI-M2 system, respectively.

As for the open-loop combustion controller, the tests are specifically designed to verify the reliability and the accuracy of the novel adaptive strategies. Two different types of fuel (with different RON) and different spark advance maps are alternated to stress and validate the specific self-learning algorithms.

3. Adaptive Strategies

The model-based algorithm converts target damage speed into the MFB50 target through the implementation of the analytical knock and Pmax models. The calibrated analytical MFB50 model calculates the ΔSA to reach the target combustion phase. The analytical knock model [2] includes the influence of the fuel RON for the accurate estimation of the engine knock tendency. In the pure open-loop controller described in the first part of this activity, the fuel RON is manually set by the operator at the test cell. Moreover, the same MFB50 and knock model are used to provide the cyclic values of MAPO and the combustion phase to the control strategy.

Considering the availability of an on-board system for the MAPO and MFB50 estimation, the previous algorithm can be modified replacing the MFB50 and knock models with the measured indexes to automatically learn the fuel RON and update the mathematical relationship between the MFB50 and the ΔSA, for each operating condition. The pure open-loop scheme evolves into that shown in Figure 2: the text in red highlights the indexes that come from the on-board indicating system and that, during this activity, are provided by OBI-M2. Such block scheme anticipates the resulting open-loop chain, defining the focus of this work (i.e., self-learning strategies and the PI controller). The engine load is expressed as the Stoichiometric Trapped Air Mass (STAM).

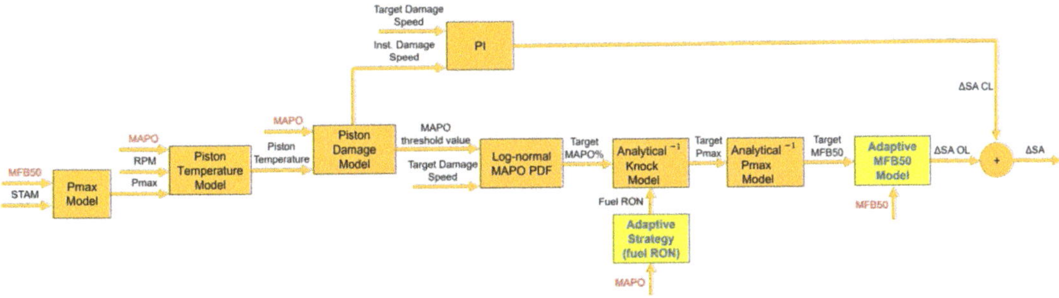

Figure 2. Open loop chain with measured MFB50 and MAPO, instead of modeled values, and PI controller. The bold text highlights the blocks in which the adaptive strategies are implemented. The exponent −1 highlights reversed models. The text in red highlights the indexes that come from the on-board indicating system.

The piston temperature [15,16] and the damage models [17,18] are discussed in detail in the previous works of the authors, and in the Part 1 of the present work. Instead, the Part 2 focuses on the development and the validation of the self-learning strategies. In this section the analytical knock intensity, Pmax and MFB50 model are briefly described, highlighting those features that have been further developed for implementing the adaptive algorithms. Moreover, the reliability of the analytical approach is demonstrated by collecting the equation coefficients that result from the calibration for two different engines. The novel algorithms are implemented in the final controller, to account for variables that can change during the vehicle use and that cannot be predicted during the calibration process, such as the fuel quality. The analytical formulation of main models is the key feature that allows to directly update some model parameters by solving a simple equation, deduced by reversing the model function. In this way, the adaptive strategies were easily implemented in the control algorithm.

3.1. *Analytical Knock Model Validation and Fuel RON Learning Strategy*

The analytical knock model is used to convert the target MAPO98 into the corresponding mean Pmax [2]. The choice of 98th percentile is arbitrary. However, it represents a good compromise between the numerical stability and the faithful representation of the highest instantaneous MAPO values.

Figure 3 shows the surfaces that derive from the model equation for the reference operating conditions for both the engines (Figure 3a for the 8-cylinder engine and Figure 3b for the 4-cylinder engine), while Table 3 shows the numerical values of the model coefficients.

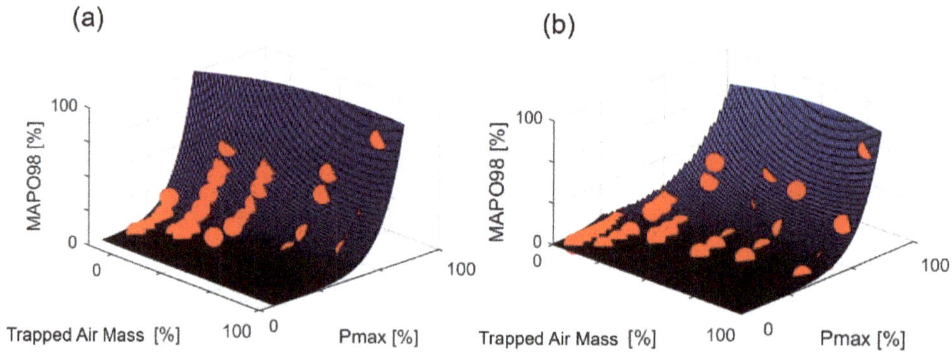

Figure 3. Normalized MAPO98 model for reference conditions of lambda, intake manifold air temperature (Tman), and fuel RON on the Trapped Air Mass-Pmax domain. The blue surfaces represent the output of the model equation while the red dots identify experimental data. The graph on the left (**a**) refers to the 8-cylinder engine, whilst the plot on the right (**b**) refers to the 4-cylinder engine.

Table 3. Knock model coefficients for the 8-cylinder and the 4-cylinder engines.

Model Coefficients	8-Cylinder Engine	4-Cylinder Engine
a	1.7×10^{-4}	2.6×10^{-4}
b	−8.4	−11.4
c	3.8	4.3
d	10.8	10.2
lm	20.2	18.4
Tm	2.2	3.5
RONm	−1.6	−1.8
ESm	1.1×10^{-4}	1.3×10^{-4}

The final formulation presented in the previous paper is reported by Equation (1), that shows the analytical equation of the model for a standard GDI, TC engine:

$$\begin{aligned} \mathrm{MAPO98} = {} & (a\,\mathrm{STAM}^{b})\,(\mathrm{Pmax} + ((\lambda - \lambda_{\mathrm{ref}})\lambda\mathrm{m}) + ((\mathrm{T_{man}} - \mathrm{T_{man\_ref}})\mathrm{Tm}) \\ & +(\mathrm{RONm}(\mathrm{RON_{ref}} - \mathrm{RON_{fuel}})))^{(c\ \mathrm{STAM}+d)} + \mathrm{RPMESm} \end{aligned} \tag{1}$$

where:

- MAPO98 is the MAPO percentile that has to be calculated
- STAM is the Stoichiometric Trapped Air Mass, which is equal to the trapped air mass (TAM) per cycle per cylinder when the mixture is stoichiometric or rich, and the ratio between TAM and lambda when the mixture is lean
- a, b, c, and d are the four calibration parameters of the model that allow to define the MAPO98 for the reference conditions (reference lambda value and air temperature in the intake manifold);
- λ_{ref} represents the reference lambda value, with which the model coefficients (a, b, c, and d) have been calibrated. In this case it is equal to 1;
- λ is the current lambda value;
- λm is the lambda multiplier, the calibratable coefficient that converts the lambda numerical difference into a ΔPmax;
- T_{man} is the current air temperature in the intake manifold;
- T_{man_ref} is the reference temperature in the intake manifold. In this case it is equal to 40 °C;
- Tm is the intake air temperature multiplier, a calibratable coefficient that converts the temperature numerical difference in a ΔPmax;

- RONref is the reference fuel RON value, used during tests that generates the database with which the parameters *a*, *b*, *c* and *d* are identified;
- RONfuel is the RON number of the current fuel;
- RONm is the fuel RON multiplier, the calibratable coefficient that converts the RON numerical difference in a ΔPmax, as well as the λm;
- ESm is the Engine Speed Multiplier, and it includes the effect of the combustion noise as a function of the engine speed (RPM)

As described by the authors in detail in [2], the effects on the knock intensity of the mixture enrichment, the cooling of the intake air temperature and the increasing fuel RON are modeled as an additive contribution to the Pmax value. The result can be seen as a mere translation of MAPO98 curve along the Pmax axis (Figure 3). In other words, for fixed engine speed, load, and fuel, when the mixture is enriched, the intake air temperature is lower than the reference or the fuel RON is higher than the reference and the same Pmax is achieved with a lower MAPO98 value.

Equation (2) shows how to convert a target MAPO98 into a Pmax value and it is thus implemented in the block that includes the reversed analytical knock model:

$$\begin{aligned} \mathrm{Pmax}_{tgt} = \sqrt[(c\ \mathrm{STAM}+d)]{\frac{(\mathrm{MAPO98_{tgt}} - \mathrm{RPM\ ESm})}{(a\ \mathrm{STAM}^{b})}} + (\lambda_{\mathrm{ref}} - \lambda)\lambda\mathrm{m} + (\mathrm{T_{man\_ref}} - \mathrm{T_{man}})\mathrm{Tm} \\ + \mathrm{RONm}(\mathrm{RON_{fuel}} - \mathrm{RON_{ref}}) \end{aligned} \tag{2}$$

In which $\mathrm{MAPO98_{tgt}}$ is the target value of MAPO98 determined by converting the target piston damage and the MAPO threshold into a MAPO PDF, and $\mathrm{Pmax_{tgt}}$ is the corresponding Pmax value, for any given operating condition. As discussed in Part 1, the target MAPO98 does not derive from a statistical calculation, but it is applied just as an index for the conversion of the MAPO threshold into a proper input for the reversed analytical knock model, that is used to convert the latter into a target Pmax value. The algebraic step that allows to achieve the desired model without requiring a new calibration process of the model parameters (*a*, *b*, *c*, *d*, ESm and so on) is a demonstration of the possibilities that derive from the analytical approach proposed by the authors. This means that once the main model parameters are determined, the analytical approach allows to use the model regardless of what are the independent variables and the output of the function. The native formulation of knock model provides the value of MAPO98 as the output [2], while the implementation of the same model in the presented controller needs to have the Pmax as the dependent variable. If a map-based approach would have been used, a new calibration process would have been needed, when the relationship between STAM, Pmax and MAPO98 is reversed. Instead, the Equation (2) can be directly reversed and used to calculate one of the independent variables as the output of the model. Indeed, when the main independent variables of the knock model are stored, $\mathrm{RON_{fuel}}$ coefficient can be directly calculated by solving Equation (2).

The engine operating range is discretized into a speed-load grid, and all the points (defined as a combination of speed and load) are numbered. In this way, variables needed to univocally identify an engine operating condition are halved, due to the possibility to indicate the corresponding point number. The fuel RON self-learning algorithm is not applied to all the engine points. Indeed, data are stored only for the operating conditions characterized by the highest load values (the threshold is defined in terms of Pmax), hence avoiding the risk of assessing the fuel RON by including the combustion noise only. The self-learning algorithm can be summarized in the following steps.

MAPO index offset by the engine speed contribution (RPM*ESm) is stored in a 2-D matrix whose rows identify the enumerated engine point, and columns are the elements of a buffer (in this case 40) that is updated every cycle, following a First-In-First-Out logic. For each matrix row (i.e., engine point), it is then possible to calculate the MAPO98 net of engine speed effect (calculated with the term RPM*ESm). Thus, $\mathrm{MAPO98_{net}}$ for each row indicates the percentile value for any given operating conditions (given engine speed and load). Equation (3) defines $MAPO98_{net}$:

$$MAPO98_{net} = MAPO98 - RPM\ ESm \tag{3}$$

The Pmax index is offset by lambda and the intake air temperature contributions $((\lambda_{ref} - \lambda)\lambda m + (T_{man\_ref} - T_{man})Tm)$ and stored in a 1-D column vector, in which each row represents the engine point. In this case the mean value is considered, and each vector element is updated by computing the weighted average between the stored value and the new one and by giving a higher weight to the old number (such as a first order low-pass filter). After each cycle, the vector is updated and it contains the mean Pmax value for each engine point net of lambda and the intake air temperature contribution, indicated as $Pmax_{net}$:

$$Pmax_{net} = Pmax - (\lambda_{ref} - \lambda)\lambda m - (T_{man\_ref} - T_{man})Tm \tag{4}$$

The STAM values are stored in a column vector by applying the same method used for the Pmax ones. The RON_{fuel} parameter can be then calculated with Equation (5):

$$RON_{fuel} = \frac{\widehat{PMAX}_{net} - \sqrt[\left(c\,\widehat{STAM}+d\right)]{\frac{\left(\widehat{MAPO98}_{net}\right)}{\left(a\,\widehat{STAM}^{b}\right)}}}{RONm} + RON_{ref} \tag{5}$$

$Pmax_{net}$, $MAPO98_{net}$ and STAM are upper marked to highlight that they are vector quantities. The calculated fuel RON is determined by solving a linear system of equations and the system degree is equal to the number of engine points. This means that the fuel RON value results as the best value for all the considered operating conditions characterized by high loads. It is important to highlight that the matrix of MAPO and the vector of Pmax contain the values for each engine point with which the operating field has been discretized. The index of each row represents an engine point number that identifies a couple of fixed engine speed and load (that are always the same) and this is the reason why the values contained in each row of the vector or matrix change slowly and typically only when fuel quality varies. The execution frequency of the fuel RON learning algorithm can be decided by the engine manufacturer: RONfuel parameter can be updated every cycle, every n cycles, once per vehicle use, and so on. The solution of Equation (5) does not introduce any kind of delay and this feature makes it particularly suitable for the real-time execution. Only the calculation of MAPO98 and the moving average applied to update the Pmax vector introduce a filtering of the measured indexes. The moving average of Pmax for each engine point is applied to remove cycle-to-cycle variation effects. The calculation of the fuel RON helps to update such value in a very fast way, and this feature is important to calculate the proper Pmax target values when the intake air temperature or the fuel is changing rapidly (for instance when the vehicle tank is empty, and the type of gasoline used to refill it is different from the previous one).

The fuel RON value is then used by the reversed knock model, which converts the target MAPO98 into the corresponding Pmax value. Figure 4 shows the open-loop controller with all the inputs needed by the fuel RON learning strategy.

3.2. *Analytical Pmax Model Validation*

As for the knock model, the simple analytical formulation of the Pmax model is applied to the 4-cylinder engine, to demonstrate the reliability of the proposed approach, despite its very simple formulation. Additionally, in this case, the calibrated coefficients are very similar for both engines. Figure 5 shows the trends of model equation on the MFB50-STAM domain and Table 4 collects the numerical values of the Pmax model coefficients for the 8-cylinder and the 4-cylinder engine.

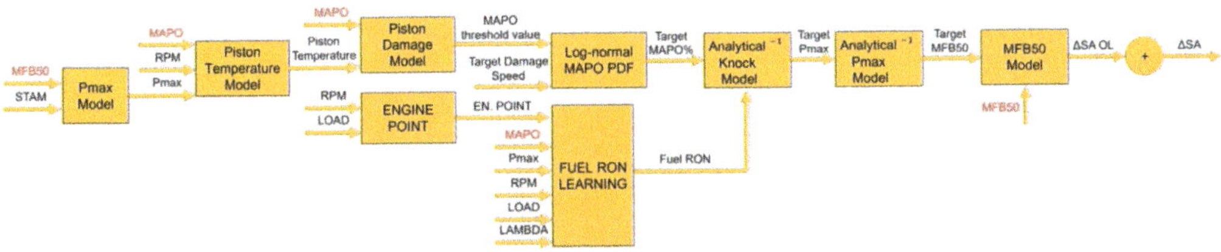

Figure 4. Controller layout, including the fuel RON adaption algorithm.

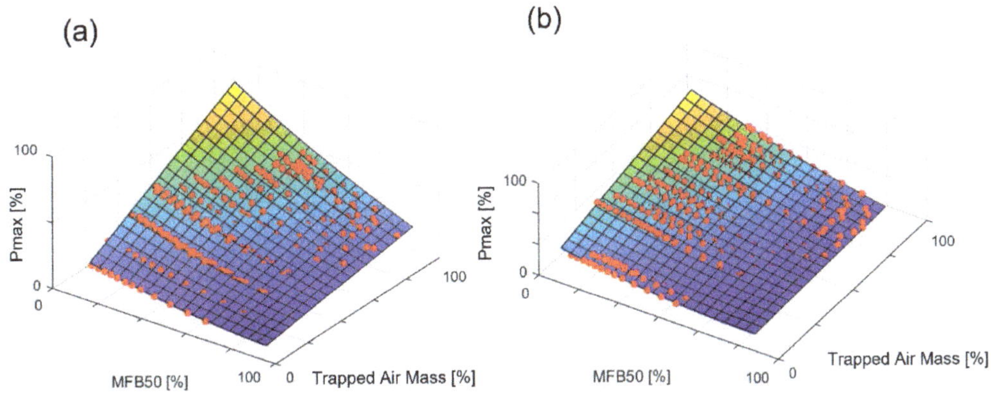

Figure 5. Normalized Pmax model on the MFB50-Trapped Air Mass domain. The surfaces represent the output of the model equation while the red dots are the experimental data. The graph on the left (**a**) refers to the 8-cylinder engine, whilst the plot on the right refers to the (**b**) 4-cylinder engine.

Table 4. Pmax model coefficients for the 8-cylinder and the 4-cylinder engines.

Model Coefficients	8-Cylinder Engine	4-Cylinder Engine
$p00$	−16.21	−6.63
$p10$	−0.19	−0.27
$p01$	148.54	145.3
$p20$	0.02	0.01
$p11$	−2.92	−2.89

As shown in Figure 4, the implementation of the reversed Pmax model in the open-loop chain of combustion controller allows the determination of the target combustion phase that achieves either target damage speed, or the best efficiency.

3.3. Adaptive Combustion Model

The final output of the open-loop chain is the ΔSA (with respect to the mapped values) that achieves the target MFB50. The MFB50 model analytically defines the relationship between the actuated spark timing and the combustion phase [3]. Figure 6 represents the block scheme of the controller open-loop chain, including the adaptive MFB50 model. The dotted line (the fuel RON learning strategy) represents the algorithm described above.

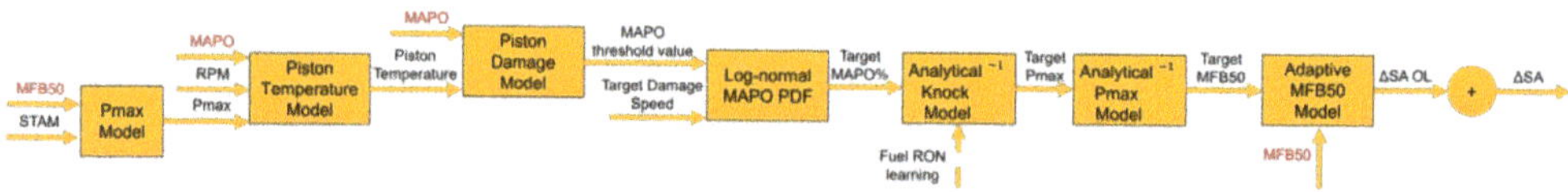

Figure 6. Controller layout, including the adaptive MFB50 model.

Such model is compatible with the adaptive strategy that updates the parabolic relationship between MFB50 and corresponding spark timing, for any given engine speed and load. For each engine point (fixed speed and load), the relationship between the mean MFB50 and the actuated SA is characterized by a parabolic trend, which can be described with a second order polynomial. Such trend derives from a variable MFB50 sensitivity to the SA: the higher the MFB50, the lower the curve slope. Figure 7 shows an example of what has been asserted, collecting data for a spark sweep recorded for given engine speed and load. SA is reported as the difference with the calibrated value (ΔSA) and MFB50 is normalized with respect to the maximum value.

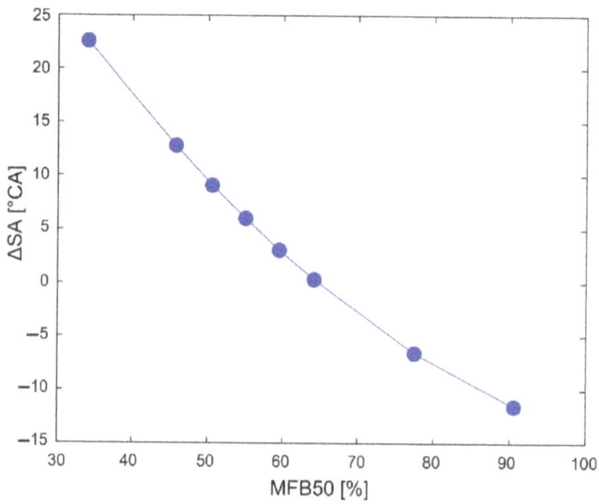

Figure 7. Experimental ΔSA-MFB50 points recorded for fixed engine speed and load. The MFB50 is normalized with respect to the maximum value.

Thus, for each engine point (defined as a couple of the engine speed and load) a second order polynomial can be calculated. Equation (6) defines the kind of equation used to fit the ΔSA on the MFB50 axis, for each engine point:

$$\mathrm{MFB50} = A\,\Delta\mathrm{SA}^2 + B\,\Delta\mathrm{SA} + C \tag{6}$$

where A, B and C are the equation parameters.

The mean MFB50 values are stored in a 2-D matrix, where rows represent the numbered engine point, and the columns are the ΔSA steps, equally distributed between positive and negative values. A unitary spark advance step corresponds to a certain number of crank angle (CA) degrees. Limitations on ΔSA are typically imposed by maximum and minimum spark timing that can be actuated by the ECU. For instance, if the controller can communicate a ΔSA step within the range ±20 (which in turn corresponds to a certain range of CA degrees), the 2-D matrix has 41 columns. Each value in the matrix is updated by computing the weighted average between the MFB50 stored value and the new one that works as a moving average. Initial values of the 2-D matrix are defined with the MFB50 model executed offline. For every execution time-step, the stored MFB50 value for current operating condition is updated with the more recent MFB50, through the moving average operator. For each matrix row, a parabolic equation can be determined by fitting the stored values with respect to the fixed ΔSA vector that determines the number of matrix columns. The identification process of the optimal values of the function parameters for the current engine point can be performed with the combustion frequency, by applying the least squares method, for the equation degree equal to 2 [19]. Considering the example described above, in which the admissible ΔSA steps go from −20 to +20, the 2-D matrix has 41 columns and the associated system of m (in this case 41) equations of degree n (with $m > n$) can be defined. The least squared method allows to directly determine the coefficients vector of $n + 1$ elements that guarantee the optimal approximation of the available data. Even in this case, the choice of the least squared method allows to perform a direct calculation of polynomial coefficients, and it instantaneously updates the parabolic relationship between the combustion phase and the spark advance. Nevertheless, differently from the fuel RON learning strategy that uses the 98th percentile of MAPO, a moving average (a low-pass filter) is applied to neglect the cycle-to-cycle variation (CCV, that normally affects combustion indexes).

For every engine cycle, the MFB50 matrix is updated including the logged MFB50 for the current engine operating conditions (speed, load, and ΔSA), and the parabolic equation is directly calculated to accurately determine the ΔSA that achieves the target MFB50. Thus, the controller layout in Figure 8 includes data the adaptive algorithm needs as input.

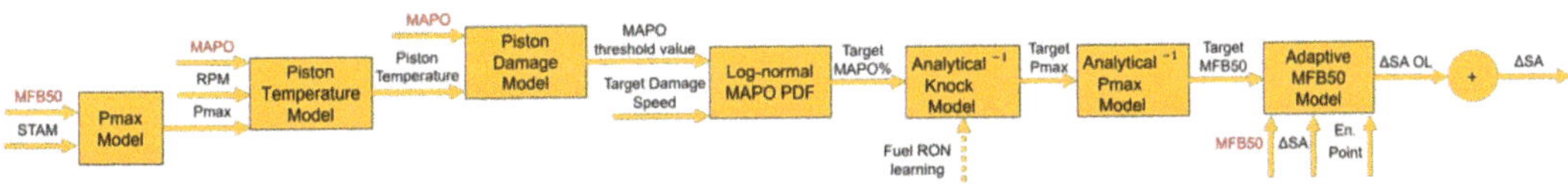

Figure 8. Controller layout, including the adaption algorithm for ΔSA-MFB50 analytical equation.

The methodology with which the vector and the matrix of combustion index and the engine load (STAM) are updated depends on the discretization of the engine speed-load operating field. On one hand, if the breakpoints of such grid are sufficiently dense, the algorithm can assign to a certain operating condition the closest engine point, rounding the speed and load with respect to the breakpoints resolution. On the other hand, once the breakpoints of the engine speed-load grid are not so dense, the weighted update of the four breakpoints between which the generic operating condition falls is performed. In other words, combustion indexes stored in the four breakpoints (two engine speed and two engine loads) that include the current operating conditions are updated by distributing the logged number on each of such breakpoints, by defining a weight that is inversely proportional to the distance between the current operating condition and each discretized engine point. Both strategies were tested during the development process of the described combustion controller, but the first solution (based on the updating of the closest breakpoint value, with respect to the current condition) was finally adopted. Indeed, a sufficiently dense discretization of the engine speed-load operating field was reached, due to a numerical optimization process of stored data for adaptive strategies that allowed to minimize the impact of matrices and vectors on the volatile memory of the RCP device. Such solution was preferred for the lower computational power needed, due to the simpler algorithm and because the performance of both solutions was practically the same for the considered grid that discretizes engine operating field.

4. Combustion Controller Validation

Differently from the pure open-loop controller described in the Part 1, piston damage is calculated using the measured MAPO as input, instead of the calculated cyclic value with the analytical knock model. Thus, a closed-loop chain based on a Proportional Integral (PI) controller is implemented for a component-protection objective. Indeed, PI controller reacts only when the instantaneous damage speed exceeds of a certain value the target (with a gain scheduling approach) and an additional integral contribution is added once a safety threshold that corresponds to a very high knock intensity level is crossed. This is because the aim is of controlling combustion phase entirely with the model-based open-loop chain in normal conditions and allowing the PI controller intervention only to the spark retardation when needed. The PI reacts with a typical sawtooth behavior that has been accurately calibrated to have a response fast enough to avoid the knock intensity remaining abundantly lower than the target. Figure 9 shows the complete combustion controller scheme. The text in red identifies inputs that come from the on-board in-cylinder pressure sensing system (MAPO and MFB50 for each combustion and for each cylinder). Excluding the adaptive algorithms described above, the open-loop strategy is identical to the one presented in Part 1. As highlighted in the previous paper, all the algebraic steps implemented in such scheme are direct calculations, and there are no operators that could induce delays in the algorithm.

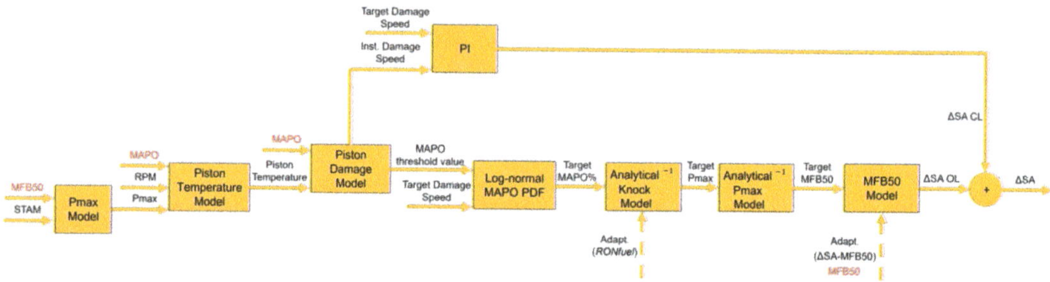

Figure 9. Complete controller scheme. The text in red indicates the combustion indexes estimated from the on-board in-cylinder pressure sensing system, i.e., MAPO and MFB50.

The combustion indexes are published once per cycle, while the combustion controller code is executed at 1 kHz to be faster that the engine cycle on the whole engine operating range. A complete controller is implemented for each cylinder, except for the fuel RON learning algorithm which is averaged between cylinders. While the open-loop chain uses instantaneous engine operating parameters (speed, load, lambda), the actuated ΔSA, and the MFB50 and MAPO recorded during the current cycle to calculate ΔSA for the following cycle, the adaption algorithms and the closed-loop chain receive the same data after the combustion. In other words, the open-loop calculations are performed before the engine cycle for which target damage speed is defined. On the contrary, the adaption algorithms and the PI controller are executed after the engine cycle for which the ΔSA is determined.

The engine at the test bench is operated with the speed and load profiles recorded during real vehicle maneuvers. The tested profiles consist of complete track laps, with both very steep transients and semi steady-state conditions. In this way, the controller is stressed with realistic engine speed and load profiles, and the results demonstrate the high reliability of the proposed tool. Moreover, some tests are carried out with a positively offset SA map, in order to validate the performance and the robustness of the self-learning algorithm that updates the mean MFB50 values and the ΔSA-MFB50 functions. The fuel RON is also changed during a portion of these tests, and the reliability of knock model is then demonstrated.

5. Results and Discussion

The main objective of the tests is to validate the algorithms developed above. Figure 10 displays a portion of the engine speed and pedal profiles tested during the validation procedure, normalized with respect to the maximum values of the engine operating range. The entire engine operating range is thus covered. The detailed results are shown only for cylinder 1, for a more concise and clearer representation. All the following figures were obtained by plotting data recorded with the RCP and the indicating system. The

RCP system provides those variables that are received by the ECU and that are calculated by the controller, while OBI-M2 allows having all the combustion indexes, such as the instantaneous and the mean MFB50, Pmax, MAPO and MAPO98.

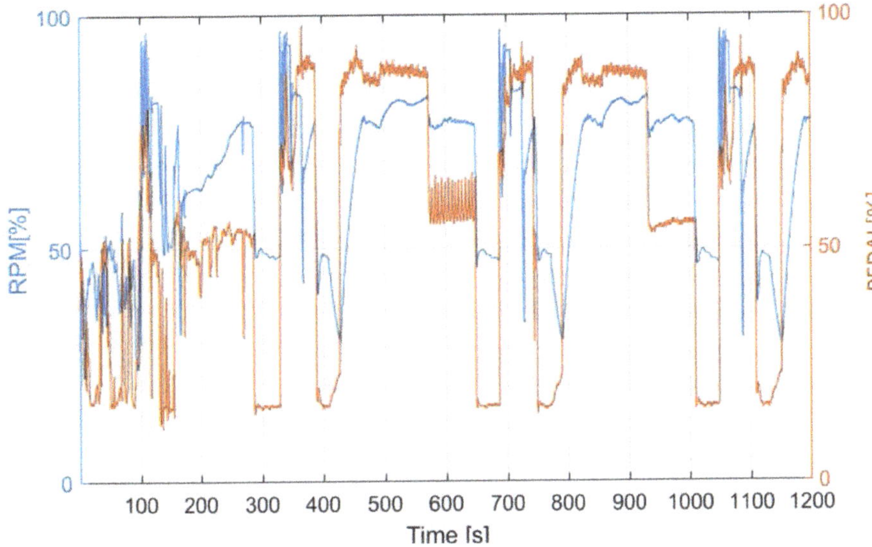

Figure 10. Normalized engine speed and load profiles related to the results presented below.

Figure 11 shows the MFB50, the Pmax, the final ΔSA and the closed-loop contribution (ΔSA CL), MAPO and its 98th percentile, and the damage speed. All the quantities reported in Figure 11 are normalized with respect to the maximum value of y-scale chosen to include all the recorded values. Instead, the values of ΔSA are reported as °CA. For this test, the matrices and the vectors included in the adaption algorithms contain initial calibrated values. As anticipated in Part 1, both the MAPO98 and the mean MFB50 shown in Figure 11 are calculated by the indicating system with a buffer of 200 cyclic values. However, such values do not have a role in the open-loop algorithm. They are just evaluated to be compared with the corresponding target value. This means that, excluding moving averages of adaptive strategies used to filter the CCV, there is not any kind of filter in the open-loop calculation chain to avoid the introduction of a delay with respect to the current combustion indexes (MFB50 and MAPO provided by the indicating system). The filtered MFB50 is always on the target, except when the PI controller reacts. This condition can be easily identified when ΔSA calculated by closed-loop is negative. As for the pure open-loop controller, the target MFB50 is defined as the value that achieves a given damage speed value, saturated at the MBT. The MBT value is the lowest MFB50 targeted by the controller. Due to the actuated SA that is generally higher than the calibrated value (as observable in the third plot of Figure 11 that reports the actuated ΔSA, ΔSA Act.), the engine operating field for which the MBT can be reached is wider than that resulting from the standard calibration. This is due to the higher MAPO that is considered as admissible (especially beyond a certain value of STAM). As for the pure open-loop controller, the most important consequence of the application of the proposed controller is a significant increase of the combustion efficiency, that can be spent with an increase of the maximum torque or a reduction of the fuel consumption. The Pmax reaches the corresponding target values only when the target MFB50 is higher than the MBT. At the same time, during such portions of the test, the MAPO98 oscillates around its target value and the average damage speed is close to the corresponding target value. The average damage speed shown in the last graph of Figure 11 is calculated as the cumulative damage (evaluated in bars) divided for the whole engine life (evaluated in seconds), and for this reason it varies slowly. Even if the controller manages the instantaneous damage speed, the mean value evaluated in such way is the most meaningful index, because it represents the average damage speed on the entire engine life and it is used to estimate the cumulative piston damage at the end of the test period. The results demonstrate that even the adaptive MFB50 model is reliable and allows to directly calculate the ΔSA needed to reach the target MFB50. As visible in the third graph of Figure 11, the PI controllers react only with small SA corrections (lower than 2 °CA) when the instantaneous damage speed exceeds a certain threshold value. Moreover, the calculated cyclic Pmax is very close to the experimental data, and this confirms the accuracy and robustness of the proposed analytical model. A more quantitative evaluation of the controller accuracy is reported in Figure 12, that shows the error between the MFB50 and the Pmax, and the corresponding target values, for the analyzed test. The statistical indexes (the mean value and the standard deviation) are reported for each plot. On the one hand, the error between the target and the filtered MFB50 has to be evaluated excluding engine cycles in which PI controller reacts and consequently the MFB50 cannot be close to the target value. On the other hand, all the engine cycles in which the target MFB50 is saturated to the MBT have to be excluded too, because the target Pmax cannot be achieved. However, since the objective of this work is to develop a combustion controller that manages the combustion phase on the entire engine operating field, the error between MFB50 and the corresponding target value is evaluated on the whole test, and not only for the knock-limited operating conditions. The error is mainly included within the range ±0.4 °CA and ±5 bar for the MFB50 and the Pmax, respectively. These values, together with the achievement of the damage speed target, is an extremely relevant result. Nevertheless, it is important to mention that the average damage speed is close to target value already at the beginning of the test and this is because the engine speed and pedal profiles shown by Figure 10 are reproduced several times in a row. In this way, the controller can reach the target damage speed value.

Figure 13 shows the calculated fuel RON value, that is practically constant during the test. In this case, standard RON 95 fuel is used. The statistical indexes values are reported in the title of the figure. As discussed above, the fuel RON value is calculated only when the Pmax exceeds a calibrated threshold, to ensure excluding the engine points in which knock does not occur. For such reason, during some portions of the test, calculated fuel RON assumes constant values.

Both the adaptive strategies are validated with specific tests. The algorithm for the ΔSA-MFB50 relationship self-learning is validated by applying an offset to the whole spark advance map, so that the parabolic function that defines the combustion phase with respect to the SA correction needs to be updated. In this way, the condition that leads to save the incorrect MFB50 values (for instance due to a broken sensor) is reproduced. In this case, the test was carried out on a stationary engine operating point, by forcing a spark timing correction on the SA map. Thus, it is possible to quantify the time needed to update the stored MFB50 values and calculate the correct parabolic function. Figure 14 displays the results in terms of MFB50 and ΔSA for the cylinder 1. A vertical dashed line indicates the moment in which the offset is applied onto the spark timing map. The adaptive strategy can recognize and calculate the new ΔSA-MFB50 parabolic function by updating the value saved in the MFB50 matrix. The error returns close to 0 in 7 s, and this kind of test demonstrates the robustness of such algorithm. The implementation of an additional closed-loop contribution to the control the MFB50 on the target value is not needed as it would introduce an additional error, especially under transient operating conditions. Moreover, in this case, the close-loop chain based on the damage speed does not react, as expected, because the SA map is modified with more retarded values, for a not knock-limited engine point.

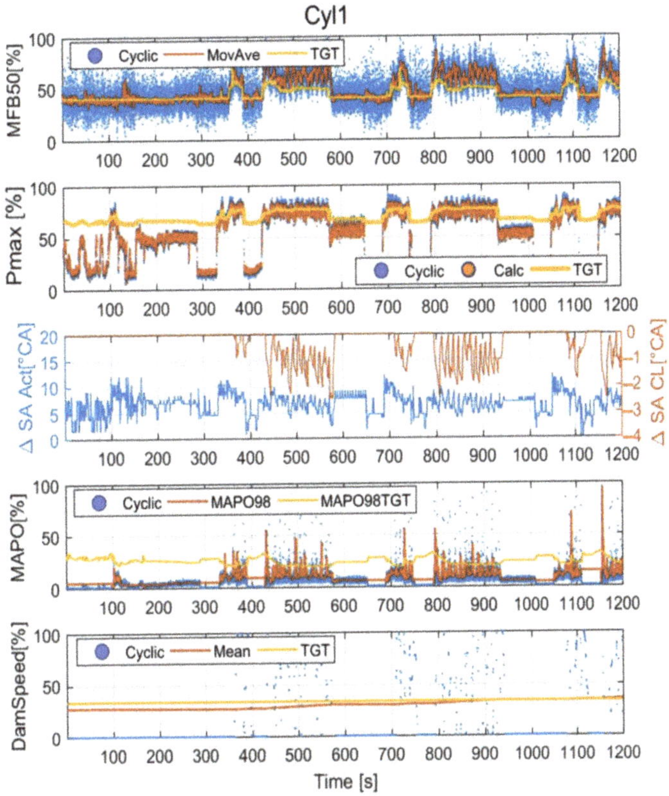

Figure 11. Target and experimental values of MFB50, Pmax, MAPO, and damage speed, for cylinder 1. In particular, the third graph represents the spark timing correction of the closed-loop chain, and the final output of the controller.

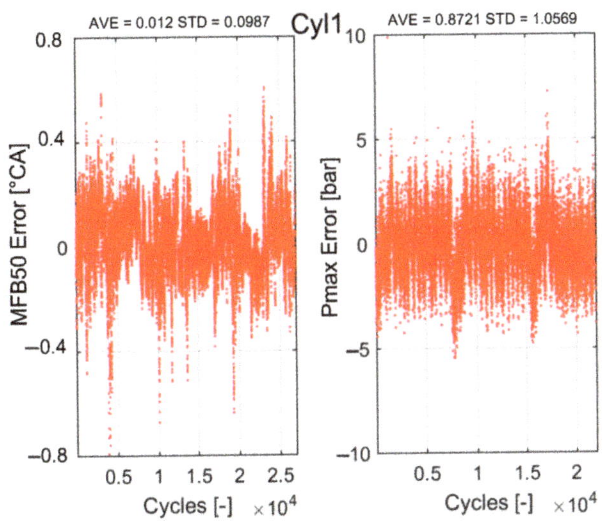

Figure 12. Error between the target and the experimental MFB50 and Pmax indexes, for cylinder 1.

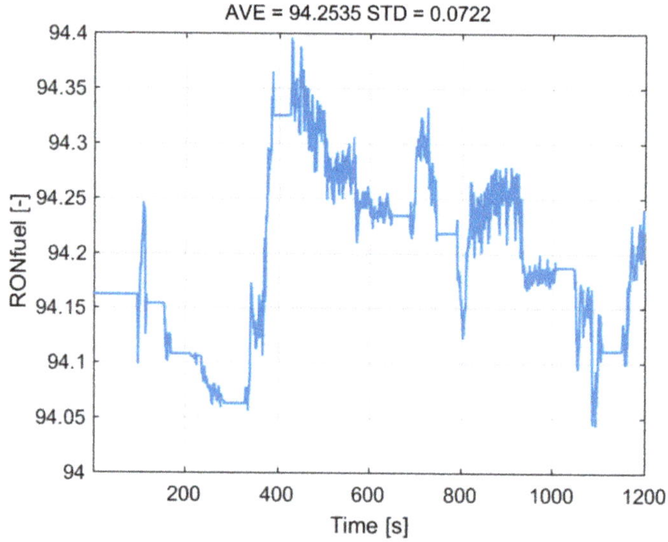

Figure 13. Calculated fuel RON. The very narrow vertical scale proves that albeit calculations are run, no modification in the RON number is adapted, as expected.

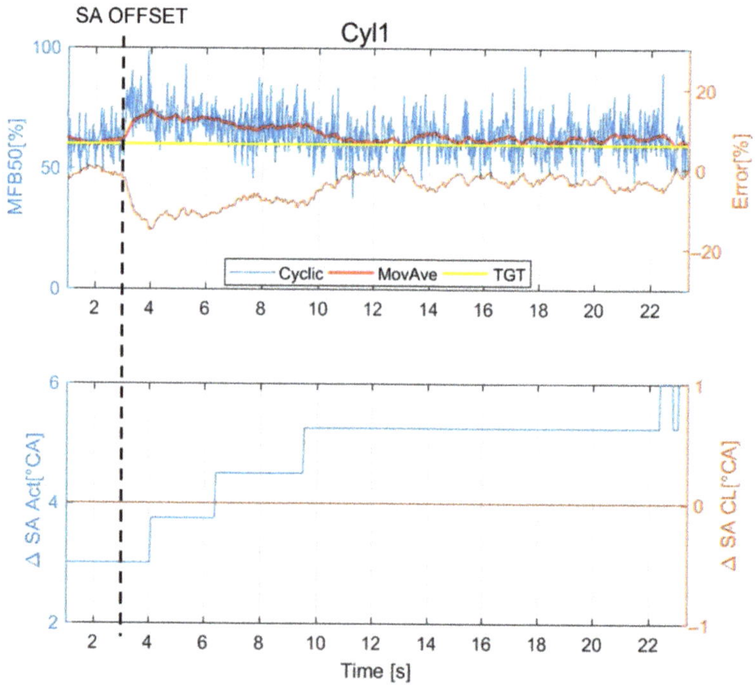

Figure 14. MFB50 and ΔSA for cylinder 1 in the test used to verify the celerity of the adaption algorithm. The vertical dotted line indicates the time an offset on the SA map is applied. The red line represents the MFB50 moving average while the orange one the error with respect to the target value. The error is very close to 0 after 7–8 s.

Lastly, the fuel RON learning algorithm is validated. A long test was carried out at the bench by reproducing the same engine speed and load profiles shown in Figure 10 and switching between two different types of fuel coming from two tanks that supply the engine test bench fuel line (with RON95 and RON100 gasoline, respectively). The switches between fuel lines are performed during the test. Even in this case, the results demonstrate the high reliability of the proposed control strategy. Figure 15 shows the trend of the calculated fuel RON value during the test. The vertical dashed lines indicate the moments in which the fuel quality is switched. As mentioned above, the RONfuel coefficient update is performed when the Pmax exceeds a calibrated threshold, and therefore the displayed value is constant during some portions of the test. The calculated RON clearly follows the nominal fuel octane number.

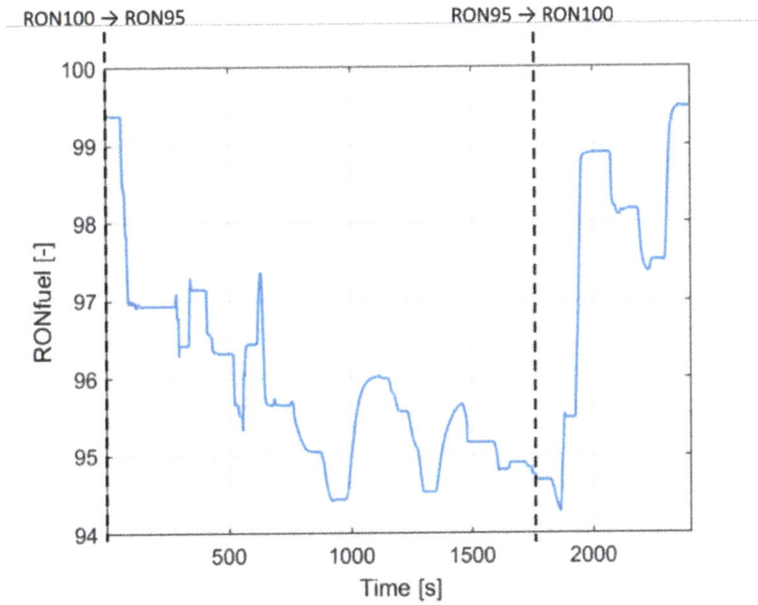

Figure 15. Calculated fuel RON for the validation test of the fuel RON adaptive strategy.

Figure 16 shows the trend of the main combustion indexes, the spark timing corrections, and the damage speed, for the cylinder 1. The fuel RON adaption algorithm allows to properly account for the effects of the fuel quality and calculate the maximum in-cylinder pressure that ensures the target MAPO98 value for the current type of fuel, on the entire engine operating range. The average damage speed remains very close to the target, as well as the MFB50 and the Pmax, when the closed-loop contribution is null and target combustion phase is not saturated to the MBT value, respectively. It can be noted that the PI controller reacts especially when the fuel RON is overestimated (for instance during the learning transient), but such corrections are null when RON value is correctly evaluated or underestimated, as expected. This is a further confirmation of the robustness of the open-loop chain, that makes the feedback controller actions very limited. Figure 17 shows the error between the target and the moving average of the cyclic MFB50 and Pmax.

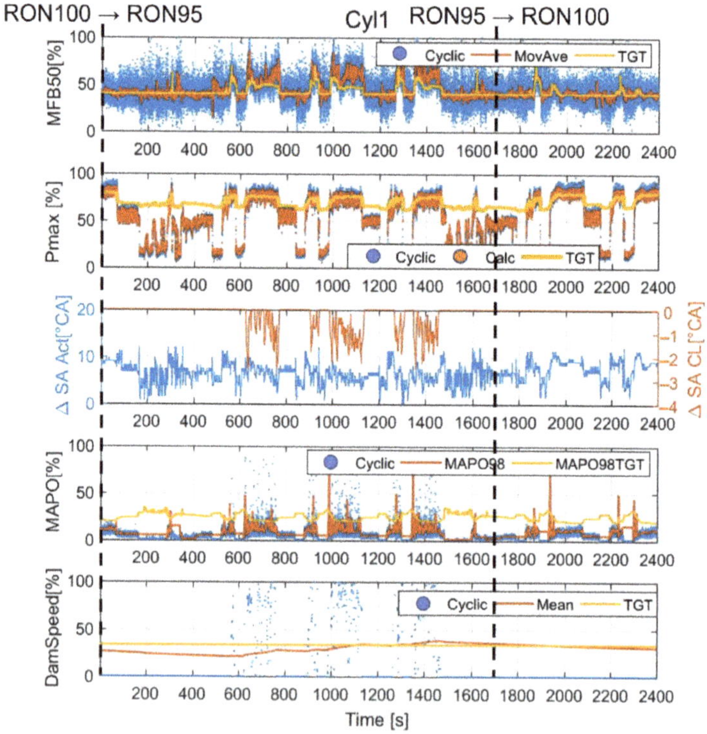

Figure 16. Target and experimental values of MFB50, Pmax, MAPO and damage speed for cylinder 1. The third graph represents the spark timing correction of the closed-loop chain and the final output of the controller, for the fuel RON adaptive strategy validation test.

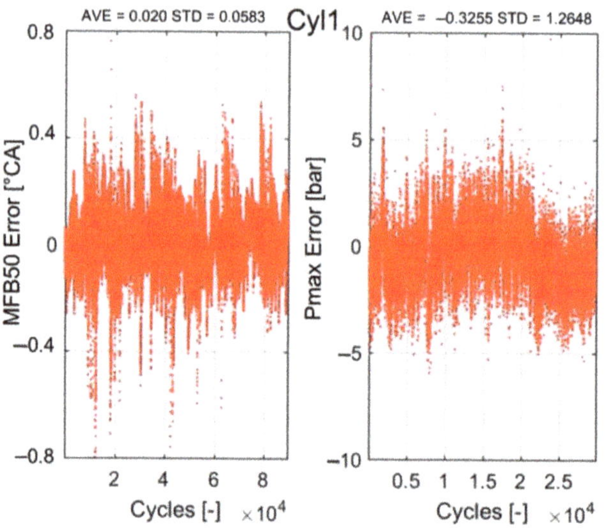

Figure 17. Error between target and experimental MFB50 and Pmax indexes recorded during the fuel RON adaptive strategy validation test, for cylinder 1.

The validation tests demonstrate the high accuracy of the model-based open-loop controller, and the reliability of the implemented adaptive strategies under real on-vehicle operating conditions.

6. Conclusions and Future Works

In this work, the innovative model-based combustion control system described in the Part 1 was further extended and validated. In this work the algorithm was developed considering the availability of an on-board indicating system that provides a knock intensity and a combustion phase index (MAPO and MFB50, respectively). The novel adaption algorithms based on the analytical formulation of the knock and the MFB50 models are now implemented in the controller, for the live adaption of fuel RON, and the relationship between the combustion phase and ΔSA. In this way, the fuel RON and the coefficients of the analytical MFB50 model that in the first version of the controller were manually set by the user during the calibration process, are now estimated automatically and continuously updated. The final controller is then implemented in an RCP system. The proposed controller needs MAPO and MFB50 indexes, that can be obtained by the on-board systems for the in-cylinder pressure signal estimation, such as the accelerometers, the ionization currents or the piezoelectric components for engine head that can sense the forces induced by the combustion process. Nevertheless, the tests shown in this section are carried out with an engine equipped with standard piezoelectric transducers and the combustion indexes are calculated with a commercial indicating system.

The analytical knock model was calibrated for both an 8- and 4-cylinder engine to demonstrate the robustness and the general validity of the analytical approach. However, the final controller was developed and calibrated for the 8-cylinder engine. The main benefits of the simple analytical formulation were shown, explaining how the proposed approach allows to directly implement an algorithm for the calculation of the fuel RON value. The analytical Pmax and MFB50 models were briefly retrieved and the implementation of the adaptive strategy for the ΔSA-MFB50 parabolic function update was discussed. The matrices that contain stored data needed for the self-learning algorithm were numerically optimized to minimize their impact on the volatile memory of RCP system, and to make the controller suitable for the following implementation in the production ECU. The validation tests were performed by reproducing at the bench the engine speed and load profiles logged during the real vehicle use on track, to validate combustion controller also under very fast transient conditions. Moreover, specific tests were performed to validate the adaption algorithms, by reproducing at the test bench some real conditions, such as a wrongly calibrated MFB50 matrix (it was simulated by offsetting the SA map of a certain value) and a fast fuel quality switch (for example a fuel tank refill with a different fuel).

The results demonstrate both the high accuracy of the complete open-loop chain, and the reliability of the self-learning algorithms. The errors between the target and the experimental indexes were low and this further confirms that the novel approach to the combustion control can be considered as a viable and robust innovation in the knock control. The resulting error was mainly included within the range ± 0.4 °CA and ± 5 bar for MFB50 and Pmax, respectively. These values, together with the achievement of the damage speed target, can be considered as a relevant result.

This combustion controller will be further validated with the use of a production-oriented, on-board-compatible indicating system that can provide MAPO and MFB50 indexes. Moreover, it will be further developed with an exhaust gas temperature model to manage even the lambda target and it will be finally implemented in a prototyping ECU for the on-vehicle tests. Modern ECUs are characterized by high computational power and their main limitation is related to the number of analog and digital channels, rather than the CPU power. Even if the computational effort strongly depends on the type of the RT machine used to perform tests, the authors confirm the proposed controller is suitable for the implementation in a production ECU.

Author Contributions: Formal analysis, M.C. and N.S.; supervision, N.C., N.R., J.M., E.C. and D.M.; writing–original draft, A.B. All authors have read and agreed to the published version of the manuscript.

Funding: This work was supported by 1 research grants founded by the Department of Industrial Engineering of the University of Bologna (grant number 2616) and the collaboration with Ferrari S.p.a. The external funding is indicated in the Acknowledgments section.

Acknowledgments: The authors gratefully acknowledge financial support from "Fondi europei della Regione Emilia-Romagna", within program POR FSE 2014–2020 in the framework of the "Accordo Regionale di insediamento e sviluppo delle imprese—progetto STEP".

Conflicts of Interest: The authors declare no conflict of interest.

Abbreviations

CAN	Controller Area Network
CCV	Cycle-to-Cycle Variation
ECU	Engine Control Unit
GDI	Gasoline Direct Injection
IMEP	Indicating Mean Effective Pressure
ION	Ionization Current
MAPO	Maximum Amplitude of Pressure Oscillation
MAPO98	98th percentile of MAPO
MBT	Maximum Brake Torque
MFB50	The 50%-of-Mass-of-Fuel-Burned
PDF	Probability Density Function
PF	Probability Function
PI	Proportional Integral controller
PID	Proportional Integral Derivative controller
Pmax	Maximum in-cylinder pressure
RCP	Rapid Control Prototyping
RON	Research Octane Number
RT	Real-Time
SA	Spark Advance
STAM	Stoichiometric Trapped Air Mass
TAM	Trapped Air Mass
TC	Turbo-Charged
VVT	Variable Valve Timing

References

1. Cavina, N.; Rojo, N.; Ceschini, L.; Balducci, E.; Poggio, L.; Calogero, L.; Cevolani, R. Investigation of Knock Damage Mechanisms on a GDI TC Engine. SAE Technical Paper 2017-24-0060. In Proceedings of the 13th International Conference on Engines & Vehicles, Capri, Italy, 10–14 September 2017. [CrossRef]
2. Cavina, N.; Brusa, A.; Rojo, N.; Corti, E. Statistical Analysis of Knock Intensity Probability Distribution and Development of 0-D Predictive Knock Model for a SI TC Engine. SAE Technical Paper 2018-01-0858. In Proceedings of the WCX World Congress Experience, Warrendale, PA, USA, 2 May 2018. [CrossRef]
3. Brusa, A.; Cavina, N.; Rojo, N.; Cucchi, M.; Silvestri, N. Development and Validation of a Control-Oriented Analytic Engine Simulator. SAE Technical Paper 2019-24-0002. In Proceedings of the 14th International Conference on Engines & Vehicles, Capri, Italy, 15–19 September 2019. [CrossRef]
4. Mahdisoozani, H.; Mohsenizadeh, M.; Bahiraei, M.; Kasaeian, A.; Daneshvar, A.; Goodarzi, M.; Safaei, M.R. Performance Enhancement of Internal Combustion Engines through Vibration Control: State of the Art and Challenges. *Appl. Sci.* **2019**, *9*, 406. [CrossRef]
5. Sateesh, K.A.; Yaliwal, V.S.; Soudagar, M.E.M.; Banapurmath, N.R.; Fayaz, H.; Safaei, M.R.; Elfasakhany, A.; El-Seesy, A.I. Utilization of biodiesel/Al2O3 nanoparticles for combustion behavior enhancement of a diesel engine operated on dual fuel mode. *J. Therm. Anal. Calorim.* **2021**, 1–15. [CrossRef]
6. Wategave, S.; Banapurmath, N.; Sawant, M.; Soudagar, M.E.M.; Mujtaba, M.; Afzal, A.; Basha, J.; Alazwari, M.A.; Safaei, M.R.; Elfasakhany, A.; et al. Clean combustion and emissions strategy using reactivity controlled compression ignition (RCCI) mode engine powered with CNG-Karanja biodiesel. *J. Taiwan Inst. Chem. Eng.* **2021**, *124*, 116–131. [CrossRef]

7. Businaro, A.; Cavina, N.; Corti, E.; Mancini, G.; Moro, D.; Ponti, F.; Ravaglioli, V. Accelerometer Based Methodology for Combustion Parameters Estimation. *Energy Procedia* **2015**, *81*, 950–959. [CrossRef]
8. Ranuzzi, F.; Cavina, N.; Scocozza, G.; Brusa, A.; De Cesare, M. Experimental Validation of a Model-Based Water Injection Combustion Control System for On-Board Application. SAE Technical Paper 2019-24-0015. In Proceedings of the 14th International Conference on Engines & Vehicles, Capri, Italy, 15–19 September 2019. [CrossRef]
9. Cavina, N.; Rojo, N.; Poggio, L.; Calogero, L.; Cevolani, R. Investigation on Pre-Ignition Combustion Events and Development of Diagnostic Solutions Based on Ion Current Signals. *SAE Int. J. Engines* **2017**, *10*, 1518–1523. [CrossRef]
10. Cavina, N.; Sgatti, S.; Cavanna, F.; Bisanti, G. Combustion Monitoring Based on Engine Acoustic Emission Signal Processing. SAE Technical Paper 2009-01-1024. In Proceedings of the SAE World Congress & Exhibition, Detroit, MI, USA, 20–23 April 2009. [CrossRef]
11. Cavina, N.; Businaro, A.; De Cesare, M.; Paiano, L. Knock Control Based on Engine Acoustic Emissions: Calibration and Implementation in an Engine Control Unit. SAE Technical Paper 2017-01-0785. In Proceedings of the WCX™ 17: SAE World Congress Experience, Detroit, MI, USA, 4–6 April 2017. [CrossRef]
12. Cavina, N.; Businaro, A.; De Cesare, M.; Monti, F.; Cerofolini, A. Application of Acoustic and Vibration-Based Knock Detection Techniques to a High Speed Engine. SAE Technical Paper 2017-01-0786. In Proceedings of the WCX™ 17: SAE World Congress Experience, Detroit, MI, USA, 4–6 April 2017. [CrossRef]
13. Romani, L.; Lenzi, G.; Ferrari, L.; Ferrara, G. Indirect Estimation of In-Cylinder Pressure through the Stress Analysis of an Engine Stud. SAE Technical Paper 2016-01-0814. In Proceedings of the SAE 2016 World Congress and Exhibition, Detroit, MI, USA, 12–14 April 2016. [CrossRef]
14. Corti, E.; Abbondanza, M.; Ponti, F.; Raggini, L. The Use of Piezoelectric Washers for Feedback Combustion Control. SAE Technical Paper 2020-01-1146. In Proceedings of the WCX SAE World Congress Experience, Detroit, MI, USA, 5–7 April 2020. [CrossRef]
15. Ceschini, L.; Morri, A.; Morri, A.; Di Sabatino, M. Effect of thermal exposure on the residual hardness and tensile properties of the EN AW-2618A piston alloy. *Mater. Sci. Eng. A* **2015**, *639*, 288–297. [CrossRef]
16. Balducci, E.; Ceschini, L.; Morri, A.; Morri, A.; Di Sabatino, M.; Arnberg, L.; Li, Y. High Temperature Behavior of the EN AW-2618A Piston Alloy Containing 0.12wt% Zr: Influence of Heat Treatment. *Mater. Today Proc.* **2015**, *2*, 5037–5044. [CrossRef]
17. Ceschini, L.; Morri, A.; Balducci, E.; Cavina, N.; Rojo, N.; Calogero, L.; Poggio, L. Experimental observations of engine piston damage induced by knocking combustion. *Mater. Des.* **2017**, *114*, 312–325. [CrossRef]
18. Balducci, E.; Ceschini, L.; Rojo, N.; Cavina, N.; Cevolani, R.; Barichello, M. Knock induced erosion on Al pistons: Examination of damage morphology and its causes. *Eng. Fail. Anal.* **2018**, *92*, 12–31. [CrossRef]
19. Matlab Documentation, The MathWorks. 2021. Available online: https://www.mathworks.com/ (accessed on 20 April 2021).

energies

MDPI

Article

Development and Experimental Validation of an Adaptive, Piston-Damage-Based Combustion Control System for SI Engines: Part 1—Evaluating Open-Loop Chain Performance

Alessandro Brusa [1,*], Nicolò Cavina [1], Nahuel Rojo [1], Jacopo Mecagni [1], Enrico Corti [1], Vittorio Ravaglioli [1], Matteo Cucchi [2] and Nicola Silvestri [2]

1 Department of Industrial Engineering, School of Engineering and Architecture, University of Bologna, 40126 Bologna, Italy; nicolo.cavina@unibo.it (N.C.); nahuel.rojo2@unibo.it (N.R.); jacopo.mecagni2@unibo.it (J.M.); enrico.corti2@unibo.it (E.C.); vittorio.ravaglioli2@unibo.it (V.R.)

2 Ferrari S.p.A., 41053 Maranello, MO, Italy; Matteo.Cucchi@ferrari.com (M.C.); Nicola.Silvestri@ferrari.com (N.S.)

* Correspondence: alessandro.brusa6@unibo.it

Abstract: This work is focused on the development and validation of a spark advance controller, based on a piston "damage" model and a predictive knock model. The algorithm represents an integrated and innovative way to manage both the knock intensity and combustion phase. It is characterized by a model-based open-loop algorithm with the capability of calculating with high accuracy the spark timing that achieves the desired piston damage in a certain period, for knock-limited engine operating conditions. Otherwise, it targets the maximum efficiency combustion phase. Such controller is primarily thought to be utilized under conditions in which feedback is not needed. In this paper, the main models and the structure of the open-loop controller are described and validated. The controller is implemented in a rapid control prototyping device and validated reproducing real driving maneuvers at the engine test bench. Results of the online validation process are presented at the end of the paper.

Keywords: knock; combustion; efficiency improvement; CO_2 emissions; control; modeling

Citation: Brusa, A.; Cavina, N.; Rojo, N.; Mecagni, J.; Corti, E.; Ravaglioli, V.; Cucchi, M.; Silvestri, N. Development and Experimental Validation of an Adaptive, Piston-Damage-Based Combustion Control System for SI Engines: Part 1—Evaluating Open-Loop Chain Performance. *Energies* **2021**, *14*, 5367. https://doi.org/10.3390/en14175367

Academic Editor: Suhan Park

Received: 21 June 2021
Accepted: 23 August 2021
Published: 28 August 2021

1. Introduction

Knock presents a completely different challenge with respect to what is commonly faced when a standard dynamic system has to be controlled. The main reasons for this deep difference between knock events and other combustion phenomena can be synthetized as follows:

1. Lack of a robust relationship between values of knock intensity indexes and the associated damage on the combustion chamber components
2. Knock stochastic nature.

On one hand, numerous indexes for knock intensity estimation have been defined from the high-frequency content of measured or estimated in-cylinder pressure trace [1], from both high- and low-frequency ranges of such signal [2], or from high-frequency vibrations induced by abnormal combustions on the engine head. Regardless of the source of such signal and the mathematical formulation of the intensity index calculation, such indexes do not provide a direct indication about the consequences that every knocking event has on the combustion chamber, such as the erosion on the piston surface. On the other hand, the non-deterministic nature of knock forced the control system developers to define some statistical quantities, i.e., some percentiles values of certain knock index (typically higher than 90th) to manage the event rate over a given threshold [3], rather than the cyclic knock intensity, which closely approximates an independent random process. Consequently, real-time knock control has never been linked to actual component damage.

Knock intensity thresholds and criteria used for the spark advance (SA) calibration process are typically the result of the experience of the engine manufacturer and, in most cases, are defined to avoid knocking combustions at all. The highest percentiles (such as the 98th or the 99th) or even the cyclic intensity values are controlled to avoid the complete exceedance of some sort of experience-driven limit. Moreover, such thresholds may have remained the same over the years, not following the improvements in materials strength [4]. Indeed, improvements in alloys' mechanical characteristics have led to a significant increase of mean specific power, not necessarily followed by a corresponding redefinition of knock intensity thresholds. This is certainly caused by the lack of a robust quantitative relationship between calculated knock index, and corresponding induced damage.

Standard knock control strategies typically use signals coming from an on-board compatible system for knock intensity sensing (accelerometers or Ionization current-based, ION) and move spark advance at every time step. Spark timing is advanced whenever the intensity is lower than the threshold, while it is retarded when knock occurs, by applying fixed corrections [5–13]. Moreover, in some cases SA maps are calibrated to maintain knock intensity under the threshold for nominal operating conditions (environmental temperature, fuel Research Octane Number, RON, and so on), and by using the knock controller just to retard SA when the on-board system senses "high" intensity knocking events, according to defined spark timing correction maps [14]. SA retarding is then gradually reduced, reaching again the mapped value: this kind of controller is the so called "sawtooth".

More generally, traditional controllers' performance under transient conditions strongly depends on values of calibration constants [15–19]. Often, on production systems, standard controllers do not manage spark timing to reach a statistical knock index target value [20], such as a percentile of maximum amplitude of pressure oscillation (MAPO). It is more common having a controller that keeps such percentile under a calibrated threshold [21,22]. On the other hand, even under steady-state conditions, the actuated mean SA may be sub-optimal due to the high spark-timing variance. For these reasons, in most cases, such controllers are additionally calibrated with a precautionary approach, and this leads to a further combustion efficiency loss. In summary, two controlling issues can be traced in the following way:

1. The definition of a threshold for statistical knock indexes that are defined independently from the consequences induced by knocking events
2. The implementation of closed-loop controllers that have an output affected by a certain delay with respect to an open-loop-based strategy, especially when an engine runs under fast transient conditions.

The knock control algorithm proposed in this work can be considered as a solution to solve both issues, and the two key tools that allow managing combustion phasing in an innovative manner are the control-oriented piston damage model [23] and the analytical knock model [24]. Such controller is designed and developed for specific applications (racing, final vehicle development phase, test at the engine bench), or could represent the basis for introducing a closed-loop correction, in terms of adaptive capability to varying environmental conditions and aging effects (see Part II). Implemented models are developed and validated in previously cited works of the authors, but their features and latest updates are also reported in this work, which for the first time integrates the two portions and introduces a new, finite element method-based approach for the piston temperature model.

The piston damage model [23] was designed for the real-time (RT) estimation of the piston surface erosion, that is calculated as a function of its temperature and MAPO, while the knock intensity model [24] analytically defines the relationship between a calibratable MAPO percentile and the maximum in-cylinder pressure (Pmax), for any given operating condition (i.e., engine speed, load, target value of lambda and fuel quality). Both models can be considered as an innovative contribution to the field of knock modeling and control, especially because they allowed the development of the complex open-loop control chain described in this document that has the capability of converting a target piston damage into a combustion phase value, cycle-by-cycle. The 50% of mass of fuel burned (MFB50) control-oriented model already described in a previous work of the authors [25] is here reversed to use target MFB50 as input and directly calculate the corresponding spark timing to reach such combustion phase value, for any given operating condition.

In this first paper (Part 1), a pure model-based, open-loop controller is developed and validated, and all combustion indexes needed to control combustion and piston damage (MFB50 and MAPO, respectively) are provided by analytical control-oriented models. The novelty introduced by the proposed solution with respect to the cited approaches is the integration of an innovative piston damage controller within a combustion one, which directly manages the instantaneous piston damage speed used as a knock intensity index alternative to MAPO. In this way, a much lower efficiency loss occurs under knock-limited operating conditions, due to a higher actuated SA with respect to standard calibrated values. Indeed, the latter is automatically defined as the value that generates an admissible piston damage speed. Results demonstrate the open-loop chain can reach target MFB50 with an extremely high accuracy even under transient operating conditions. The instantaneous knock intensity is evaluated through a novel index (piston damage speed), and it is directly controlled without applying any statistical methodology or buffer, as performed instead by standard knock controllers. It is important to anticipate that a predicted percentile of MAPO is used to feed the analytical knock model, but it is an index evaluated through a direct calculation, and it is updated once per cycle. It is not calculated with a buffer and thus, no delays are introduced in the control algorithm.

With the knock control strategy proposed in this work, the authors are thus proposing a solution for both issues listed above:

1. The target value of MAPO percentile derives from the admissible piston damage in a certain amount of time (i.e., a target piston damage speed), and this means it is defined according to the existing relationship between knock intensity and induced damage. Pmax and MFB50 models convert such value in a target combustion phase
2. With the open-loop chain the target MFB50 can be achieved by calculating the required SA. The controller is thought to be cycle-resolved, which makes such an algorithm suitable to accurately manage the combustion phase cycle-by-cycle even under transient conditions.

Outside knock limited operating range, the MFB50 model continues to calculate SA to maintain the combustion phase close to the target value. When the target MFB50 (the value that guarantees the target piston damage) is lower than such calibrated value, the MBT combustion phase becomes the actual target. For this reason, such a controller efficiently integrates both knock controller and combustion phase management.

A detailed description of model development and an assessment of model accuracy is reported in previous works of the authors [23–25].

However, it is important to mention the described pure open-loop controller cannot estimate accurate combustion indexes when events that are not considered by models occur. For instance, engine and exhaust system aging or extreme environmental conditions could have unpredictable effects on the combustion process, and thus on the calculated combustion phase and knock intensity. However, the presented models can be further developed to introduce dependency to such kinds of events, and this becomes particularly important when the engine is installed on a production vehicle. Indeed, the proposed open-loop combustion controller can be applied to conditions where extreme environmental conditions are avoided, and the engine has a limited life (such as at the test cell or during racing events), or it should be supported by a closed-loop controller for adapting to varying external conditions. Of course, the availability of a measurement of such indexes can prevent undesirable events and guarantees having a more robust algorithm (and this is the focus of Part II of this activity) for standard applications. Nevertheless, a pure open-loop control strategy developed in this paper represents an innovative solution and it is concretely used for special applications. Results are valuable and the presented tests demonstrate the general validity of the proposed approach. Moreover, authors can anticipate that the closed-loop chain introduced with Part II of this activity is calibrated to react for protection reasons (i.e., just with negative SA corrections), and it is not active in most cases. In other words, even when the closed-loop is available, it reacts rarely and with small corrections (averagely -1 crank angle degrees of magnitude), and the controller still works mainly with open-loop chain.

The experimental setup and the database used for model calibration and validation are described, and each model is then briefly introduced below. The open-loop combustion controller is described in detail. The controller is then implemented in a rapid control prototyping (RCP) device and validated online at the engine test bench. Tests are carried out by reproducing engine speed and load profiles recorded during real vehicle maneuvers performed on track. Results are finally collected and discussed in the last part of the paper.

2. Experimental Setup

Experimental tests for model calibration and controller validation are carried out on a high-performance, 8-cylinder, Gasoline Direct Injection (GDI) and Turbo-Charged (TC) production engine. Each cylinder is equipped with a piezoelectric pressure transducer. The main engine characteristics and the layout of the development network used to test the complete controller via RCP are presented in Table 1 and Figure 1, respectively. The combustion chamber of this engine is characterized by a central spark plug and a side-mounted fuel injector. The indicating system is composed of CHAMPION charge amplifiers and OBI-M2 indicating system, both provided by Alma Automotive S.r.l. (Bologna, Italy), that are used for signal conditioning and acquisition, respectively. The indicating system acquires in-cylinder pressure signals at 200 kHz. Calculation of MFB50, indicated mean effective pressure (IMEP) and Pmax combustion indexes use the low-pass filtered pressure trace, while the same signal is windowed and band-pass filtered (specific cut-off frequencies and angular window cannot be disclosed for confidentiality reasons) to calculate MAPO, as defined by Equation (1):

$$\text{MAPO} = \max\left(|p_{filt}|\right) \tag{1}$$

where p_{filt} is the band-pass filtered, in-cylinder pressure signal. In the following text, the nomenclature "MAPO" or "knock intensity" indicates cyclic values of such an index, while, if it refers to a percentile value, it is always specified or it appears as "MAPO98" or "MAPO99.5" (for the 98th and the 99.5th percentile, respectively) or as "statistical knocking level".

Table 1. Eight-cylinder engine characteristics.

Displaced Volume	3.9 L (8 Cylinder)
Stroke	82 mm
Bore	86.5 mm
Connecting rod	143 mm
Compression ratio	9.45:1
Number of valves per cylinder	4

As mentioned above, the combustion controller has been tested through its deployment in an RT machine for RCP application that communicates with the ECU. The controller receives via the ECU the main engine parameters (such as speed, load, etc.), and it calculates for each cylinder a SA correction with respect to the mapped value. The models implemented in the open-loop controller calculate the combustion and knock indexes needed to determine spark advance corrections, and the indicating system is used to monitor the combustion process and to log in-cylinder pressure curves.

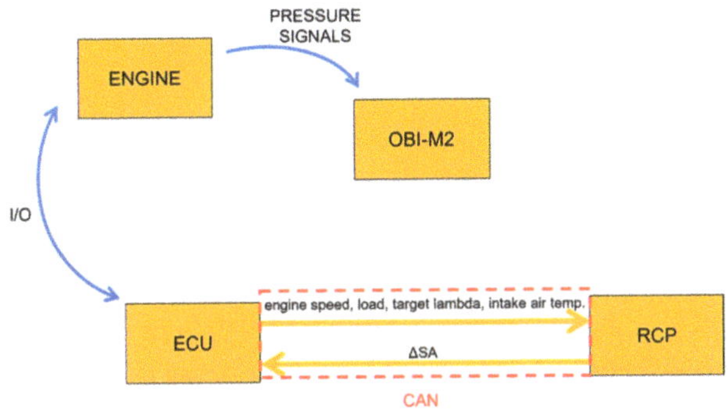

Figure 1. Functional layout of communication loop.

Data used for the controller validation were recorded during tests that reproduced different, real-use engine speed and load profiles from in-vehicle acquisitions. On the other hand, the database used for calibration of the analytical models (such as knock, Pmax, and MFB50 models) comes from an engine experimental campaign, needed to calibrate some ECU control strategies. In other words, such combustion data were not logged for this specific modeling activity but were already available. The piston damage model was calibrated instead with data recorded with the same 8-cylinder engine during a dedicated experimental campaign, described below. In the following sections, the main engine variables have been normalized with respect to their maximum value for confidentiality reasons.

3. Implemented Models

In this section, the piston damage model, the analytical knock intensity model, Pmax, and MFB50 models are described, highlighting those features that have to be considered to understand the development of the combustion controller. The results presented below are only a portion of the complete campaign carried out to develop each model, described in the previous works by the authors [23–25]. In this case, analytical models (knock, Pmax and MFB50) are specialized for each cylinder. This approach is needed to accurately control damage speed and combustion phase also including the cylindrical variability that is always present in multi-cylinder engines.

Knock, Pmax, and MFB50 models were developed to analytically describe the effects of engine operating conditions and other influent parameters, such as lambda and fuel RON (especially for the knock model), on a certain MAPO percentile, on maximum in-cylinder pressure, and combustion phase, respectively. They are developed by applying the separation-of-effects method [26] and only the engine parameters that significantly change when engine runs and that effectively affect such combustion indexes are included in these models. Valve timing, for instance, is an engine actuation that has a strong impact on the combustion process, but the variable valve timing (VVT) map is one of those calibrations that rarely changes, and it can be considered frozen after ECU calibration. Spark timing and lambda are varied by specific strategies that react to manage, for instance, catalyst heating, turbine inlet gas temperature, as well as traction of the vehicle and so on, and for this reason, the effect of lambda and SA changes must be counted by the models. On the contrary, fuel quality and intake air temperature are uncontrollable variables that undoubtedly change during vehicle life. These variables affect both knock tendency and SA-MFB50 relationship, and for this reason, RON of fuel available has to be correctly provided by the operator by calibrating the related variable, while air temperature is communicated to the control algorithm via CAN by ECU. Nevertheless, adaptive strategies are developed and presented in the second paper focused on this topic and they lead to automatically calculating current fuel RON so as to include its effects on the calculation of the final SA. The proposed approach is focalized to include only the parameters that can be significantly changed during the engine operation. Such a method for the prediction of the effect of operating conditions variation on combustion indexes can be considered robust and reliable, especially when the final purpose of the modeling activity is that of developing control-oriented algorithms for a production engine.

Analytical models need to be calibrated when they have to be applied to a different engine, but as mentioned above, the data used to perform such a process come from a standard engine characterization database. Variables needed for model calibration are engine speed, load (evaluated as stoichiometric trapped air mass, (STAM)), SA, target lambda, fuel RON, MFB50, Pmax and MAPO and the process (that takes a couple of minutes) is based on the execution of a script that identifies the optimal coefficients of each model. Indeed, the formulation of each model is pre-defined, and only some specific coefficients need to be recalibrated for a different engine. On the other hand, piston temperature and damage models are developed following a physical approach. The piston temperature model can therefore be easily recalibrated by updating the piston geometry. As described below, the chosen piston finite element method (FEM) model approximates the piston shape as a flat plate, and thus only the thickness must be updated. Instead, the piston damage model is strongly related to the alloy with which the piston is made. For this reason, it can be applied to a different piston by comparing hardness curves of the two different alloys and properly scaling the corresponding coefficients.

Figure 2 shows the scheme of the complete algorithm, to anticipate the role of each model within the control strategy. The model-based controller converts the target damage speed into a target MFB50, and it is solved once per cycle, for each cylinder. The MFB50 and knock model provide combustion phase and knock intensity index to the piston temperature and damage models, respectively. The target damage speed is then converted into a target 98th percentile of MAPO, to properly feed the inverted knock model (the model inversion is highlighted with exponent -1 in the scheme). Finally, the inverted Pmax model outputs the target value of the combustion phase, and the reversed MFB50 model can directly calculate the SA correction needed to reach the target MFB50. The whole algorithm is solved through a direct calculation, and each index or target value is updated every cycle. The models implemented in such controllers are described in this section.

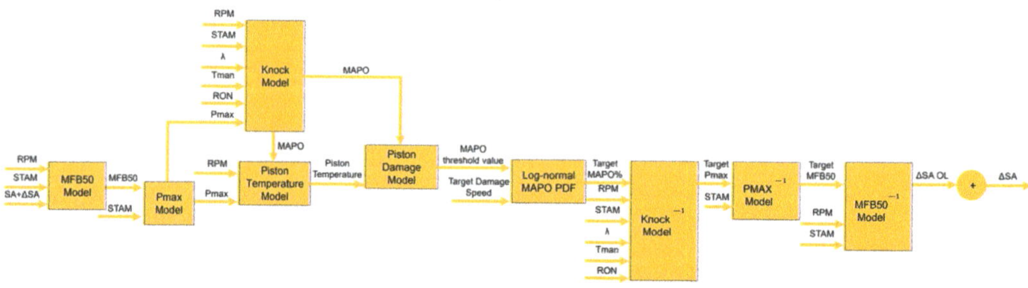

Figure 2. Complete open-loop controller scheme. All inputs needed by each model are reported.

3.1. Damage Model

The piston damage [23] has to be evaluated by counting the effect of piston temperature, that changes the piston alloy resistance which is then damaged by knock-induced mechanical stress [27–30]. Figure 3 shows the decay of aluminum alloy hardness as a function of exposure time at different mean temperatures. Such a figure represents a material characteristic and Brinell hardness (HB) is reported on the y-axis.

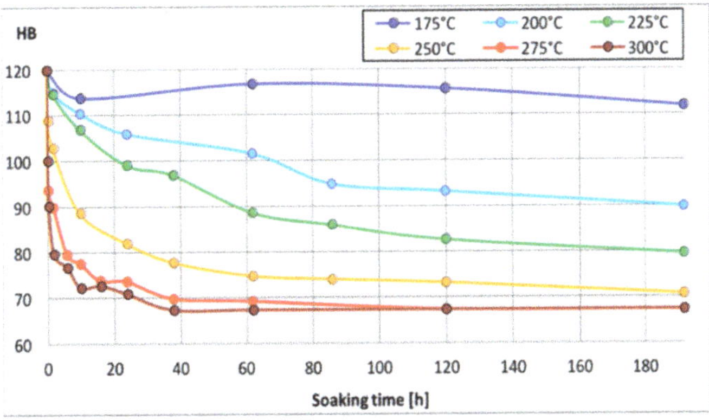

Figure 3. Hardness-temperature-time (HTT) curves for the aluminum alloy.

Development and calibration of a physical control-oriented piston temperature model for aluminum pistons (those mounted in the 8-cylinder engine considered in this work) precedes the discussion of the damage model. Tests for identification and calibration of both models are carried out on-engine in steady-state conditions, at 6000 RPM and 19.8 bar of brake mean effective pressure (BMEP), targeting a fixed knock level, (defined as a value of a MAPO percentile). Only 4 cylinders at a time are operated, and knock intensity is controlled independently for each cylinder. The tests were performed with RON 95 (cylinders 1–4) and RON 91 (cylinders 5–8) gasoline, and for different time intervals for the two engine banks (7.5 h for pistons 5–8 and 10 h for pistons 1–4). Target levels of statistical knock intensity were different for each cylinder, and for both engine banks. In addition, cylinder 5 was operated knock-free to have a reference condition. Such tests have been performed under steady-state conditions and the statistical knocking level was defined by using the 99.5th percentile of MAPO, that was evaluated on a buffer of 200 elements.

With the analysis of both in-cylinder pressure traces recorded during tests and piston damage, a predictive control-oriented piston damage model that is compatible with RT execution was then developed. Inputs are engine speed, Pmax, and MAPO, while output is the instantaneous damage speed, calculated as the cyclic damage induced by a single knocking event divided by the engine cycle duration. Of course, the same algorithm is applied for each cylinder. The layout of the resulting damage model is shown in Figure 4.

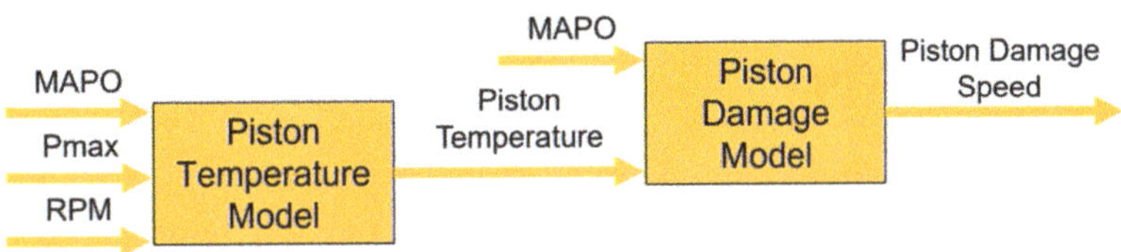

Figure 4. Scheme of piston damage model with logic blocks.

Residual hardness analysis on damaged pistons was carried out by performing micro-Vickers tests on the piston crown, considering 5 measurement locations on the intake side, and 5 on the exhaust side. Deep visual investigations were conducted with a digital microscope to determine the damage morphology in the area of the exhaust valve pockets and top-land, that are particularly sensitive to knocking erosion damage and which are where erosion was actually found. Roughness tests were also carried out in order to quantify the visible damage in the same areas.

By performing tests with different fuels (RON 91 and RON 95 gasoline) it is possible to separate the contributions of the heat transfer induced by the combustion process, and the additional amount due to knock. This evaluation is important to accurately determine the piston surface temperature, hence the piston erosion. Indeed, if two cylinders are controlled to have different knocking levels with a fixed type of fuel, they necessarily experience different maximum in-cylinder pressures (due to a different spark timing) and wall thermal load, regardless of knock intensity. Consequently, differences in measured thermal effects on pistons (residual hardness) are not fully attributable to knock, because they depend on both different Pmax and knock intensity (always evaluated as the 99.5th percentile of MAPO). Thus, combustion-related and knock-related heat transfer rates cannot be separated by moving spark timing only. The use of fuels with different RON numbers enables obtaining the same Pmax level by applying the same SA with different knock intensities, and vice versa. Thus, the proposed method is needed also to quantify and calibrate the temperature role in the damage mechanism by controlling, for example, different piston temperatures at the same knock intensity. Further details about such considerations are provided by the authors in [23].

The mean piston surface temperature can be estimated by measuring residual hardness after the tests on 5 spots on each side, and by using the hardness-temperature-time curves (Figure 3). These measurements show non-uniformity in the temperature field, both on the intake and exhaust side. Nevertheless, the measurements turned out to be symmetrical with respect to the thrust-antithrust axis of the piston surface geometry. Symmetry can therefore be used to average the measurements. In this way, as shown in Figure 5, for every piston it is possible to calculate 6 global hardness values, of which 4 are averaged, while the remaining 2 are calculated according to the thrust-antithrust axis. This approach significantly simplifies the piston temperature model, without excluding the spatial temperature range on the piston crown. Such observations are used for piston temperature model calibration.

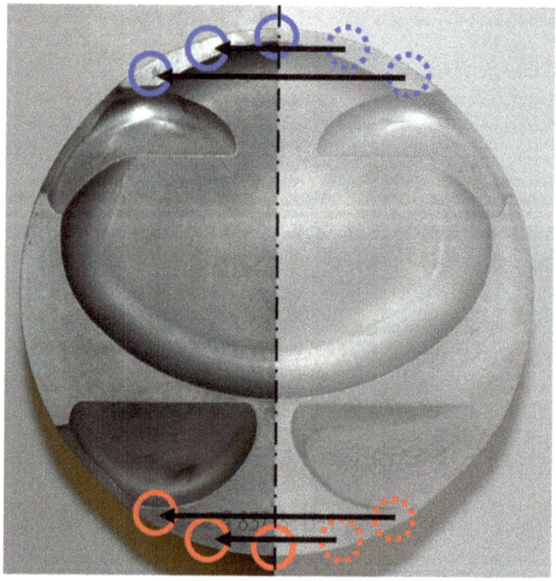

Figure 5. Hardness measurements symmetry, used to average homologous locations.

The temperature model was then developed according to the following hypotheses:

1. Even if the heat transfer rate on both piston surfaces (the combustion chamber and the sump ones) changes by orders of magnitude during the combustion cycle, the skin piston temperature can be supposed to be almost constant during the cycle, due to the high conductivity and the specific heat of the material. The model is therefore computed once per cycle.

2. The exchange condition between gas and piston crown is assumed to be uniform on the entire surface. The specific heat flux is imposed on the piston surface, which assumes that heat transfer is not dependent on surface temperature. Albeit inaccurate, this is deemed to be an acceptable simplification for calculation purposes because the piston temperature is lower than the gas space-averaged temperature.
3. Oil jets cooling is assumed to be the only active path of piston heat rejection (it represents 60% of the total heat rejection, according to [31–33]), so that the heat exchange through the ring belt and the skirt is neglected. These contributions are acknowledged not to be negligible, but their calibration would require, for example, to conduct tests with different coolant temperatures (and to perform the resulting hardness measurements). Conductive heat transfer is very complex and difficult to investigate with specific experimental tests, and for this reason, it is simplified by supposing that the piston bottom inner surface is involved in the heat exchange process and most of such process takes place in the surface directly touched by the oil jets. Figure 6 shows the method for the estimation of the distance between different locations on the piston surface and the area touched by the oil jets.
4. The ratio between gas-exposed (upper) and oil-impinged (lower) areas is unitary, which means that in stationary conditions the specific heat flux across the thickness is uniform, and the heat flux is mono-dimensional. Substantially, it is equivalent to a flat plate, but with differential thermal conductivity (because of the varying distance between the two sides in the real piston).

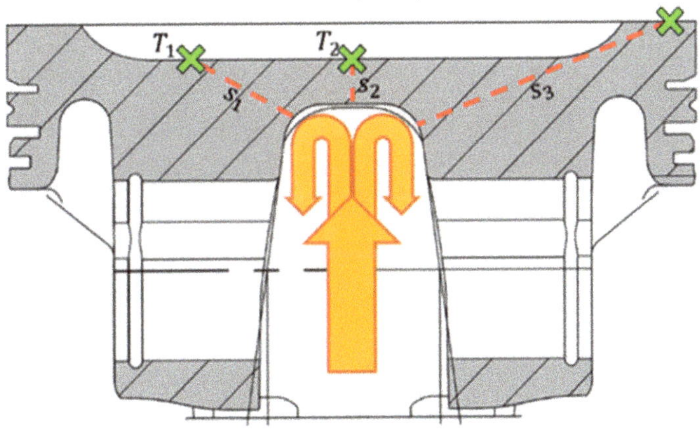

Figure 6. Qualitative oil jets position, used for the estimation of the measurement location distance from the oil-impinged surface.

The result is a non-stationary finite-element model [34,35], where nodes are distributed along the thickness of the piston, and an equivalent thickness is defining the location. Such parameter is the only one that needs to be updated for the piston temperature model recalibration when a different engine geometry is considered. The boundary conditions are the same for every measurement location (i.e., resulting thickness) considered:

- imposed heat flux on the gas side (Neumann condition)
- conductive heat transfer on the oil side (Robbins condition)

The temperatures in the nodes T_t at time t are calculated by:

$$T_t = MM \backslash FF \tag{2}$$

With MM and FF defined by Equations (3) and (4):

$$MM = \frac{MC}{ts} + \alpha\,(MK + MH) \tag{3}$$

$$FF = \left(\frac{MC}{ts} - (1-\alpha)(MK + MH)\right)T_{t-1} + F \tag{4}$$

where:

- α defines the method: 0, 1, 0.5 for explicit, implicit and Crank-Nicholson, respectively
- ts is the time-step
- T_{t-1} is the temperature vector at the previous time-step $t-1$
- MC is the capacity matrix

$$MC = \frac{\rho\, c\, l}{2}\begin{pmatrix} 1 & 0 & 0 & 0 & 0 \\ 0 & 2 & 0 & 0 & 0 \\ 0 & 0 & 2 & 0 & 0 \\ 0 & 0 & 0 & 2 & 0 \\ 0 & 0 & 0 & 0 & 1 \end{pmatrix} \tag{5}$$

where ρ, c and l are the density, the specific heat, and the elements length, respectively. MK is the conductivity matrix:

$$MK = \frac{\lambda}{l}\begin{pmatrix} 1 & -1 & 0 & 0 & 0 \\ -1 & 2 & -1 & 0 & 0 \\ 0 & -1 & 2 & -1 & 0 \\ 0 & 0 & -1 & 2 & -1 \\ 0 & 0 & 0 & -1 & 1 \end{pmatrix} \tag{6}$$

where λ is the material conductivity. MH is the convective matrix:

$$MH = H_{oil}\begin{pmatrix} 0 & 0 & 0 & 0 & 0 \\ 0 & 0 & 0 & 0 & 0 \\ 0 & 0 & 0 & 0 & 0 \\ 0 & 0 & 0 & 0 & 0 \\ 0 & 0 & 0 & 0 & 1 \end{pmatrix} \tag{7}$$

where H_{oil} is the convective coefficient at the oil-piston boundary. F is the vector of the loads:

$$F = \begin{pmatrix} q_{pist_t}\,\alpha + q_{pist_{t-1}}(1-\alpha) \\ 0 \\ 0 \\ 0 \\ H_{oil}T_{oil} \end{pmatrix} \tag{8}$$

In which:

- q_{pist_t} is the specific heat flux at the gas-piston boundary at time t;
- $q_{pist_{t-1}}$ is the specific heat flux at the gas-piston boundary at time $t-1$;
- T_{oil} is the oil temperature.

H_{oil} is an unknown model parameter, such as q_{pist}. The heat transfer problem is represented in Figure 7, and it needs to impose H_{oil} or q_{pist} to be solved. In the figure, piston and oil-piston interface are two thermal resistances connected in series.

Under steady-state conditions, H_{oil} can be calculated through Equation (9) from piston temperature, piston thickness s (evaluated by analyzing the piston geometry) and specific power q_{pist} [36]:

$$H_{oil} = \frac{1}{\frac{T_{crown}-T_{oil}}{q_{pist}} - \frac{s}{\lambda}} \tag{9}$$

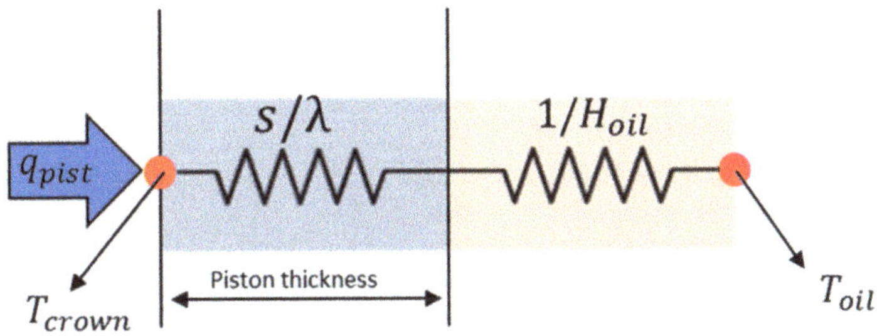

Figure 7. Schematic representation of heat transfer through thermal resistances series.

The mean value of the specific heat flux within the combustion cycle, q_{pist}, is imposed on the gas side and it is estimated by a linear empirical correlation, with the maximum pressure of the cycle and the engine speed as inputs. This is supposed to be valid in non-knocking combustions, while an additional contribution is given by knock, but through a localized effect. For this engine, piston damage was observed on the exhaust side, so it is assumed this is the location where knock occurs. Consequently, the knock-related heat transfer contribution is to be considered only on the exhaust side. Thus, the specific power model, $q_{pist} = f(\text{Pmax, RPM, MAPO})$, is defined as a linear function of Pmax and engine speed. Moreover, for the exhaust side, it is multiplied by a term linearly increasing with MAPO:

$$q_{pist} = (a\ \text{Pmax} + b\ \text{RPM} + c)K_{\text{MAPO}} \tag{10}$$

where $K_{\text{MAPO}} = 1$ for the intake side and $K_{MAPO} = \max(\text{MAPO} - 4,\ 0)d$ for the exhaust side, and a, b, c, and d have to be identified.

K_{MAPO} definition is thought of as a line with positive slope, passing through MAPO = 4 and saturating at 0. The meaning is that cycles with a MAPO value lower than 4 (an arbitrary and calibratable threshold) are not increasing the nominal q_{pist} value, while higher values of MAPO are amplifying such an amount.

In Figure 8, the calibration process is summarized. Combustion indexes (Pmax and MAPO) recorded during bench tests are vectors describing the whole pressure history (one value for every combustion cycle) experienced by the piston. The piston temperature model estimates the temperature history (one value for every cycle) for every measurement point. Then, with the HTT curves seen in Figure 3 the hardness history for every measurement point can be estimated by using data about cumulative exposure time and mean temperature. Finally, the residual hardness values are compared to the measured ones. This is applied to every tested piston. To minimize the cost function, which is defined as the sum of all the absolute errors, for every piston, for every measurement point, *fminsearch* function in MATLAB (R2018, MathWorks Inc., Natick, MA, USA) optimization toolbox [37] is used.

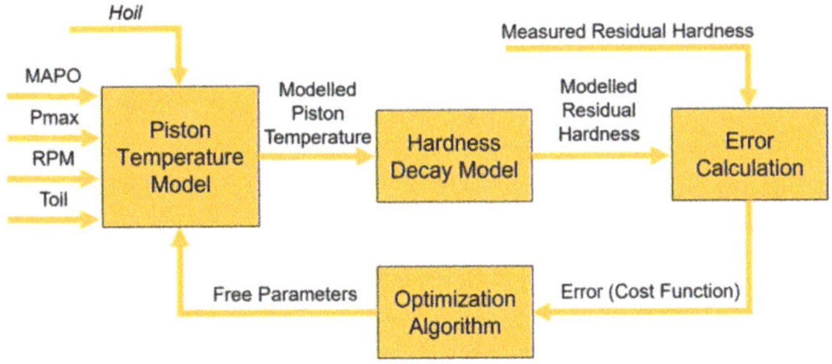

Figure 8. Calibration process for the piston temperature model.

Knock damage model inputs are thus MAPO, representing the mechanical stress acting on piston surface, and piston surface temperature, from which piston alloy resistance strongly depends (Figure 3). Some sort of linear dependence between measured knock intensity and some representative mechanical stress values is assumed since it is very difficult to accurately estimate mechanical forces on piston surface due to knock. As described by the authors in [23], the knock model is based on the consideration that there is a knock threshold, definable as a function of piston surface temperature, beyond which the material deformation falls in the alloy plastic field. Such threshold is therefore calibrated in this model.

Temperature dependence is obtained by moving plastic field boundary (i.e., knock index threshold), while the plastic slope is assumed to be constant. The total modeled damage is given by the summation of the corresponding damage of every cycle. The model calibration consists of the identification of 2 parameters (the gain and the offset of linear MAPO threshold with respect to piston temperature). The calibration algorithm is schematized in Figure 9, and it is very similar to the one adopted for the piston temperature model. Model inputs are MAPO and the piston temperature, updated with engine cycles frequency. The cost function (CF) is defined to maximize Pearson's correlation, ρ, between the measured and the modeled damage D:

$$CF = 1 - \rho(D_{meas}, D_{mod}) \tag{11}$$

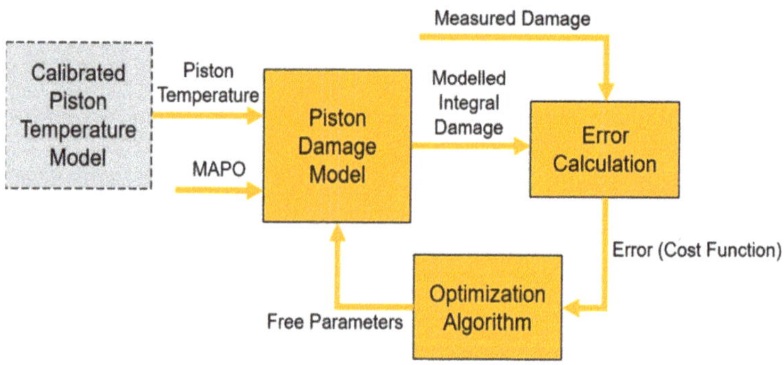

Figure 9. Damage model calibration process.

The best configuration found for the knock index threshold dependence on temperature is reported in Figure 10, which is also the adopted threshold map.

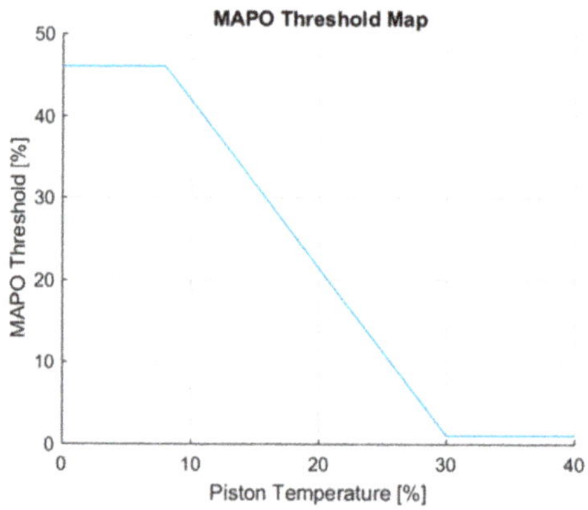

Figure 10. MAPO threshold as a function of normalized piston temperature.

Model calibrations clearly depend on piston alloy characteristics. This means that the whole approach can be easily calibrated for a different engine with the same piston alloy, by just updating piston thickness, for the area in which knock-induced erosion occurs. Once such an area is identified and the piston geometry is known, the methodology can be adapted for another engine. By comparing the online MAPO values and the calibrated threshold for given piston temperature, the excess over the threshold can be calculated for every cycle, estimating the induced damage in the piston exhaust valve pockets and top-land. The calculated damage represents the amount of pressure (MAPO) that exceeds the threshold, and it can then be expressed in *bar*.

At this point, the corresponding allowable knock intensity, and the resulting engine performance, have to be spread throughout engine life. Indeed, an accelerated damage generation and the corresponding performance increment over a short time would be a very inefficient way of using this model, and it could also result in non-uniform engine performance achievable by the driver over time. Moreover, if the target of the model would be defined as the cumulative piston damage evaluated at the end of engine life, the controller would reach too severe knocking combustions, especially after a period in which the calculated damage is zero (for instance, due to a drive with a low torque request). Of course, this is an undesirable behavior due to the high knocking levels that could cause the engine seizure. For such reason, rather than the cumulative damage at the end of engine life, the control strategy target should be a constant damage speed, which finally generates the desired damage level upon completion of the design life cycle. The acceptable piston erosion has been experimentally established by the engine manufacturer, by verifying the maximum damage that does not compromise any engine functional parameter. On the other hand, even the time of the engine life spent under knock-limited operating conditions (i.e., engine points for which it could be reached the target damage speed) has been determined by the engine manufacturer, through a market investigation that figured out the behavior of the "mean" driver of the vehicles that mount the considered engine. Thus, supposing a predefined time of the engine life spent under knock-limited operating conditions, $Life_{tgt}$ (expressed in seconds) and a target modeled damage (that derives from the maximum admissible piston damage evaluated at the end of engine life), $Damage_{tgt}$ (expressed in *bar*), a damage speed target DS_{tgt} can be defined, as described in Equation (12):

$$DS_{tgt} = \frac{Damage_{tgt}}{Life_{tgt}} \quad \left[\frac{bar}{s}\right] \tag{12}$$

Similarly, the estimated instantaneous damage speed can be defined as:

$$DS_{est} = \frac{Damage_{est|P}}{P} \quad \left[\frac{bar}{s}\right] \tag{13}$$

where P is the considered period and $Damage_{est|P}$ is the damage per cycle estimated by the model. For the model implementation in the open-loop-based combustion controller, the period P is the engine cycle time, that for a 4-stroke engine can be calculated with Equation (14):

$$P = \frac{2 * 60}{RPM} \quad \left[\frac{s}{cycle}\right] \tag{14}$$

The piston temperature model coupled with the damage model calculates the instantaneous damage speed, using engine speed, Pmax and MAPO as inputs. These models are the solution proposed by the authors for the first issue discussed in the Introduction. Indeed, they represent a validated and reliable control-oriented tool for the conversion of the cyclic MAPO value into the corresponding piston surface erosion. Thus, such an approach replaces the traditional knock intensity index used in calibration with a more significant and tangible damage index. However, MAPO remains a knock intensity index that can be directly measured through the in-cylinder pressure sensing system, while the piston surface erosion can be only calculated online through the model. In this work, the cyclic MAPO value is calculated by the knock model, as described by the authors in [24]. For given operating conditions, the complete probability density function (PDF, see the Appendix of [24]) is determined and the cyclic MAPO value is extracted. Indeed, MAPO is needed by the piston damage model to calculate the piston temperature. Considering the final purpose of the feed-forward combustion controller (i.e., converting the target damage speed into a target combustion phase), the translation of the target piston surface erosion into a combustion index correlated with MFB50 is needed. The analytical knock model previously developed by the authors [24,25] becomes, thus, a useful tool to perform such a conversion, and for this reason, it is briefly described below. Its simple analytical formulation makes it particularly suitable to be implemented in a combustion controller for a production ECU. A MAPO percentile becomes the input and Pmax becomes the output, simply by reversing the standard model equation. Such percentile (that can be arbitrarily chosen) needs to be fed to the reversed analytical knock model, and it depends on MAPO threshold and, thus, on piston temperature.

Depending on both operating conditions and oil temperature, the FEM model described above estimates the piston temperature and based on this, the MAPO threshold between safe and damaging intensities is defined. Hence, for given MAPO threshold and for given target damage per cycle, the MAPO log-normal PDF that produces such damage can be univocally defined. The cyclic damage is calculated multiplying the damage speed with the time (expressed in seconds) of an engine cycle. It is then possible to calculate the cyclic knock damage as the integral of MAPO values over such threshold. Consequently, given the instantaneous estimated piston temperature and the target damage speed, the corresponding MAPO PDF (within the ∞^1 possible distributions, depending on operating condition) can be identified and converted into the value of a given MAPO%. For a certain MAPO PDF, the related damage per cycle is graphically represented in Figure 11 and it can be calculated through Equation (15):

$$\textit{Damage per cycle}\ [bar] = \int_{THR}^{+\infty} (\text{PDF}\ \Delta\text{MAPO})\ \text{dMAPO} = \int_{THR}^{+\infty} \text{PF dMAPO} \tag{15}$$

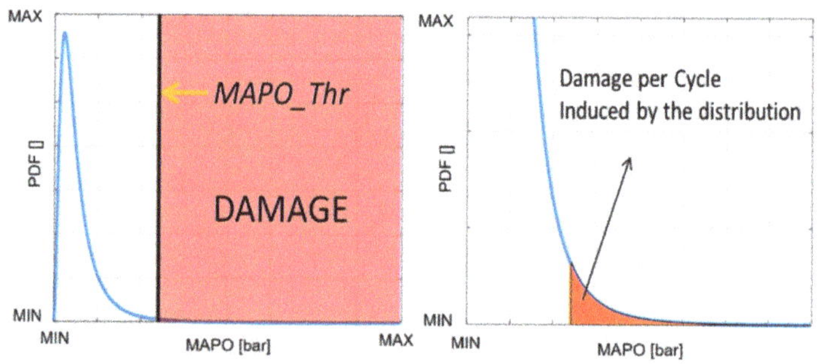

Figure 11. Damage per cycle associated to the MAPO distribution for the given MAPO threshold.

In which ΔMAPO is the interval used to discretize the MAPO domain and PF is the probability function.

An iterative calculation of the targeted PDF (and corresponding MAPO percentile) is implemented in the control strategy. As mentioned above, once the log-normal PDF that fits the integral value over the MAPO threshold calculated by the knock model is identified through Equation (15), the corresponding value of a MAPO percentile (98th in this case) can be used by the knock model as input. Such algorithm uses the linear function between two percentiles of a log-normal distribution introduced in the Appendix in [11] and reported in Equation (16):

$$\text{MAPO50} = g\ \text{MAPO98} \tag{16}$$

Based on the authors' experience, g is fixed on the entire engine operating field, and it can be univocally determined by analyzing the ratio between MAPO50 and MAPO98. In other words, it is identified when the analytical knock model is calibrated for the considered engine and then it is kept fixed. In this way, it is possible to halve the independent variables that the iterative calculation has to change because once the value of MAPO98 is defined, MAPO50 can be calculated via Equation (16). Of course, the approximation of the MAPO for fixed operating conditions with a log-normal PDF introduces an error, but, as described in [24,38], such kind of function is one of the most accurate ways to describe the distribution of MAPO probability. The error introduced by such approximation is evaluated in detail in [24]. In this work, the focus is the development of the combustion controller and in the last part of this paper, the overall error (that incudes even the error introduced by each model) between targeted and measure values is reported. At the same time, when two percentiles are known (and one of these is 50th), the mean value (μ) and the standard deviation (σ) can be directly calculated by using the properties of the log-normal PDF:

$$\mu = \ln(\text{MAPO50}) \tag{17}$$

$$\sigma = \frac{\ln(\text{MAPO98}) - \ln(\text{MAPO50})}{2.0057} \tag{18}$$

In other words, the procedure varies the highest percentile in a certain calibrated range (in that case 98th, but it can be arbitrarily chosen as a percentile higher than the 50th) and 50th percentile (directly deductible with Equation (17)) to find the MAPO PDF for which the integral value of knock index values over the threshold (defined by damage model) coincides with the cyclic target damage. This can be calculated by reversing Equation (13):

$$Damage_{tgt} = DS_{tgt}\ Life_{tgt}\ [bar] \tag{19}$$

Figure 12 shows the algorithm scheme, while Equation (20) indicates the mathematical expression of such an algorithm.

$$\text{MAPO98}_i = \text{MAPO98}_{i-1} + \left(Damage_{tgt} - \int_{THR}^{+\infty} (\text{PDF}_{i-1}\ \Delta\text{MAPO})\ \text{dMAPO} \right) \tag{20}$$

where i and $i-1$ indicate the current and the previous iteration of such iterative calculation. However, an iterative procedure is not compatible with RT execution of the complete controller because it needs multiple iterations that must be completed between two consecutive engine cycles. Thus, this procedure was executed offline for a reasonable range of target damage rates and for several MAPO threshold values (Figure 10, and a 2-D lookup table was then calibrated. A direct calculation of target MAPO98 for a given MAPO threshold would be possible, but it would contain mathematical operations between values with different orders of magnitude and this would be numerically unsuitable for the final implementation in a production hardware. Thus, considering the limitations imposed by the final implementation on the ECU, such a kind of operations should be converted into a Lookup Table (LUT) that contains the result. For such reasons, the authors preferred to simplify the whole calculation of the target MAPO98 by introducing a map. The target damage speed and MAPO threshold are the inputs, while the target MAPO98 value is the output. At this point, it is clear how such MAPO98 is not calculated using a buffer (that would inevitably introduce a delay in the calculation), but it is applied just as an index for the conversion of the current MAPO threshold into a proper input for the reversed analytical knock model (that is briefly introduced in the following section). The whole control algorithm is thus characterized by direct calculations and there is not any kind of filter (buffers, moving averages, and so on). Even such a feature can be considered as a significant novelty in the field of knock control.

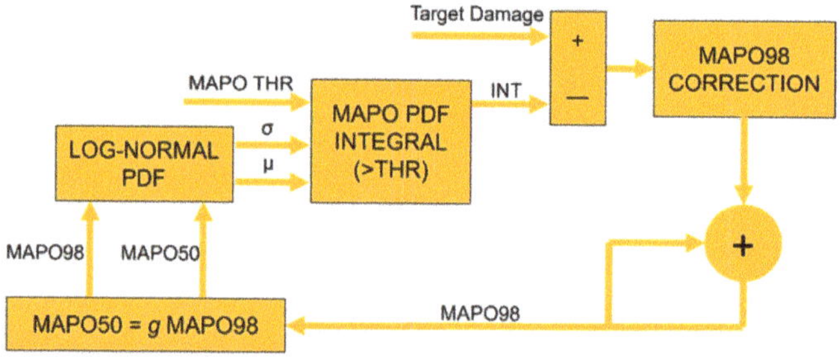

Figure 12. Block scheme of the algorithm used to find the MAPO PDF that achieves the target cyclic piston damage.

3.2. Analytical Knock Model

The analytical knock model is used to convert the target MAPO98 value into corresponding Pmax [24]. The choice of 98th percentile is arbitrary. It was chosen for the knock model calibration and, consequently, as the output of the iterative procedure described above because it represents a good compromise between numerical stability and sound representation of high-end MAPO values. Equation (21) shows the model analytical formulation for a standard GDI, TC engine:

$$\begin{aligned}\mathrm{MAPO98\%} = {} & (a\,\mathrm{STAM}^{b})\,(\mathrm{Pmax} + ((\lambda - \lambda_{\mathrm{ref}})\lambda\mathrm{m}) + ((\mathrm{T_{man}} - \mathrm{T_{man\_ref}})\mathrm{Tm}) \\ & + ((\mathrm{RON}_{ref} - \mathrm{RON}_{fuel})\mathrm{RON}m))^{(c\,\mathrm{STAM}+d)} + \mathrm{RPM\ ESm}\end{aligned} \tag{21}$$

where:

- MAPO% is the MAPO percentile to be calculated
- a, b, c, and d are the four calibration parameters of the model that allow the definition of MAPO% for the reference conditions (reference lambda value and air temperature in the intake manifold)
- STAM is the stoichiometric trapped air mass, which is equal to the trapped air mass (TAM) per cycle per cylinder when the mixture is stoichiometric or rich, and the ratio between TAM and lambda when the mixture is lean
- λ is the current lambda value
- λ_{ref} represents the reference lambda value, with which model coefficients (a, b, c, and d) are calibrated. In this case, λ_{ref} is equal to 1
- λm is the lambda multiplier, a calibratable coefficient that converts the lambda numerical difference in ΔPmax
- $\mathrm{T_{man}}$ is the current air temperature in the intake manifold
- $\mathrm{T_{man\_ref}}$ is the reference temperature in the intake manifold. In this case it is equal to 40 °C
- Tm is the intake air temperature multiplier, a calibratable coefficient that converts the temperature numerical difference in ΔPmax
- RON_{ref} is the reference fuel RON value, used during tests which generated the database with which parameters a, b, c and d were calibrated
- RON_{fuel} is the RON number of the current fuel
- RONm is the fuel RON multiplier, a calibratable coefficient that converts RON numerical difference into ΔPmax, as well as the λm.
- ESm is the engine speed multiplier, and it includes the effect of combustion noise as a function of engine speed.

As observable in Equation (21), the effects of the mixture enrichment and intake air temperature on MAPO are modeled as an additive contribution to Pmax value. The result can be seen as a mere translation of MAPO98% curve along the Pmax axis. In other words, for fixed engine speed, load, and fuel, when the mixture is enriched, or the intake air temperature is lower than reference, the same Pmax is achieved with a lower MAPO98 value. Such phenomenon is visible in Figure 13, in which normalized MAPO98% curves on Pmax axis are reported for three spark sweeps performed with the same engine speed, load (7000 RPM, 1800 mbar of intake manifold pressure) and type of fuel, but for different lambda values. Knock index is offset by the modeled engine speed contribution. Legend includes the lambda value, the intake air temperature, and the fuel RON, respectively. Of course, the same behavior can be observed when the intake manifold pressure varies.

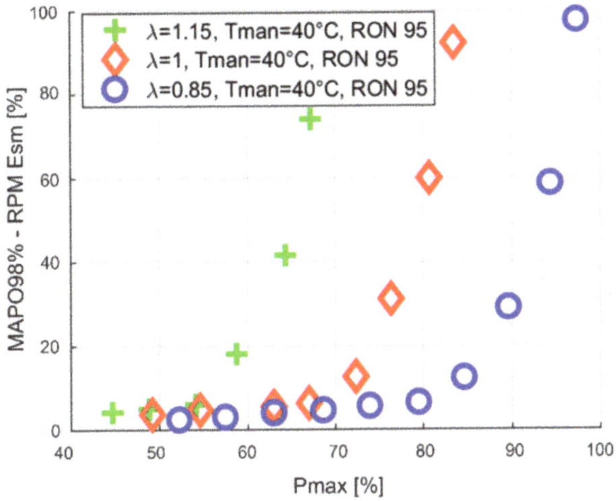

Figure 13. Normalized and offset MAPO98% curves for fixed operating conditions (7000 RPM, 1800 mbar of intake manifold pressure) but for different lambda values.

The same approach used to model the influence of lambda and intake air temperature on statistical MAPO98 is applied to include fuel RON variations. Indeed, as shown in Figure 14, a fuel RON increase acts as a mere translation of MAPO98 curves along Pmax axis, showing the same qualitative trend observed for mixture enrichment.

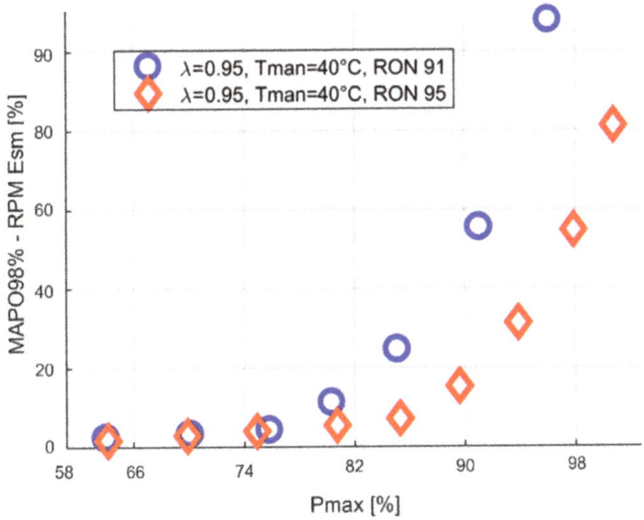

Figure 14. Normalized and offset MAPO98% curves for fixed operating conditions (7000 RPM, 1800 mbar of intake manifold pressure) for two fuel RON values.

Model equation is then reversed to convert a target MAPO98% into a Pmax value. Analytical formulation of such model is shown by the following Equation (22):

$$\mathrm{Pmax}_{tgt} = \sqrt[(c\ \mathrm{STAM}+d)]{\frac{(\mathrm{MAPO98\%_{tgt}} - \mathrm{RPM\ ESm})}{(a\ \mathrm{STAM}^{b})}} + (\lambda_{\mathrm{ref}} - \lambda)\lambda\mathrm{m} + (\mathrm{T_{man\_ref}} - \mathrm{T_{man}})\mathrm{Tm} + \mathrm{RONm}\left(\mathrm{RON_{fuel}} - \mathrm{RON}_{ref}\right) \tag{22}$$

where MAPO98$\%_{tgt}$ is the target MAPO98 value determined by converting target piston damage and MAPO threshold into a MAPO PDF, and Pmax$_{tgt}$ is the corresponding Pmax value, for fixed operating conditions. The model inversion can be performed without carrying out a new calibration process of model parameters (a, b, c, d, ESm and so on) and this is a further demonstration of the possibilities that derive from the analytical approach proposed by the authors.

3.3. Analytical Pmax Model

The control-oriented Pmax model was described and validated in a previous work of the authors [25]. The reversed analytical formulation of such a model can be easily implemented in the combustion controller to convert Pmax value into the corresponding MFB50. Equation (23) represents the standard model formulation. It is a first-order polynomial with respect to engine load (evaluated as STAM) and a second-order polynomial with respect to MFB50. This means that two possible solutions exist by reversing such an equation to calculate MFB50. Nevertheless, only the positive one defines the physical value of combustion phase. Thus, Equation (24) shows the proper formulation of the reversed Pmax model.

$$\mathrm{Pmax} = p_{00P} + p_{10P}\mathrm{MFB50} + p_{01P}\mathrm{STAM} + p_{20P}\mathrm{MFB50}^2 + p_{11P}\mathrm{MFB50\ STAM} \tag{23}$$

$$\mathrm{MFB50} = \frac{-(p_{11P}\mathrm{STAM} + p_{10P}) + \sqrt{(p_{11P}\mathrm{STAM} + p_{10P})^2 - 4\,p_{20P}(\mathrm{Pmax} + p_{00P} + p_{01P}\mathrm{STAM})}}{2\,p_{20P}} \tag{24}$$

The implementation of the reversed Pmax model in the open-loop combustion controller allows the determination of the target combustion phase that achieves the target damage speed. Figure 15 shows the controller layout at this point of the system development.

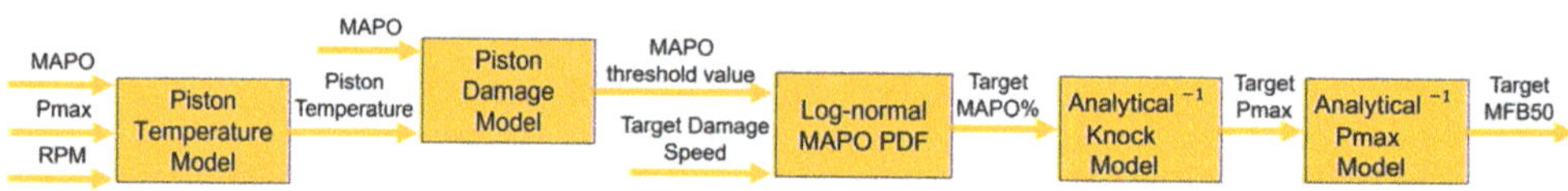

Figure 15. Controller layout including reversed Pmax model.

3.4. Analytical MFB50 Model

The final output of the open-loop chain is the ΔSA (with respect to mapped value) that achieves target MFB50. MFB50 model analytically defines the relationship between actuated spark timing and combustion phase [25]. Figure 16 represents the block scheme of the controller open-loop chain, including the reversed MFB50 model.

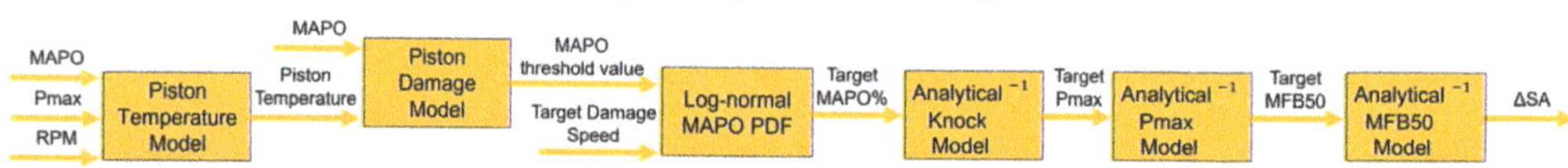

Figure 16. Controller layout including MFB50 model.

Such model is reversed with respect to what is described by the authors in [25] to calculate ΔSA. For each engine point (any given engine speed and load), the relationship between mean MFB50 and actuated SA has a parabolic trend, which can be described with a second- order polynomial. Such trend derives from a variable MFB50 sensitivity to the SA: the higher the MFB50, the lower the curve slope. Figure 17 shows an example of what has been asserted, collecting data for a spark sweep recorded for fixed engine speed and load. SA is reported as the difference from the mapped value (ΔSA) and the MFB50 is normalized with respect to maximum value.

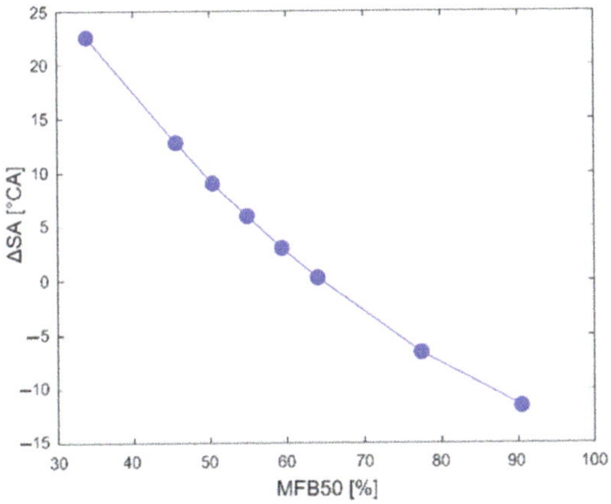

Figure 17. Experimental MFB50-ΔSA points recorded during a spark-sweep for fixed engine speed and load. MFB50 is normalized with respect to the maximum value.

Thus, for each engine point (speed and load), a second order polynomial can be calculated and the SA that univocally determines the target MFB50 value can be directly evaluated.

4. Open-Loop Combustion Controller Validation

The controller scheme shown in Figure 16 also needs cyclic Pmax and MAPO values as inputs of the piston temperature and damage model. Thus, the control system is further developed, by introducing a standard Pmax model defined by Equations (16)–(18), Equations (21) and (22), to calculate cyclic MAPO value. Such model implements the algorithm that defines the MAPO PDF as a function of engine speed, load, Pmax (and thus spark timing), lambda, intake air temperature and fuel RON. Once the PDF is known, the model extracts some values randomly but according to the probability determined by such PDF. Such operation is carried out by using the Matlab/Simulink random function and specifying the mean value and the standard deviation of the current log-normal distribution [37]. In this way, even the random behavior of MAPO is reproduced. The output of knock model block is thus the instantaneous MAPO, needed by the piston damage model to evaluate the corresponding piston temperature (and thus the target MAPO threshold and the corresponding MAPO98), calculated once per cycle. Moreover, Pmax model needs current MFB50 as input and for such reason also the MFB50 analytical model is added to the final controller scheme. Of course, in this case, the polynomial equation described by the authors in [25] is implemented to calculate the MFB50 when SA is given. The scheme shown in Figure 16 evolves into that anticipated by Figure 2. The spark advance that is used as input for the standard MFB50 model (left-hand side of Figure 2) is the value that results by adding the mapped value with the correction calculated by the controller for the previous combustion (right-hand side of Figure 2).

The resulting combustion controller is finally implemented in an RCP system. The complete open-loop controller is available for each cylinder, and it uses current engine operating parameters (engine speed, STAM, target lambda and mapped SA) communicated by ECU via CAN, and modeled combustion indexes.

The engine at the test bench is operated with speed and load profiles recorded during real vehicle maneuvers. The tested profiles consist of complete track laps and include both very steep transients and semi steady-state conditions. In this way, controller is stressed with realistic engine speed and load profiles, and results demonstrate the high reliability of the proposed tool on the entire engine operating range and for all the operating conditions. Tests were performed with RON 95 fuel, and such value is manually tuned and kept fixed.

Figure 18 shows the engine speed and pedal traces of an exemplary profile tested during the validation procedure, normalized with respect to maximum values of the engine operating range. Thus, the entire engine operating field is covered with these tests. Results only for cylinder 1 are shown in detail for a more concise and clear representation.

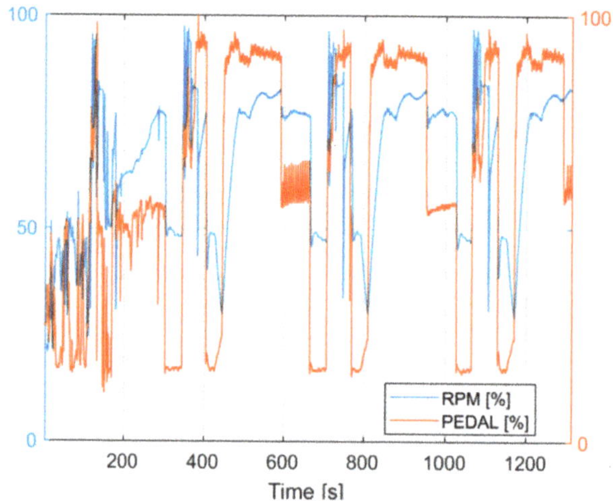

Figure 18. Normalized engine speed and pedal profiles related to results presented below.

During the tests, the in-cylinder pressure curves and combustion indexes measured by the indicating system have been recorded to verify the accuracy of modeled values and the robustness of the control strategy. On one hand, measured MFB50, Pmax, cyclic MAPO and ΔSA determined by the controller are shown in Figure 19. On the other hand, even the calculated indexes are reported (i.e., instantaneous Pmax, target MAPO98, target Pmax, target MFB50 and damage speed values). Rather than the MAPO threshold of piston damage model, the target MAPO98 is reported because, as described above, these two indexes are equivalent (they are two different values of the same MAPO PDF) and, at the same time, the instantaneous MAPO threshold would provide the same indication of the instantaneous damage speed (already shown in the last graph of Figures 19 and 20). All the quantities reported in Figures 19 and 20 are normalized with respect to the maximum value of y-scale, which is chosen to include all the recorded values, and the MAPO graph is zoomed to optimize the data visualization. Instead, the values of ΔSA are reported as °CA. The shown MFB50 and MAPO are the measured ones to demonstrate the accuracy of the control strategy. As anticipated above, the MAPO98 shown in Figure 19 is calculated by the indicating system with a buffer of 200 cyclic values. However, such value has not a role in the management of piston damage speed. It is just evaluated to be compared with the target MAPO98 calculated by the control algorithm. In the same way, mean MFB50 is calculated by the indicating system and it is not an index included in the controller scheme. This means that there is not any kind of filter in the controller calculation chain to avoid the introduction of a delay with respect to the current combustion indexes. The target MFB50 is defined as the value that determines a given damage speed value, saturated at the maximum combustion efficiency value (MBT). Consequently, the MBT value is the lowest MFB50 targeted by the controller. Of course, the actuated SA is generally higher than the calibrated value (as observable in the third plot of Figure 19 that reports the ΔSA) and the engine operating field for which the MBT can be reached by the controller is wider than that resulting from the standard calibration. This is due to the higher MAPO that is considered admissible (especially beyond a certain value of STAM). The most important consequence of the application of the proposed controller is a significant increase in the combustion efficiency that can be spent as a torque increase or a fuel consumption reduction. When the combustion phase value given by the conversion of the target damage speed is lower than the MBT value, the resulting target saturates to the calibrated maximum combustion efficiency value. For this reason, Pmax and MAPO98 reach the corresponding target values only when target MFB50 is higher (i.e., more retarded) than the MBT, and such cases identify also the knock-limited operating conditions. Indeed, during such portions of the test, the measured Pmax and MAPO98 are very close to the target, allowing to achieve the damage speed target, as observable in the related graph. Indeed, despite the high variability of MAPO98, the actuated SA achieves a value of such an index that oscillates around the target value. The MAPO values reported in the last graph of Figure 19 show a low variability. This is because a portion of the highest values is covered by the MAPO98 line, but also because such values are approximated by a log-normal distribution. However, the error introduced by such approximation is low and acceptable as demonstrated by the authors in [24].

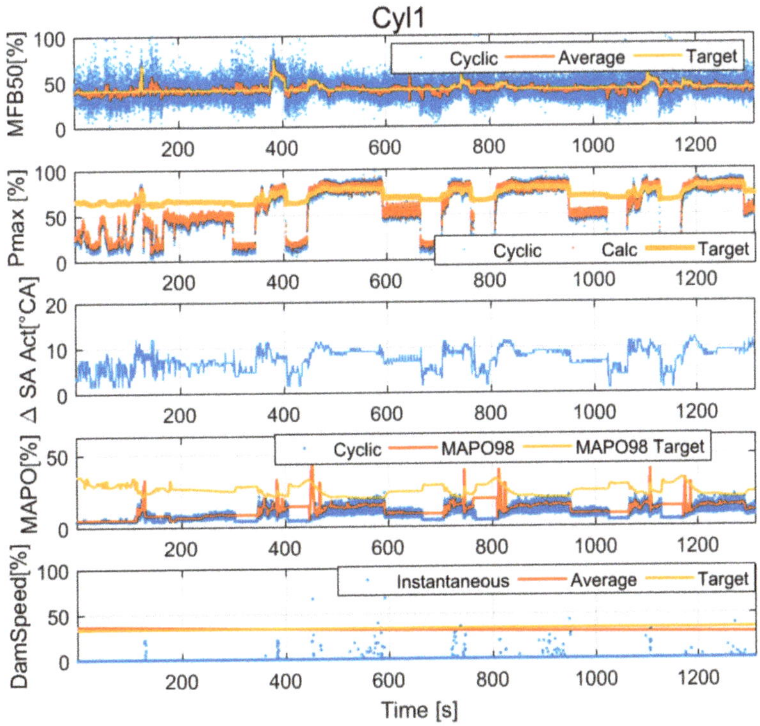

Figure 19. The target and the experimental values of MFB50, Pmax, MAPO, and the calculated damage speed for the cylinder 1. The second graph from the top also includes the calculated Pmax.

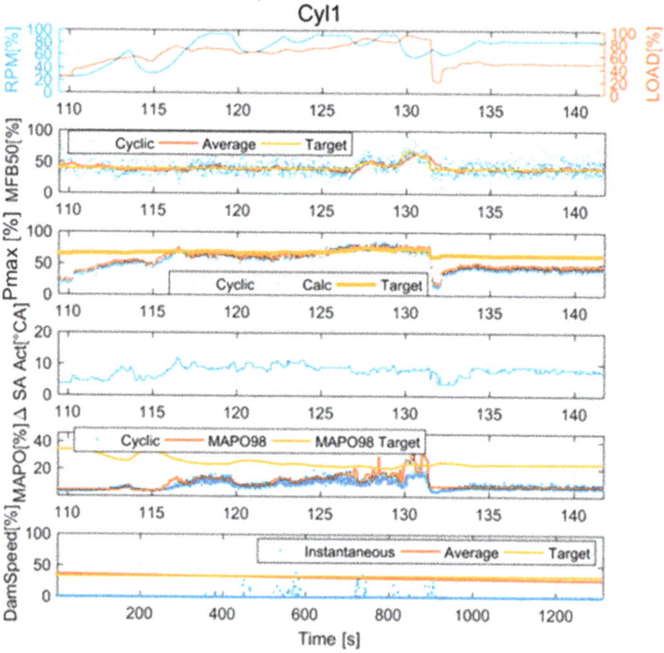

Figure 20. Zoom of Figure 19 between the seconds 110 and 140.

It is important to mention that the average damage speed is close to target value already at the beginning of the test and this is because the engine speed and pedal profiles shown in Figure 18 are reproduced several times in a row. In this way, the controller can reach the target damage speed value. The average damage speed shown in the last graph of Figure 19 is calculated as the cumulative damage (evaluated in bars) divided by the whole engine life (evaluated in seconds), and for this reason, it varies slowly. Even if the controller manages the instantaneous damage speed, the mean value evaluated in such a way is the most meaningful index because it represents the average damage speed on the entire engine life and the most proper value to estimate the cumulative piston damage at the end of the test period. In other words, the average damage speed calculated in this way is the index used to further validate the robustness of the piston damage model at the end of the test period. Indeed, it is directly used to estimate the resulting cumulative damage that is compared with the experimental one. At the same time, it is important to mention that the average damage speed does not have a role in the control strategy, which uses only its instantaneous value. This result demonstrates that the model-based, open-loop chain dynamically estimates the combustion indexes and converts the target damage speed into the corresponding combustion phase that achieves a desired piston surface erosion. The reliability of all the described models is therefore confirmed.

Figure 20 shows a zoom of Figure 19 between seconds 110 and 140. The first plot reports the engine speed and load in such time window (they are the corresponding values taken from Figure 18) and their trends are characterized by variations that cover the entire engine operating field. Results demonstrate the controller can keep combustion indexes close to the target values, even under rapid transient conditions. Only the moving average of MFB50 provided by the indicating system shows a slight delay with respect to the instantaneous value and the target one, but this is the common result of applying a filter. Instead, cyclic MFB50 follows with high accuracy the target (as quantified by Figure 21). It can be noted that the instantaneous damage does not only depend on the cyclic MAPO value but also on the piston temperature and thus on the engine speed and Pmax. Indeed, the piston damage can be higher than zero (but not higher than the target) even when the MAPO98 (that is not a directly controlled index) is slightly lower than the target, depending on piston temperature values. However, when the engine speed and load conditions allow reaching target Pmax, also MAPO98 goes close to the corresponding target value. In some cases, the instantaneous damage speed exceeds the threshold, but, as it is clearly visible in the last graph of Figure 20, such exceedance is limited and always acceptable. Even if in the Part II a PI controller is introduced in the control strategy, its actions are very limited (the calculated ΔSA is always lower than 2°CA), and they react only when the instantaneous damage speed exceeds a safety threshold. At the same time, the authors can confirm that it would be inactive during the test shown in Figures 19 and 20, due to the damage speed values that remain always under the mentioned safety threshold. Hence, Figure 20 demonstrates the accuracy of the complete open-loop controller even under transient conditions.

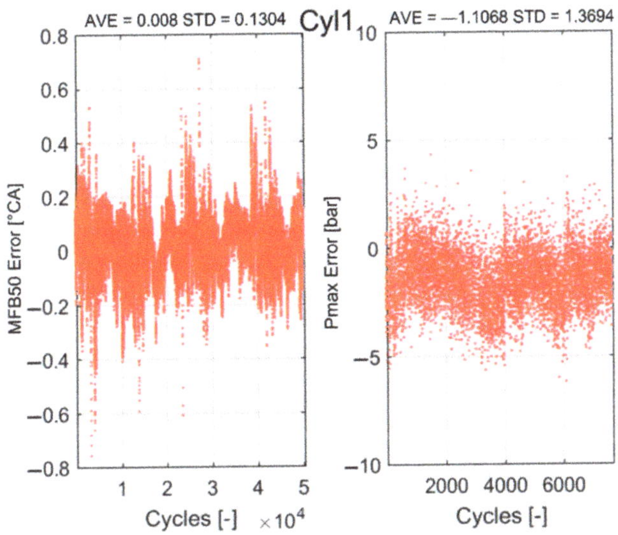

Figure 21. Error between the target and the experimental MFB50 and Pmax indexes, for the cylinder 1. The mean value (AVE) and the standard deviation (STD) are indicated in the title of each graph.

The calculated cyclic Pmax is often overlaid to the experimental one and this result confirms the accuracy and the robustness of the proposed analytical models that calculate the MFB50 and the maximum in-cylinder pressure. It is possible to state that the analytical MFB50 and Pmax models are extremely accurate, and they introduce negligible errors. However, a quantitative evaluation of the model accuracy is provided in Figure 21. Indeed, Figure 21 shows the percentage error between the MFB50 and Pmax, and the corresponding target values. The mean value and the standard deviation of the error are even reported. The proper way to calculate error between the target and the experimental Pmax is excluding all the engine cycles in which target MFB50 is saturated to the MBT because the target Pmax cannot be achieved. On the contrary, since the objective of this work is that to develop a combustion controller that manages the combustion phase on the entire engine operating field, the error between the MFB50 and the corresponding target value is evaluated for the whole test. For such indexes, the error is mainly within the range +−0.6 °CA and +−5 bar, respectively. These values, together with the achievement of the damage speed target, can be considered as a further confirmation of the controller accuracy and robustness.

5. Conclusions and Future Works

In this work, a novel model-based combustion control system is described. The proposed algorithm completely redefines the traditional knock control approaches, especially for applications in which feedback is not required, providing a solution to efficiently integrate the knock intensity management with the combustion phase control. This is possible by converting the target knock intensity values into the corresponding target MFB50, and this is realized through the implementation of a chain of models that describe the trends of main combustion indexes in a very accurate and simple way. Target knock intensity is defined as a piston erosion level, that represents the effective and tangible consequence of knocking combustion, and it is controlled through an open-loop chain that guarantees the expected behavior even under fast transient conditions. The models are developed for a designed and calibrated 8-cylinder engine. Indeed, the final controller is implemented in an RCP system that communicates the calculated spark timing corrections via CAN to the ECU, that applies such corrections to the mapped SA, for any given operating condition.

At first, a description of the main models presented in previous works of the authors is reported, deepening the latest updates and the modifications carried out for their implementation in the controller. The development and the calibration of the piston damage model is presented, underlining the novel improvements and the methodology applied for the conversion of target damage speed into a target MAPO percentile. The analytical knock model is calibrated for the 8-cylinder engine used for this work and it is reversed to convert the target MAPO percentile into the corresponding Pmax value. The main benefits of the simple analytical formulation are thus shown. The analytical Pmax and MFB50 models are briefly retrieved and the implementation of the ΔSA-MFB50 parabolic function is discussed. The model-based open-loop chain converts the target piston damage speed into a target combustion phase, saturated to the MBT.

The validation is performed by reproducing at the engine bench the speed and load profiles logged during real, on-track vehicle maneuvers, to test the combustion controller under fast transient conditions.

The results demonstrate both the robustness of the complete open-loop chain and the high accuracy of the combustion models, returning very small errors in estimating main combustion indexes. However, as already mentioned, it is important to highlight the described pure open-loop controller cannot estimate accurate combustion indexes when events that are not considered by the models occur, such as the engine or the exhaust system aging. However, the presented models can be further developed to introduce the dependency to such kinds of events. Of course, the availability of a measurement of the MFB50 and MAPO can prevent undesirable events. Nevertheless, the results achieved with the open-loop controller are valuable (error of +−0.6 °CA on MFB50 and +−5 bar on Pmax) and the presented tests demonstrate the general validity of the proposed approach. More generally, a SA higher than the mapped value is applied by the controller and a significant combustion efficiency increase is achieved.

Such controller is concretely used for special applications, and it will be further developed with an exhaust gas temperature model to manage even the lambda target and then it will be implemented in a prototyping ECU. Indeed, modern ECUs are characterized by high computational power and their main limitation is related to the number of analog and digital channels, rather than the CPU power. Even if the computational effort strongly depends on the type of RT machine used to perform the tests, the authors confirm that the proposed controller is suitable for the implementation in a production ECU.

In the second paper (Part 2) focused on this topic, the open-loop combustion controller is further developed by introducing a PI controller and model-based adaptive strategies that automatically determine the fuel RON value and update the analytical relationship between the ΔSA and the MFB50. Of course, such adaptive strategies need the measurement of the knock intensity and of the combustion phase, but this can be considered as a viable solution also for on-board applications, due to the availability of commercial systems that can estimate such indexes. Such development is needed when the controller has to be applied to production systems. The resulting controller will be directly deployed on a prototyping ECU and validated at the engine test bench.

Author Contributions: Formal analysis, M.C. and N.S.; Supervision, N.C., N.R., J.M., E.C. and V.R.; Writing—original draft, A.B. All authors have read and agreed to the published version of the manuscript.

Funding: This work was supported by 1 research grants founded by the Department of Industrial Engineering of the University of Bologna (grant number 2616) and the collaboration with Ferrari S.p.a. The external funding is indicated in the Acknowledgments section.

Acknowledgments: The authors gratefully acknowledge financial support from "Fondi europei della Regione Emilia-Romagna", within program POR FSE 2014–2020 in the framework of the "Accordo Regionale di insediamento e sviluppo delle imprese—progetto STEP".

Conflicts of Interest: The authors declare no conflict of interest.

Abbreviations

BMEP	Brake Mean Effective Pressure
CAN	Controller Area Network
ECU	Engine Control Unit
FEM	Finite Element Method
GDI	Gasoline Direct Injection
HB	Brinell-Hardness
HTT	Hardness-Time-Temperature curves
IMEP	Indicating Mean Effective Pressure
ION	Ionization Current
MAPO	Maximum Amplitude of Pressure Oscillation
MAPO98	98th percentile of MAPO
MAPO99.5	99.5th percentile of MAPO
MBT	Maximum Brake Torque
MFB50	The 50%-of-Mass-of-Fuel-Burned
PDF	Probability Density Function
PF	Probability Function
PID	Proportional Integral Derivative controller
Pmax	Maximum in-cylinder pressure
Pmax90	90th percentile of Pmax
RCP	Rapid Control Prototyping
RON	Research Octane Number
RT	Real-Time
RzD	Roughness Depth index
SA	Spark Advance
STAM	Stoichiometric Trapped Air Mass
TAM	Trapped Air Mass
TC	Turbo-Charged
VVT	Variable Valve Timing

References

1. Wang, Z.; Liu, H.; Reitz, R. Knocking combustion in spark-ignition engines. *Prog. Energy Combust. Sci.* **2017**, *61*, 78–112. [CrossRef]
2. Corti, E.; Forte, C. *Statistical Analysis of Indicating Parameters for Knock Detection Purposes*; SAE Technical Paper; SAE: Warrendale, PA, USA, 2009. [CrossRef]
3. Spelina, J.; Peyton Jones, J.; Frey, J. Recent advances in knock analysis, simulation, and control. *SAE Int. J. Engines* **2014**, *7*, 947–955. [CrossRef]
4. Gao, J.; Yao, A.; Feng, L.; Xu, H.; Yao, C. *Experimental Investigation on the Failures of Engine Piston Subjected to Severe Knock*; SAE Technical Paper; SAE: Warrendale, PA, USA, 2019. [CrossRef]
5. Samimy, B.; Rizzoni, G.; Leisenring, K. *Improved Knock Detection by Advanced Signal Processing*; SAE Technical Paper; SAE: Warrendale, PA, USA, 1995. [CrossRef]

6. De Cesare, M.; Ravaglioli, V.; Carra, F.; Stola, F. *Review of Combustion Indexes Remote Sensing Applied to Different Combustion Types*; SAE Technical Paper; SAE: Warrendale, PA, USA, 2019. [CrossRef]
7. Moro, D.; Cavina, N.; Ponti, F. In-cylinder pressure reconstruction based on instantaneous engine speed signal. *ASME J. Eng. Gas Turbines Power* **2002**, *124*, 220–225. [CrossRef]
8. Romani, L.; Bianchini, A.; Vichi, G.; Bellissima, A.; Ferrara, G. Experimental assessment of a methodology for the indirect in-cylinder pressure evaluation in four-stroke internal combustion engines. *Energies* **2018**, *11*, 1982. [CrossRef]
9. Shimasaki, Y.; Kobayashi, M.; Sakamoto, H.; Ueno, M.; Hasegawa, M.; Yamaguchi, S.; Suzuki, T. *Study on Engine Management System Using In-cylinder Pressure Sensor Integrated with Spark Plug*; SAE Technical Paper; SAE: Warrendale, PA, USA, 2004. [CrossRef]
10. Franke, A.; Reinmann, R.; Larsson, A. *The Role of the Electrodes for the Ionization Sensor Signal*; SAE Technical Paper; SAE: Warrendale, PA, USA, 2003. [CrossRef]
11. Panousakis, D.; Gazis, A.; Patterson, J.; Chen, R. *Analysis of SI Combustion Diagnostics Methods Using Ion-Current Sensing Techniques*; SAE Technical Paper; SAE: Warrendale, PA, USA, 2006. [CrossRef]
12. Cavina, N.; Poggio, L.; Sartoni, G. Misfire and partial burn detection based on ion current measurement. *SAE Int. J. Engines* **2011**, *4*, 2451–2460. [CrossRef]
13. Tong, S.; Li, H.; Yang, Z.; Deng, J.; Hu, Z.; Li, L. *Cycle Resolved Combustion and Pre-Ignition Diagnostic Employing Ion Current in a PFI Boosted SI Engine*; SAE Technical Paper; SAE: Warrendale, PA, USA, 2015. [CrossRef]
14. Peyton Jones, J.; Muske, K.; Frey, J.; Scholl, D. *A Stochastic Knock Control Algorithm*; SAE Technical Paper; SAE: Warrendale, PA, USA, 2009. [CrossRef]
15. Spelina, J.M.; Peyton Jones, J.C.; Frey, J. Characterization of knock intensity distributions, Part I: Statistical independence and scalar measures. *Proc. Inst. Mech. Eng. Part D J. Auto Eng.* **2017**, 117–128. [CrossRef]
16. Spelina, J.M.; Peyton Jones, J.C.; Frey, J. Characterization of knock intensity distributions, Part II: Parametric models. *Proc. Inst. Mech. Eng. Part D J. Auto Eng.* **2013**, *227*, 1650–1660. [CrossRef]
17. Peyton Jones, J.C.; Spelina, J.M.; Frey, J. Likelihood-based Control of Engine Knock. *IEEE Trans. on Control Sys. Tech.* **2013**, *1*. [CrossRef]
18. Thomasson, A.; Shi, H.; Lindell, T.; Eriksson, L.; Shen, T.; Jones, J.C.P. Tuning and Experimental Evaluation of a Likelihood-Based Engine Knock Controller. In Proceedings of the 2013 IEEE 52nd Conference on Decision & Control, Firenze, Italy, 10–13 December 2013. [CrossRef]
19. Spelina, J.M.; Peyton Jones, J.C.; Frey, J. Stochastic simulation and analysis of a classical knock controller. *Int. J. Engine Res.* **2014**, *16*, 461–473. [CrossRef]
20. Frey, J.; Peyton Jones, J.C. Stochastic simulation of a cumsum knock controller. *IFAC-PapersOnLine* **2016**, *49–11*, 210–216. [CrossRef]
21. Shayestehmanesh, S.; Peyton Jones, J.; Frey, J. *Stochastic Characteristics of Knock and IMEP*; SAE Technical Paper; SAE: Warrendale, PA, USA, 2018. [CrossRef]
22. Peyton Jones, J.; Frey, J. performance assessment of knock control systems subject to disturbances. *SAE Int. J. Engines* **2021**, *14*. [CrossRef]
23. Cavina, N.; Rojo, N.; Ceschini, L.; Balducci, E.; Poggio, L.; Calogero, L.; Cevolani, R. *Investigation of Knock Damage Mechanisms on a GDI TC Engine*; SAE Technical Paper; SAE: Warrendale, PA, USA, 2017. [CrossRef]
24. Cavina, N.; Brusa, A.; Rojo, N.; Corti, E. *Statistical Analysis of Knock Intensity Probability Distribution and Development of 0-D Predictive Knock Model for a SI TC Engine*; SAE Technical Paper; SAE: Warrendale, PA, USA, 2018. [CrossRef]
25. Brusa, A.; Cavina, N.; Rojo, N.; Cucchi, M.; Silvestri, N. *Development and Validation of a Control-Oriented Analytic Engine Simulator*; SAE Technical Paper; SAE: Warrendale, PA, USA, 2019. [CrossRef]
26. Abgrall, R.; Shu, C.-S. *Handbook of Numerical Methods for Hyperbolic Problems*; Springer: Berlin/Heidelberg, Germany, 2017; ISBN 9780444639103.
27. Ceschini, L.; Morri, A.; Balducci, E.; Cavina, N.; Rojo, N.; Calogero, L.; Poggio, L. Experimental observations of engine piston damage induced by knocking combustion. *Mater. Des.* **2017**, *114*, 312–325. [CrossRef]
28. Balducci, E.; Ceschini, L.; Rojo, N.; Cavina, N.; Cevolani, R.; Barichello, M. Knock induced erosion on Al pistons: Examination of damage morphology and its causes. *Eng. Fail. Anal.* **2018**, *92*, 12–31. [CrossRef]
29. Ceschini, L.; Morri, A.; Morri, A.; di Sabatino, M. Effect of thermal exposure on the residual hardness and tensile properties of the EN AW-2618A piston alloy. *Mater. Sci. Eng. A* **2015**, *639*, 288–297. [CrossRef]
30. Balducci, E.; Ceschini, L.; Morri, A.; Morri, A.; di Sabatino, M.; Arnberg, L.; Li, Y. High temperature behavior of the EN AW-2618A piston alloy containing 0.12wt% Zr: Influence of heat treatment. *Mater. Today Proc.* **2015**, *2*, 5037–5044. [CrossRef]
31. DIN 4768. *Determination of Values of Surface Roughness Parameters Ra, Rz, And Rmax Using Electrical Contact (Stylus) Instruments Concepts and Measuring Conditions*, 1990th ed.; Deutsches Institut für Normung: Berlin, Germany, 1990.
32. Cantore, G.; Giacopini, M.; Rosi, R.; Strozzi, A.; Pelloni, P.; Forte, C.; Achilluzzi, M.; Bianchi, G.M.; Ceschini, L.; Morri, A. Validation of a Combined CFD/FEM Methodology for the Evaluation of Thermal Load Acting on Aluminum Alloy Pistons through Hardness Measurements in Internal Combustion Engines. In *Metallurgical Science and Technology*; Public Knowledge Project: Vancouver, BC, Canada, 2011.
33. Mahle, G.M.B.H. *Piston and Engine Testing*; Springer: Berlin/Heidelberg, Germany, 2012.
34. Cesari, F. *Calcolo Matriciale delle Strutture 2*; Pitagora Editrice Bologna: Bologna, Italy, 1997.

35. Cesari, F. *Meccanica delle Strutture. Metodo degli Elementi Finiti. 120 Problemi Risolti*; Pitagora Editrice Bologna: Bologna, Italy, 2011.
36. Timoshenko, S.P. *Theory of Elasticity*, 3rd ed.; McGraw-Hill: New York, NY, USA, 1970.
37. Matlab Documentation. The MathWorks. 2021. Available online: https://www.mathworks.com/ (accessed on 20 April 2021).
38. Ghandhi, J.; Kim, K. *A Statistical Description of Knock Intensity and Its Prediction*; SAE Technical Paper; SAE: Warrendale, PA, USA, 2017. [CrossRef]

energies

MDPI

Article

Analysis of the Use of Fatty Acid Methyl Esters as an Additive to Diesel Fuel for Internal Combustion Engines

Łukasz Muślewski [1], Marietta Markiewicz [1,*], Michał Pająk [2], Tomasz Kałaczyński [1] and Davor Kolar [3]

1 Faculty of Mechanical Engineering, Bydgoszcz University of Science and Technology, Al. Prof. S. Kaliskiego 7, 85-796 Bydgoszcz, Poland; lukasz.muslewski@pbs.edu.pl (Ł.M.); tomasz.kalaczynski@pbs.edu.pl (T.K.)
2 Faculty of Mechanical Engineering, University of Technology and Humanities in Radom, Stasieckiego 54, 26-600 Radom, Poland; m.pajak@uthrad.pl
3 Faculty of Mechanical Engineering and Naval Architecture, University of Zagreb, Ivana Lučića Street 5, 10002 Zagreb, Croatia; davor.kolar@fsb.hr
* Correspondence: marmar000@pbs.edu.pl

Abstract: This study presents pro-ecological activities focused on an analysis of the use of biofuels as an environmentally friendly fuel. The research objects were different concentration mixtures of diesel fuel and fatty acid methyl esters, that is, transesterified plant oils. The tests involved an assessment of the performance parameters of a drive unit in a vehicle powered by diesel fuel for different mixtures of diesel oil and fatty acid methyl esters in the following proportions: 10%, 30%, 50% and 50% with chemical additives. The tests were comparative and were conducted for 'pure' diesel (ON). The study presents test results of selected performance parameters of the analyzed power unit. The object of the tests was a self-injection engine with a maximum power of 81 kW. The main tests which were most important for assessment of the mixtures, from the point of view of their effect on the analyzed performance parameters, involved measuring power and torque, and the toxic components of exhaust gases. Based on the obtained results, a statistical analysis was carried out, and a model for the evaluation of how the research object functions when fed with different fuel mixtures was developed. The research found which mixture can be considered the most optimal, and what the influence of individual fuel mixtures is on the analyzed performance parameters of the tested power unit.

Keywords: transport; engines; diesel oil; biocomponents; environment; sustainable development

Citation: Muślewski, Ł.; Markiewicz, M.; Pająk, M.; Kałaczyński, T.; Kolar, D. Analysis of the Use of Fatty Acid Methyl Esters as an Additive to Diesel Fuel for Internal Combustion Engines. *Energies* **2021**, *14*, 7057. https://doi.org/10.3390/en14217057

Academic Editor: Dimitrios C. Rakopoulos

Received: 30 September 2021
Accepted: 23 October 2021
Published: 28 October 2021

1. Introduction

Due to rapid technological and economic development, many countries have introduced economic changes to improve efficiency of production and management, focusing on the ecological character of legislation and the relevant actions to be taken. The use of renewable energy sources to produce transport fuel for combustion engines is dictated by a high demand for fuel and the depletion of natural resources. The structures of combustion engines enable their modernization and, subsequently, a return to the idea of plant fuels. The exhaustion of petroleum fuels is an argument for turning to the concept of plant fuels. Plant oils are considered to be the best for this purpose. The most common biofuels include rape oil, soya oil, sunflower oil, arachis oil and animal fats. The above-listed plant fuels, referred to as biofuels, must undergo chemical processing to achieve physical-chemical properties similar to those of diesel oil. Rape oil is most frequently used for economic, technical, structural and technological reasons. The results of research provided by many scientific centers prove that the physical-chemical properties of rape biofuel are the most similar to diesel, which makes it the number one biofuel when compared to other biofuels. The content of oil and hydrogen in rape oil is lower than in traditional diesel, whereas its density and viscosity is higher [1]. The application of rape oil involves the presence of sulfur in exhaust emission, and a smaller number of solid particles such as hydrocarbons and carbon monoxides. The combustion of conventional fuels in self-ignition engines generates a big amount of sulfur oxide and nitric oxide. Powering self-ignition engines with fatty acid methyl esters causes a reduction in toxic emissions of exhaust gases such as carbon monoxides, hydrocarbons and solid particles [2–7]. Nitric oxides are the exception, as they increase by several percent when compared to pure diesel oil [8]. It needs to be highlighted though, that the most commonly used indicators of combustion engines are: power, torque, fuel consumption, noise emission and the composition of exhaust fumes and exhaust smoke. Tests conducted by other authors have provided 7% lower values in engine performance parameters and a lower power reduction compared to diesel, depending on the combustion system and characteristics of the injection unit [8,9]. According to the literature, using renewable energy sources to power combustion engines reduces the emission of pollution of such exhaust components as: carbon monoxide, hydrocarbons and particulates [5,7–17]. Nitric oxides, which cause an increase of several percent compared to diesel, are the exceptions. The use of biofuels combined with low temperature combustion can be a way to improve the economy of fuel management and emission reduction. However, research centers and the industry are working on innovative combustion systems to reduce CO2 emission and improve NOx-Soot trade-offs, such as additive manufacturing, bowl design, innovative fuel injection systems, injection strategy and EGR systems [18,19].

The goal of this study is to analyze the advisability of the application of one type of alternative fuel, mixtures of diesel and fatty acid methyl esters, as well as evaluate their impact on the selected performance parameters that characterize the operation of the research object—the power unit of a transport means. The main subject of this work is to find out which mixtures of fatty acid methyl esters with diesel oil would have the performance parameters of diesel, as well as assess their environmental impact. A multi-criteria and comprehensive analysis of the impact of the use of biofuels on the change of the basic performance parameters of drive units is a new approach to the subject.

2. The Research Materials

The research objects were mixtures with different concentrations of diesel fuel and fatty acid methyl esters, that is, plant oils subjected to transesterification—a process of exchanging glycerin bonded in a triacyloglicerole molecule with methyl alcohol catalyzed by the addition of base or acid catalyst—commonly known as a biocomponent.

In the main tests, the performance parameters of a transport-means drive unit, powered by mixtures of diesel fuel with fatty acid methyl esters, were measured according the following proportions:

- 10% of fatty acid methyl esters and 90% of diesel oil,
- 30% of fatty acid methyl esters and 70% of diesel oil,
- 50% of fatty acid methyl esters and 50% of diesel oil,
- 50% fatty acid methyl esters with chemical additives and 90% diesel oil.

A chemical additive introduced to the mixture of 50% diesel fuel with 50% fatty acid methyl esters was a substance which increased the lubrication properties of the fuel and its cetane number. Table 1 presents a comparison of the results for diesel oil, biodiesel and plant oil.

Table 1. Physical chemical properties of fuels used in combustion engines.

Property	Diesel Oil	Biodiesel	Plant Oil
Viscosity $[mm^2 \ s^{-1}]$	2.0–4.5	3.5–5.5	7.2
Sulfur content $[mg{\cdot}kg^{-1}]$	≤35	≤10	no
Density $[g{\cdot}cm^{-3}]$	0.82–0.45	0.86–0.90	0.88
Cetane number	≥51	≥47	>40
Calorific value [MJ/kg]	43	41	37.6
Ignition point [°C]	≥55	≥101	≥220

The object of the tests was a self-injection engine with a maximum power of 81 kW. The power unit is considered to be representative as it has a wide application in many vehicles, with a maximum authorized weight up to 3.5 t, a 1.6 HDi engine with a Common Rail direct fuel injection system, and the use of electromagnetic injectors. The characteristics of the factory settings of one of the test engines are presented in Figure 1.

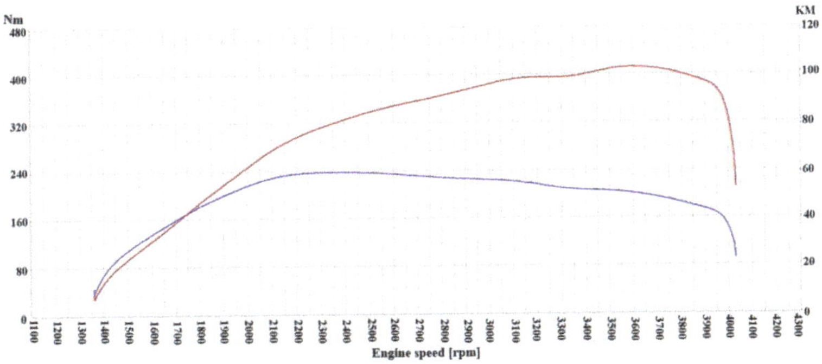

Figure 1. Characteristics of the test engine.

3. The Research Methods

The tests were divided into two parts, the initial and main tests, which are presented in Figure 2. The initial tests involved the determination of the value of the energy value and the cetane number of the mixtures of diesel fuel and fatty acid methyl esters. The tests were repeated and carried out in order to determine the content of the mixtures of diesel fuel with a biocomponent. The tests were of comparative character, and were performed for 'pure' diesel (ON). The notion 'pure' diesel fuel refers to a fuel without fatty acid methyl esters. The tests were performed on a chassis dynamometer, simulating road conditions. The following parameters were tested: power, torque, exhaust gas content and content of particulates. The emission of noise generated by the power unit was also assessed.

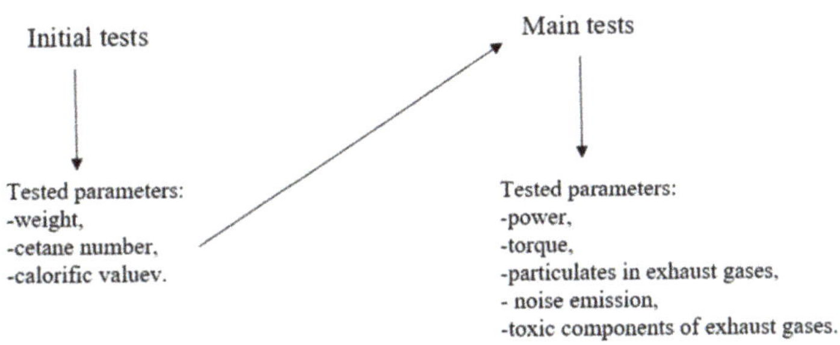

Figure 2. Stages of the tests.

The initial tests, whose aim was to identify the tested material (of particular mixtures), included:

- The measurement of fuel energy parameters—specific heat and calorific value—with the use of a colorimeter;
- A viscosity test—the measurement of the liquid viscosity coefficient—with the use of a rotational viscosimeter;
- the measurement of the cetane number—with the use of a calibrated model engine according to ASTM D613 and PN-EN ISO norms.

The main tests that played the most important role in the assessment of the mixtures, from the point of view of their effect on the analyzed performance parameters, included measurements of:

- Power and torque—using a load bearing chassis dynamometer with a DynoTech electromechanical brake with a valid manufacturers certificate. Measurements included the impact of air moisture, pressure changes, changes in air temperature and atmospheric pressure, whose values were normalized. The measurement results were calculated onto values corrected in accordance with the applicable standard (25 °C and 1000 hPa) norm [20];
- The toxic components of exhaust gases—using an exhaust fume analyzer MGT-5, which along with a periodically validated measuring probe, complied with the requirements provided in 22/2004WE directive. The experimental tests allowed the definition of the values of emitted compounds, which are presented in Table 2;
- The content of solid particles in exhaust gases—the measurement of the emission of solid particles with dimensions exceeding 100 mm was performed measured with MPM-4 analyzer of Maha company. The analyzer had a certificate of periodic validation;
- The value of noise emission—the test was conducted with the use of a noise level meter with an embedded spectrum analyzer with a validation certificate complying with the requirements of IEC 61672-1:2002 and IEC 60651 norms. The orientation method was used to define the standard power of the engine noise acoustics by calculating the level of power corrected in frequency bands based on the acoustic pressure level in the frequency bands, and the level of noise measured by means of a correction filter.

This section presents a description of the proposed assessment method and an assessment model for the evaluation of the influence of fatty acid methyl ester additives to diesel fuel on the values of selected performance parameters of the analyzed research object.

Table 2. The emitted compounds that were tested.

Emitted Compounds	Symbol
Hydrocarbons	HC
Oxygen	O_2
Carbon oxide	CO
Carbon dioxide	CO_2
Nitric oxides	NO_2
Excess air coefficient	λ

3.1. Description of the Assessment Method

It was established that assessment should be carried out based on tests of the impact of the mixtures of diesel fuel with fatty acid methyl esters on selected measurable performance parameters of the analyzed power unit. The assessment covered parameters that, due to their importance and impact on the test results, were accepted in a mathematical description [21,22]. The assessment process involved the identification of the parameter set {Xi} i = 1, 2, ... , p (characteristics), describing the tested engines from the point of view of the research purpose. Next, the value of each accepted parameter was measured. At the same time, it must be remembered that the considered set of parameters accepted for assessment consists of two subsets. One of them is a subset whose values are determined based on the measurements, and these are of constant character. The second subset, however, consists of characteristics assessed in a discrete manner. Based on this, the analyzed characteristics are approached in terms of a fuzzy representation [23]. When their values are determined from the measurements, the obtained value is to be defined with accuracy, taking into consideration the measurement error [24]. Thus, only the interval containing a given value can be defined, but not the value itself:

$$X_0 \in \langle X_p - \delta_u, X_p + \delta_u \rangle \tag{1}$$

where:

X_0—calculation value of the characteristics;
X_p—measuring value of the characteristics;
δ_u—measuring point of the device.

In the case of direct measurement, the value is additionally burdened with an error of the measurement method. This involves the necessity of accepting a tolerance interval, which needs to be considered when analyzing the results.

In the case of measurements containing insensitivity zones, the tolerance interval was modelled as a fuzzy set of type Π:

$$FS_{\Pi}(x) = \begin{cases} \frac{x - lrs}{lrk - lrs} \\ \frac{rrs - x}{rrs - rrk} \end{cases} \tag{2}$$

where:

$FS_{\Pi}(x)$—a function of affinity for a fuzzy set of type Π;
lrk—the lowest value belonging to the nucleus of fuzzy set;
lrs—the lowest value belonging to the fuzzy set medium;
rrk—the highest value belonging to the fuzzy set nucleus;
rrs—the highest value belonging to the medium of fuzzy set.

In the remaining cases, the tolerance interval was modeled in the form of a fuzzy set of type Λ:

$$FS_{\Lambda}(x) = \begin{cases} \frac{x - lrs}{lrk - lrs} \\ \frac{rrs - x}{rrs - rrk} \end{cases} \tag{3}$$

where:

$FS_{\Lambda}(x)$—function of affinity for the fuzzy set of type Λ;
lrk—the lowest value belonging to the nucleus of fuzzy set;
lrs—the lowest value belonging to the medium of fuzzy set;
rrs—the highest value belonging to the fuzzy set medium.

The value of a fuzzy set modal was accepted as the measurement value, whereas its medium is equal to the accepted tolerance interval.

In the case of characteristics that make up the second identified set, the assessment can be of subjective character. At the same time, for the assigned discrete scale to be used for grading the respective criterion fulfilment degree, the assessment is determined with approximation, which results from the applied discretization. The above inaccuracy can be modeled by means of fuzzy sets of type Λ. A modal value is accepted to be the criterion fulfilment degree for the considered set, whereas twice the distance between the discrete values of the rating scale is accepted to be the medium.

3.2. Construction of an Assessment Model for the Impact of a Fatty Acid Methyl Esters Additive to Diesel Fuel on the Values of Performance Parameters of the Considered Means of Transport

Let X_i, i = 1, 2, ... , p, stand for a feature which is a random variable representing the assessment of a transport means, functioning from the point of view of it being powered by a given fuel mixture. A vector of features:

$$X = \langle X_1, X_2, \ldots, X_p \rangle \tag{4}$$

is considered.

The component X_i, i = 1, 2, ... , p, of vector X is a one-dimensional random variable in R space, describing the i-th feature, and representing the value of one of the parameters accepted for the evaluation. Vector X is a p–dimensional random variable providing a full (collective) assessment of the research object in space R^p. Then, expression:

$$X \times \Omega \rightarrow R^p \tag{5}$$

means that for each elementary (ω), where $\omega \in \Omega$, X(ω) is a p–dimensional vector, its components are real numbers expressing the analyzed parameters of the mixture fueling in the transport-means power unit, where:

X—p–dimensional assessment vector;
Ω—set of elementary events;
ω—elementary event;
R^p—p–dimensional space made up of vectors (x1, x2, ... , xp);
x_i—p–element sequences; x_i R, i = 1, 2, ... , p.

For a given research object, the random variable is defined in the form:

$$Z_X = \sum_{i=1}^{p} \alpha_i X_i \tag{6}$$

where:

$$\alpha_i \geq 0, \qquad \sum_{i=1}^{p} \alpha_i = 1, \tag{7}$$

α_i, i = 1, 2, ... , p—denote the values of weights for each parameter;
Z_X—a random variable being a finite mixture of variables X_i, i = 1, 2, ... , p.

For the mean value, it can be:

$$EZ_X = \sum_{i=1}^{p} \alpha_i EX_i \tag{8}$$

Mean value EZX is a linear combination of mean values EX_i, i = 1, 2, ... , n. The formula finds application even if variables X_i, i = 1, 2, ... , n are dependent ones.

With these considerations, the aim of the study is to compare the fueling of the propulsion unit with different mixtures in relation to the standardized mixture, or adopted as a reference point ('pure' diesel), and, based on this, determine the impact of each mixture on the respective performance parameters. It should to be noted that the values of the parameters describing the test samples may be less, equal or higher than the values of the parameters that are the point of reference. Thus, in a graphic interpretation (Figure 3), the length of the components of vector W^B (representing particular values of the parameters describing a given mixture) analogically, can be different or equal to the components of vector (W^B) (representing values of parameters for 'pure' diesel), and a comprehensive assessment will result from the total values of the particular parameters and assign to them weights.

A multi-criteria optimization analysis was used in the assessment process, in consideration of the above case. This involves arranging the assessed variants in a specific order and, if possible, determining the overall quality of each variant by assigning to them appropriate degrees and results. The application of this analysis allows:

- The determination of the criteria that make up the multi-criteria optimization assessment system—quantitative and qualitative criteria; the common method to be used for assigning a grade to each variant, according to each criterion being an object of the same domain;
- The determination of a system of weights for each criterion;
- The determination of a system of weights for a particular decision makers' assessments;
- The implementation of the calculation process for an overall assessment of a given variant;

- The interpretation of the overall assessment provided.

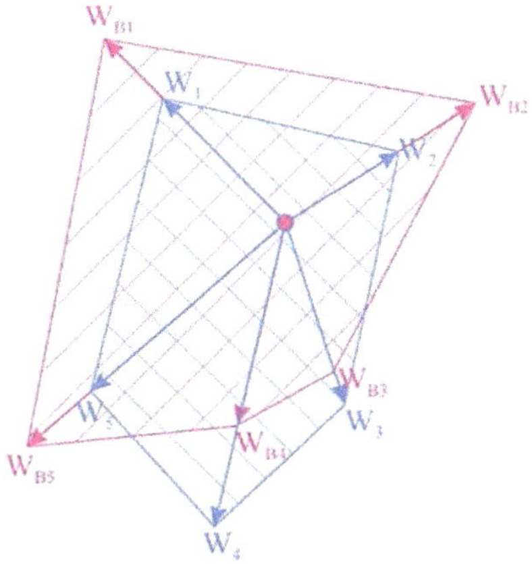

Figure 3. Test stages.

The MOA SMART and AHP methods are proposed for the multi–criteria optimization analysis [25,26]. In the process of assessment, the AHP (Analytic Hierarchy Process) is used for the determination of the weights of particular criteria [27–29]. It enables the specification of the expert's knowledge in a way that is natural for a human mind. A scale is provided to assign numerical values to verbal comparisons. It is a method of comparing parameters that can be represented in the form of a matrix:

$$q = \begin{bmatrix} 0 & q_{1,2} & \cdots & q_{1,n} \\ 0 & 0 & \cdots & \vdots \\ 0 & 0 & 0 & q_{n-1,n} \\ 0 & 0 & 0 & 0 \end{bmatrix} \quad (9)$$

The determination of the values of grades defining the significance of the analyzed parameters is carried out based on a seven-degree grading scale, described in Table 3. The grading scale was selected according to the AHP method.

Table 3. Grading scale used in a comparative assessment of the parameters.

Comparison of Variant *a* with Variant *b*	Relative Quality of Variant *a* as Compared to *b*	Assigned Value q_{ab}
a much better than *b*	Strong preference of *a*	6
a better than *b*	Preference of *a*	4
a a little better than *b*	Peak preference of *a*	2
a equally good as *b*	No preference	0
a slightly worse than *b*	Weak preference of *b*	−2
a worse than *b*	Preference of *b*	−4
a much worse than *b*	Strong preference of *b*	−6

According to the presented method, the values of the weights are determined based on the dependence:

$$w_{wj} = \frac{1}{n_w} \sum_{k=1}^{n_w} q_{wjk} \quad (10)$$

where:

q_{wjk}—value of the preference criterion with index w_j, in relation to a criterion with index k;
n_w—number of criteria;
w_{wj}—value of weight for the w_j-th criterion.

4. Results

4.1. Initial Tests

The tests (anti-anti-anti-anti) involved the use of different test methods. The samples were observed in order to define the negative phenomena of delamination, gelation and others. The samples were put into containers for protection against light and air for approximately 72 h, at a constant temperature of 210 °C. The solutions were examined after 12 h. The properties affecting the engine operation were identified in the next stage of the tests. Calorific value and the mass of ash left after burning the fuel, cetane number and viscosity were tested. An assessment of calorific value of the samples was performed by means of a KL-12Mn calorimeter by burning 1 g of fuel in the atmosphere of oxygen, in a glass pot. After each test, the mass of ash from burning was defined. Tests were repeated three times for each fuel. A measurement of the viscosity coefficient was performed with the use of a rotating viscometer, at a temperature of 300 °C, three times. Measurements of the cetane number were made with the engine method. Mean quantities of energetic value, cetane number and viscosity is presented in Table 4.

Table 4. Mean quantities of energetic value, cetane number and viscosity.

	Calorific Value [MJ·kg^{-1}]	Cetane Number	Viscosity [cSt]
Mixture I	43.096	53.50	6.47
Mixture II	43.197	53.99	6.37
Mixture III	41.957	54.97	7.80
Mixture IV	40.588	55.95	8.57
Mixture V	37.903	58.40	1.87

Based on the observations carried out in the first stage of the tests, it was found that the color of samples III, IV and V had already changed after the first 12 h (the samples darkened slightly). However, neither the delamination nor coagulation mentioned by other authors [8] were observed. The calorific value of the tested fuels, the ash mass and viscosity did not change statistically significantly for biocomponent content 0 or 10%. For a biocomponent content of 30% and 50%, a drop in calorific value by approximately 3.4% was observed, compared to samples I and II. However, for sample V, the drop was 12.1%. The course of change for the fuel calorific value and viscosity is demonstrated in Figures 2 and 3. The ash mass is different for fuels with a biocomponent content higher than 30% (samples III, IV i V) and is lower by about 63% for fuels with a higher content of biocomponent.

4.2. Construction of the Assessment Model

In this study, performance parameters were assessed for their impact on the quality of the analyzed object functioning in the environment. Moreover, the physical-chemical properties of the tested mixtures, such as viscosity, cetane number, calorific value and water content (initial tests), were tested. Fuel consumption tests were neglected, as this subject has already been profoundly analyzed by many authors [7,30–33]. However, the overall costs to be borne in connection with the use of biofuels are the company internal costs and the costs of materials. The costs of materials include diesel fuel and methanol with a catalyst. The catalyst is potassium hydroxide or possibly sodium hydroxide. A ready-to-use product (RME biodiesel and technical glycerin) are the profits. This is regardless of the fatty acid methyl ester production costs, which depend on the current prices according to the stock exchange listing in Rotterdam [34].

In the proposed model, X denotes one-dimensional vectors, which are accepted as random variables. The analyzed parameters ($X_1 \div X_{10}$) represent the evaluation of transport means powered with mixtures of diesel oil and a variable content of fatty acid methyl esters. Then, the considered vector assumes the following form:

$$Xi = \langle X_1, X_2, X_3, X_4, X_5, X_6, X_7, X_8, X_9, X_{10} \rangle \quad (11)$$

where the vector components are:

X_1—power;
X_2—torque;
X_3—emission of sound generated by the engine;
X_4—content of particulates in exhaust gases;
X_5—carbon monoxide;
X_6—carbon dioxide;
X_7—oxygen;
X_8—nitric oxides;
X_9—air excess coefficient lambda;
X_{10}—hydrocarbons.

A random variable in the form of:

$$Z_x = \alpha_{1\times1} + \alpha_{2\times2} + \alpha_{3\times3} + \alpha_{4\times4} + \alpha_{5\times5} + \alpha_{6\times6} + \alpha_{7\times7} + \alpha_{8\times8} + \alpha_{9\times9} + \alpha_{10\times10} \quad (12)$$

where α denotes the values of weights (determined by the AHP method) for particular parameters, which are presented in Table 5.

Table 5. Values of weights for particular parameters [35–37].

Vector Component Denotation	Weight Denotation	Explanation	Weight
X_1	α_1	Power [kW]	0.2616
X_2	α_2	Torque [Nm]	0.45547
X_3	α_3	Emission of sound generated by the engine [dB]	0.01239
X_4	α_4	Content of particulates in exhaust gases [ppm]	0.2616
X_5	α_5	Carbon monoxide [% vol.]	0.00044
X_6	α_6	Carbon dioxide [% vol.]	0.00235
X_7	α_7	Oxygen [% obj.]	0.00046
X_8	α_8	Nitric oxides [ppm]	0.00539
X_9	α_9	Air excess coefficient lambda	0.00019
X_{10}	α_{10}	Hydrocarbons [ppm]	0.00011

4.3. Results of Experimental Tests

This study presents a detailed comparative analysis of the values of the performance parameters of a transport-means power unit fueled with mixtures of diesel oil and fatty acid methyl esters.

A comparison of the tested mixtures made it possible to refer to the values obtained for particular performance parameters to a 'pure' diesel, which was used as a reference for the needs of this study. The assessment process is considered to be an example of a multi-criteria optimization analysis (MOA). The assessment is expressed as an arithmetical mean, being the most efficient, unburdened estimator of an unknown expected value [38] for particular values of the performance parameters of the transport-means power unit (five mixtures of diesel fuel and fatty acid methyl esters), and is presented in Table 6.

Table 6. Assessment of the values of performance parameters for particular mixtures of diesel fuel and fatty acid methyl esters.

Mixture Number	Power [kW]	Torque [Nm]	Noise [dB]	Solid Particles [ppm]	Carbon Oxide [% obj.]	Carbon Dioxide [% obj.]	Oxygen [% obj.]	Nitric Oxides [ppm]	Lambda	Hydrocarbons [ppm]
I	0.8482	0.8698	0.6826	0.5164	0.7232	0.3464	0.5200	0.4740	0.6048	0.6482
II	0.8888	0.8564	0.7966	0.5008	0.7944	0.4206	0.6706	0.4630	0.6280	0.8122
III	0.9326	0.9282	0.6774	0.4916	0.7230	0.4190	0.6210	0.4794	0.6056	0.6666
IV	0.8378	0.9494	0.6110	0.4534	0.8058	0.4218	0.6428	0.4746	0.5990	0.6248
V	0.9500	0.9226	0.5982	0.4432	0.7400	0.132	0.6236	0.3670	0.6090	0.6350

Tests of a transport-means power unit fueled with different mixtures of diesel and fatty acid methyl esters were conducted every 24 h, with 10 repetitions for each parameter. Based on the tests results, values were determined for each parameter in a given time moment (intervals) ti, i = 1, 2, 3. Based on this, the mean values of grades were also determined for particular measurement groups. The values determined for the parameters were recorded so that the minimum values would reflect the worst level, whereas the maximum values would reflect the value desired for particular parameters.

The obtained results allowed the definition of the size of the components of particular vectors. The determination of these components enables the geometric interpretation of the performance parameter mean values for a transport-means power unit fueled with mixtures of diesel and fatty acid methyl esters. For transparency and unambiguity of the results, the values of the relevant parameters from the analyzed set were normalized onto interval <0 ÷ 10>, using the following dependence:

$$10 \times \frac{(X_i - X_{min})}{(X_{max} - X_{min})} \tag{13}$$

Furthermore, variability intervals were determined for the set of relevant parameters, which are presented in Table 7.

Table 7. Variability intervals for the set of relevant parameters.

Parameter	Minimal Value	Maximal Value
Power [kW]	100.0	110.0
Torque [Nm]	230.0	250.0
Noise [dB]	118.0	126.0
Particulates [ppm]	87.0	110.0
Carbon oxide [% obj.]	0.05	0.2
Carbon dioxide [% obj.]	11.0	12.8
Oxygen [% obj.]	7.0	9.0
Nitric oxides [ppm]	230.0	281.0
Lambda	1.3	1.6
Hydrocarbons [ppm]	10.0	60.0

First, the components of the vector which was accepted to be the point of reference, the 'pure' diesel fuel (mixture I), were analyzed. Next, a comparative analysis was performed for the remaining four mixtures. Below are the component values of the parameter measurement vector for mixture V (50% diesel fuel and 50% fatty acid methyl esters with a chemical additive) are presented in Table 8.

Table 8. Average measurement results for mixture V.

Parameter	Measurement Number			Mean Value
	1	2	3	
Power [kW]	104.786	104.7718	104.8774	104.8117
Torque [Nm]	247.024	246.834	246.598	246.8187
Emission of sound generated by the engine [dB]	121.062	121.122	121.102	121.0953
Content of particulates in exhaust gases [ppm]	88.932	88.61	89024	88.85533
Carbon monoxide [% vol.]	0.1584	0.0652	0.0574	0.093667
Carbon dioxide [% vol.]	11.9966	12.216	12.186	12.13287
Oxygen [% obj.]	9.1444	8.2088	8.2538	8.535667
Nitric oxides [ppm]	256.32	265.28	268.56	263.3867
Air excess coefficient lambda	1.53238	1.52244	1.52428	1.526367
Hydrocarbons [ppm]	27.28	24.46	27.2	27.31333

The results obtained for each parameter were presented in a geometric interpretation (Figure 4) for mixture V, in comparison with the vector, which is the point of reference.

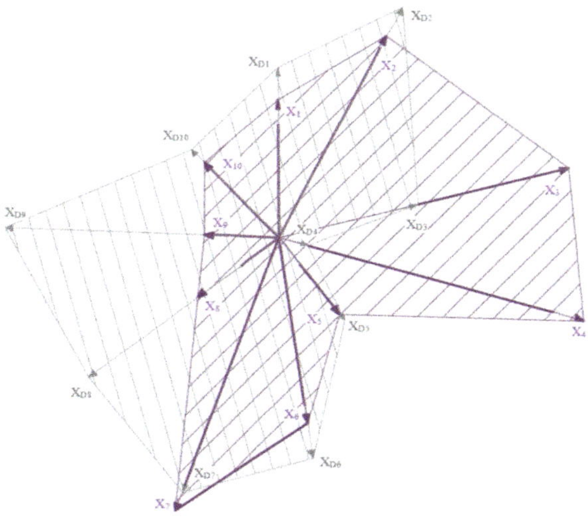

Figure 4. Graphic interpretation of the vector components for mixture V.

Based on the obtained test results, the shapes of random variables Zx were determined for particular mixtures.

- for mixture I

$$Z_x = 1.0218 + 2.9138 + 0.1006 + 2.2601 + 0.0012 + 0.0139 + 0.0038 + 0.0147 + 0.0004 + 0.0003$$

$$Z_x = 6.3306$$

- for mixture II

$$Z_X = 0.8389 + 2.7469 + 0.1006 + 2.0277 + 0.0005 + 0.0113 + 0.0038 + 0.0070 + 0.0008 + 0.0002$$

$$Z_X = 5.7377$$

- for mixture III

$$Z_X = 0.6828 + 1.8569 + 0.0415 + 2.1449 + 0.0008 + 0.0151 + 0.0026 + 0.0157 + 0.0010 + 0.0001$$

$$Z_X = 4.7614$$

- for mixture IV

$$Z_X = 1.1322 + 3.7144 + 0.0469 + 1.8758 + 0.0013 + 0.0168 + 0.0015 + 0.0525 + 0.0011 + 0.0009$$

$$Z_X = 6.8461$$

- for mixture V

$$Z_X = 1.2588 + 3.3309 + 0.0479 + 0.2111 + 0.0013 + 0.0148 + 0.0035 + 0.0353 + 0.0014 + 0.0004$$

$$Z_X = 4.9054$$

Figure 5 shows standardized results for random variable Z_X in the form of diagrams for each mixture.

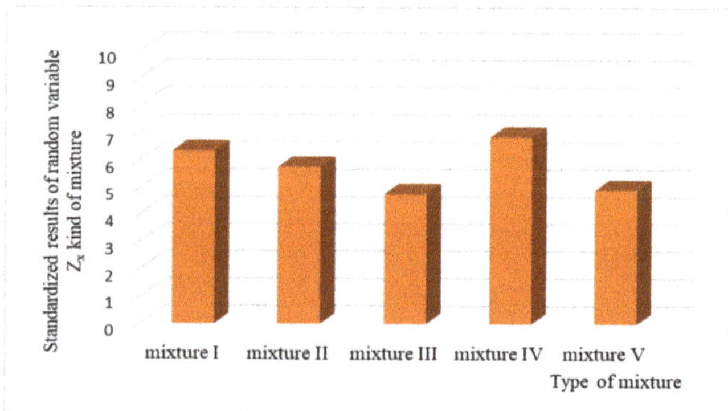

Figure 5. Graphic interpretation of the value of random variable Z_X.

The values of random variable Z_X for particular fuel mixtures (ON, BIO10, BIO30, BIO50 i BIO50+) represent a sum of the vector components and assigned weights to them. The reference point was mixture ON ('pure' diesel). As the diagram shows, the determined value of the random variable for mixtures II, III and V was lower than the reference point. Whereas, the value reached for the random variable for mixture BIO50 was higher by almost 8% than that of the 'pure' diesel. Therefore, based on the analysis, and using the accepted assessment parameters, the application of mixture IV—BIO50 was found to be advisable, as it was characterized by the most beneficial performance properties for both the engine operation and the environment.

4.4. Statistical Analysis of the Results

The experiment results underwent statistical analysis to provide information on the studied relationships. For this purpose, a comparison of the mixture mean values was carried out.

Five mixtures of diesel fuel and fatty acid methyl esters were compared. Each of the presented mixtures (I, II, III, IV, V) was tested five times. In these tests, 10 performance parameters of the drive unit were measured. The hypothesis of means/medians equality was accepted. Significance levels are presented in Table 9 for the mean value comparative test.

Table 9. Verification of the hypothesis for the analyzed parameters [6].

Parameters	I	II	III	IV	V
Power [kW]	0.0001	0.0001	0.0001	0.0001	0.0001
Torque [Nm]	0.0001	0.0001	0.0001	0.0001	0.0001
Emission of sound generated by the engine [dB]	0.0001	0.0001	0.0006	0.0001	0.5171
Content of particulates in exhaust gases [ppm]	0.0001	0.0001	0.0001	0.0001	0.0001
Carbon monoxide [% vol.]	0.1331	0.0001	0.1015	0.0039	0.0003
Carbon dioxide [% vol.]	0.0415	0.3460	0.0802	0.0381	0.0302
Oxygen [% obj.]	0.0001	0.0001	0.0001	0.7957	0.0002
Nitric oxides [ppm]	0.1172	0.3548	0.3714	0.0009	0.0001
Air excess coefficient lambda	0.0427	0.8826	0.0006	0.0009	0.0001
Hydrocarbons [ppm]	0.0001	0.0001	0.0001	0.0001	0.0001

The table includes the results of the significance level, where $p = 0.0001$ means that the actual significance level was smaller. In eight cases, the value exceeded the established significance level. For these cases, there were no grounds to reject the equality of the means/medians hypothesis.

In the next stage, the relationships between the performance parameters were analyzed, depending on the mixture content. The analyzed population was a 150-element sample n, which, for the studied parameters x and y, provided definite measurement results. The amount of fatty acid methyl ester additives to the diesel fuel was determined as X, and y was used to determine the results of the performance parameter tests on individual values for individual mixtures. Hence, in the regression analysis, the percentage of fatty acid methyl esters in the diesel fuel was an independent variable, and the dependent variables were the individual values of the engine performance parameters. In the case of engines powered with plant oil fuels, the interpretation of regression function in the statistical model is about the applicability of the variables for the prediction of the level of the dependent variables.

This work presents exemplary regression function equations for the performance parameters of a transport-means power unit. Below, there is an exemplary analysis of a test of the solid particle content in exhaust gases for mixture IV. A regression straight was obtained by marking the empirically provided points on the diagram. A diagram of the studied linear regression is presented in Figure 6.

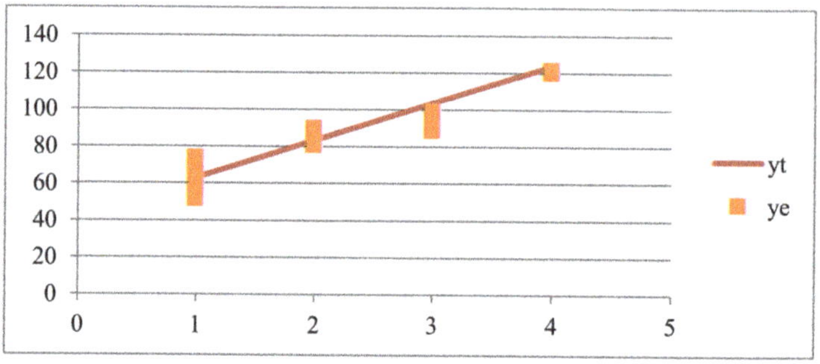

Figure 6. Diagram of linear regression of the particulate parameter for mixture IV.

The above diagram shows the empirical points marked as ye, along with a regression straight yt, determined for theoretical parameters.

The least squares method was used to determine the values of the estimators a and b. The differences between the measurement values and their mean value, as well as the values of the defined functions, were also calculated. On this basis, a simple regression was determined in the form:

$$y = 20.265x + 42.722 \quad (14)$$

For the above regression equation, the correlation coefficient was $r = 0.974$. Tests of hypothesis H0: $a = 0$ yielded $p < 0.0001$, which means that the studied dependence is statistically significant.

Analogically, an analysis was performed for each parameter of the respective mixtures.

5. Discussion

The whole discussion included in this study is supposed to show the impact of a biofuel component additive to diesel fuel as an alternative fuel from the perspective of sustainable development policy, particularly in the automotive industry. These solutions need to be consistent with already existing technologies (in the analyzed case, power and torque of a power unit), and take into account the implementation costs, including the prices of fuel and its consumption. On the other hand, the actions to be taken should also take into account a widely understood interest of future generations, including the exhaustion of natural resources, an assessment of the impact of given solutions on the life of people and animals, and environmental protection. When making a comparative analysis of the toxic parameter values of exhaust gases emitted by transport means powered with the so-called 'pure' diesel and a fuel with biocomponent additives, a question arises as to whether this policy is justified in terms of all the above mentioned aspects.

Emissivity reduction in a self-ignition engine while maintaining its high efficiency needs to be further explored for ecological reasons. The conducted tests have confirmed that the use of fuels from renewable sources contributes to reduction in the emissions of exhaust gases harmful to the natural environment, which has been proven by many authors involved in this subject [11,39–49]. The tests described in this study were performed for the standard settings of the vehicle deck computer. The literature does not provide information on the adjustment of the fuel injection controller, which would allow the matching of a given setting with the optimal operation of the drive unit of a given transport means powered with a mixture of diesel oil and fatty acid methyl esters. The adjustment of the vehicle deck computer would allow efficient operation of the engine (maintaining its power and torque) with the reduction in the harmful content of exhaust gases. In the next stage of the experiment, the performance of measurements of the engine fuel injection controller adjustments is planned.

6. Summary and Conclusions

The presented methodology and the proposed model enable assessment and inference regarding the applicability of a given fuel, depending on the accepted (optional) assessment parameters.

Based on the experimental tests, the following conclusions have been formulated:

- 30% and more of a fatty acid methyl ester additive to diesel fuel is a threshold value for its applicability, due to the power and torque criterion;
- The analysis of the values of the operating parameters of the drive unit of the means of transport, carried out on the basis of a random variable defined for this object, showed that the best results were obtained for mixture IV, and the worst for mixture III;
- The assessment shows that it was the criteria of power and torque that were most affected by the fuel mixture content, whereas the emission of carbon dioxide and nitric oxides were the least affected;
- The application of fatty acid methyl esters in diesel fuel had a positive impact on a reduction in the emission of noise generated by the transport-means power unit;
- The assessment shows that the composition of the fuel blend had the greatest impact on the criterion of power and torque of the drive unit of the measured transport, and the lowest emissions of carbon dioxide and nitrogen oxides;
- The performed statistical analysis makes it possible to define guidelines to be developed into decision-making strategies regarding the correct composition of the fuel mixture;
- Changes in the properties of the tested operational parameters of the drive unit of the means of transport are presented in a vector manner, which allows for a simultaneous analysis of changes in the tested parameters;
- The analysis makes it possible to set the guidelines for the development of a decision-making strategy for the application of appropriate content of the fuel mixture to be used in a car.

Author Contributions: Conceptualization: Ł.M., M.P. and M.M.; formal analysis: M.M.; investigation: M.M.; methodology: Ł.M., M.M. and D.K.; resources: M.M.; visualization: M.M. and T.K.; writing—original draft: M.M.; writing—review and editing: M.M. All authors have read and agreed to the published version of the manuscript.

Funding: The research did not receive external financing.

Institutional Review Board Statement: Not applicable.

Informed Consent Statement: Not applicable.

Data Availability Statement: Not applicable.

Conflicts of Interest: The authors declare no conflict of interest.

References

1. Struś, M. Efficiency of self ignition combustion engines powerd with bioxdiesel with fuly renewable components. *Mach. Eng.* **2014**, *14*, 29–38.
2. Dzieniszewski, G. Selected problems of biofuels for powering delf ignition engines. *Agric. Eng.* **2008**, *10*, 39–45.
3. Arapaki, N.; Bekeas, E.; Karavalakis, G.; Tzirakis, E.; Stournas, S.; Zannikos, F. Regulated and unregulated emission characteristics of diesel vehicle operating with diesel biodiesel blend. *SAE Tech. Pap.* **2007**, *1*, 1–9.
4. Bala, B.K. Studies on biodiesels from transformation of vegetable oils for diesel engines. *Energy Educ. Sci. Technol.* **2005**, *15*, 1–43.
5. Karavalakis, G.; Kousoulidou, M.; Tzamkiozis, T.; Ntziachristos, L.; Bakeas, E.; Stournas, S.; Samaras, Z. Effects of biodiesel on passenger car fuel consumption, regulated and non-regulated pollutant emissions over legislated and real-world driving cycles. *Fuel* **2009**, *88*, 1608–1617.
6. Knothe, G.; Sharp, C.; Ryan, T. Exhaust emission of biodiesel, petrodiesel, neat methyl esters and alkanes in a new technology engine. *Energy Fuels* **2006**, *20*, 403–408. [CrossRef]
7. Ozer, C.; Ozturka, E.; Solmazb, H.; Aksoyc, F.; Çinarb, C.; Yucesub, S. Combined effects of soybean biodiesel fuel addition and EGR application on the combustion and exhaust emissions in a diesel engine. *Appl. Therm. Eng.* **2016**, *95*, 115–124.

8. Azoumaha, Y.; Blinac, J.; Dahob, T. Exergy efficiency applied for the performance optimization of a direct injection compression ignition engine using biofuels. *Renew. Energy* **2009**, *34*, 1494–1500. [CrossRef]
9. Tompkins, B.; Song, H.; Bittle, J.; Jacobs, T. Efficiency considerations for the use of blended biofuel in diesel engines. *Appl. Energy* **2012**, *98*, 209–218. [CrossRef]
10. Man, X.; Cheung, C.; Ning, Z.; Wei, L.; Huang, Z. Influence of engine load and speed on regulated and unregulated emissions of a diesel engine fueled with diesel fuel blended with waste cooking oil biodiesel. *Fuel* **2016**, *180*, 41–49. [CrossRef]
11. Armas, O.; Yehliu, K.; Boehman, A. Effect of alternative fuels on exhaust emission during diesel engine operation with matched combustion phasing. *Fuel* **2010**, *89*, 438–456. [CrossRef]
12. Chauhan, B.; Kumara, N.; Chob, H.; Limc, H. A study on the performance and emission of a diesel engine fueled with Karanja biodiesel and its blends. *Energy* **2013**, *56*, 1–7. [CrossRef]
13. Hoekman, S.; Robbins, C. Review of the effects of biodiesel on NOx emissions. *Fuel Process. Technol.* **2012**, *96*, 237–249. [CrossRef]
14. Necati, A.; Canakci, O. Determination of performance and combustion characteristics of a diesel engine fueled with canola and waste palm oil methyl esters. *Energy Convers. Manag.* **2011**, *52*, 108–116.
15. Ozener, O.; Yuksek, L.; Ergenç, T.; Ozkan, M. Effects of soybean biodiesel on a DI diesel engine performance, emission and combustion characteristics. *Fuel* **2014**, *115*, 875–883. [CrossRef]
16. Xue, J.; Grift, T.; Hansen, C. Effect of biodiesel on engine performances and emissions. *Renew. Sustain. Energy Rev.* **2011**, *15*, 1098–1116. [CrossRef]
17. Rakopoulos, C.; Dimaratos, A.; Giakoumis, E.; Rakopoulos, D. Study of turbocharged diesel engine operation, pollutant emissions and combustion noise radiation during starting with bio-diesel or n-butanol diesel fuel blends. *Appl. Energy* **2010**, *88*, 3905–3916. [CrossRef]
18. Belgiorno, G.; Boscolo, A.; Dileo, G.; Numidi, F.; Concetto Pesce, F.; Vassallo, A.; Ianniello, R.; Beatrice, C.; Di Blasio, G. Experimental Study of Additive-Manufacturing-Enabled Innovative Diesel Combustion Bowl Features for Achieving Ultra-Low Emissions and High Efficiency. *SAE Int. J. Adv. Curr. Pract. Mobil.* **2020**, *130*, 672–684. [CrossRef]
19. Guido, C.; Beatrice, C.; Di Iorio, S.; Napolitano, P.; Di Blasio, G.; Vassallo, A.; Ciaravino, C. *Assessment of Closed-Loop Combustion Control Capability for Biodiesel Blending Detection and Combustion Impact Mitigation for an Euro5 Automotive Diesel Engine*; SAE 2011 World Congress & Exhibition; SAE International: Pittsburgh, PA, USA, 2011.
20. Sivaramakrishnan, K.; Ravikumar, P. Determination of cetane number of biodiesel and it's influence on physical properties. *ARPN J. Eng. Appl. Sci.* **2012**, *7*, 2.
21. Woropay, M.; Muślewski, Ł. Quality as a system on example of transport system. *J. KONES Intern. Combust. Engines* **2004**, *11*, 334–346.
22. Woropay, M.; Muślewski, Ł.; Piętak, A.; Niezgoda, T.; Żurek, J. The application example of the evaluation model in the case of the transport system operation quality. *ACSIM* **2004**, *2*, 220–226.
23. Pająk, M.; Muślewski, Ł. Fuzzy model of quality assessment of operation of a complex operation system. In Proceedings of the 10th Jubilee Congress of Technical Equipment Operation, ITE, Warsaw, Poland, 11–12 December 2005.
24. Sharma, D. Designing and Modeling Fuzzy Control Systems. *Int. J. Comput. Appl.* **2011**, *16*, 46–53. [CrossRef]
25. Broniewicz, E.; Ogrodnik, K. Multi-criteria analysis of transport infrastructure projects. *Transp. Res. Part D: Transp. Environ.* **2020**, *83*, 102351. [CrossRef]
26. Camargo Perez, J.; Carrillo, M.; Montoya-Torres, J. Multi-criteria approaches for urban passenger transport systems: A literature review. *Ann. Oper. Res.* **2015**, *226*, 69–87. [CrossRef]
27. Sawik, B.; Faulin, J.; Pérez-Bernabeu, E. Multi-Criteria Optimization for Fleet Size with Environmental Aspects. *Transp. Res. Procedia* **2017**, *27*, 61–68. [CrossRef]
28. De Luca, S. Public engagement in strategic transportation planning: An analytic hierarchy process based approach. *Transp. Policy* **2014**, *33*, 110–124. [CrossRef]
29. Duleba, S.; Moslem, S. Examining Pareto optimality in analytic hierarchy process on real Data: An application in public transport service development. *Expert Syst. Appl.* **2019**, *116*, 21–30. [CrossRef]
30. Demirbas, A. Importance of biodiesel as transportation fuel. *Energy Policy* **2007**, *35*, 4661–4670. [CrossRef]
31. Ali, O.; Mamat, R.; Abdullah, N.; Abdullah, A. Analysis of blended fuel properties and engine performance with palm biodiesel–diesel blended fuel. *Renew. Energy* **2016**, *86*, 59–67. [CrossRef]
32. Alptekin, E. Emission, injection and combustion characteristics of biodiesel and oxygenated fuel blends in a common rail diesel engine. *Energy* **2017**, *119*, 44–52. [CrossRef]
33. Wang, L.; Wang, J.; Xiao, J. Combustion and emission characteristics of diesel engine fueled with diesel/biodiesel/pentanol fuel blends. *Fuel* **2015**, *156*, 211–218.
34. E-Petrol. Available online: https://www.e-petrol.pl/quotes/biofuels (accessed on 10 January 2021).
35. Markiewicz, M.; Muślewski, Ł. Survey performance and emission parameters of diesel engine powered by diesel oil and fatty acid methyl esters using fuzzy logic techniques. *Fuel* **2020**, *277*, 118179. [CrossRef]
36. Markiewicz, M.; Muślewski, Ł.; Pająk, M. Impact of Biocomponent Additive to Diesel Oil on Values of Selected Functional Parameters of Transport Means. *Pol. J. Environ. Stud.* **2020**, *29*, 3475–3483. [CrossRef]
37. Velasquez, M.; Hester, P. An Analysis of Multi-Criteria Decision Making Methods. *Int. J. Oper. Res.* **2013**, *10*, 2.

38. Agarwal, A.K.; Srivastava, D.K.; Dhar, A.; Maurya, R.K.; Shukla, P.C.H.; Singh, A.P. Effect of fuel injection timing and pressure on combustion, emissions and performance characteristics of a single cylinder diesel engine. *Fuel* **2013**, *111*, 374–383. [CrossRef]
39. Choi, S.; Oh, Y. The emission effects with the use of biodiesel fuel. *Int. J. Mod. Phys. B* **2006**, *20*, 4481–4486. [CrossRef]
40. Lapuerta, M.; Armas, O.; Rodriguez-Fernandez, L. Effect of biodiesel fuels on diesel engine emission. *Prog. Energy Combstion Sci.* **2008**, *34*, 198–223. [CrossRef]
41. Leung, D.; Luo, Y.; Chan, T. Optimization of exhaust emission of a diesel engine fuelled with biodiesel. *Energy Fuels* **2006**, *20*, 198–223. [CrossRef]
42. Muślewski, Ł.; Pająk, M.; Landowski, B.; Żółtowski, B. A method for determining the usability potential of ship steam boilers. *Pol. Marit. Res.* **2016**, *23*, 105–111. [CrossRef]
43. Jablonicky, J.; Tkac, Z.; Majdan, R.; Mullerova, D.; Hujo, L.; Vozarova, V. *Hodnotenie Vlastnosti Biopaliw a Biomazin*; Slovak University of Agriculture in Nitra, Fakulty of Engineering: Nitra-Chrenová, Slovakia, 2012; ISBN 978-80-552-0766-7.
44. Curkowski, A.; Mroczkowski, P.; Oniszk-Ppławska, A.; Wiśnieski, G. *Agricultural Biogas Production and Utilization*; Mazowiecka Energy Agency: Warszawa, Poland, 2009.
45. Markiewicz-Patalon, M.; Muślewski, Ł.; Kaszkowiak, J.; Sójka, M. *Analysis of Efficiency of the Vehicle Transport Facilities Powered with Diesel Oil with Additive of Biocomponent*; KONMOT: Kraków, Poland, 2018; pp. 297–299.
46. Markiewicz-Patalon, M.; Muślewski, Ł.; Kaszkowiak, J. *Noise Emission Research of Diesel Engine Powered with Mixture of Diesel Oil and Methyl Esters*; KONMOT: Kraków, Poland, 2018; pp. 272–274.
47. Muślewski, Ł. The implementation of the fuzzy logic elements in the area of the transport system operation quality assessment. Maritime Industry, Ocean Engineering and Coastal Resources. Maritime Transportation. In *Balkema—Proceedings and Monographs in Engineering, Water and Earth Sciences*; Soares, C.G., Kolev, P.N., Eds.; Taylor & Francis Group: London, UK; Leiden, The Netherlands; New York, NY, USA; Philadelphia, PA, USA; Singapore, 2007; Volume 1.
48. Landowski, B.; Muślewski, Ł. Decision model of an operation and maintenance process of city buses. In Proceedings of the 58th International Conference of Machine Design Departments—ICMD 2017, Praque, Czech Republic, 6–8 September 2017; Czech University of Life Sciences Prague: Praha-Suchdol, Czech Republic, 2017; pp. 188–193.
49. Aleksandrowicz, P. Verifying a truck collision applying the SDC method. In Proceedings of the 58th International Conference of Machine Design Departments, Praque, Czech Republic, 6–8 September 2017; pp. 14–19.

MDPI
St. Alban-Anlage 66
4052 Basel
Switzerland
Tel. +41 61 683 77 34
Fax +41 61 302 89 18
www.mdpi.com

Energies Editorial Office
E-mail: energies@mdpi.com
www.mdpi.com/journal/energies